Global Forest Monitoring from Earth Observation

Earth Observation of Global Changes

Series Editor
Chuvieco Emilio

Global Forest Monitoring from Earth Observation
edited by Frédéric Achard and Matthew C. Hansen

Global Forest Monitoring from Earth Observation

Edited by
Frédéric Achard • Matthew C. Hansen

CRC Press
Taylor & Francis Group
Boca Raton London New York

CRC Press is an imprint of the
Taylor & Francis Group, an **informa** business

Joint Research Centre, Institute for Environment and Sustainability

CRC Press
Taylor & Francis Group
6000 Broken Sound Parkway NW, Suite 300
Boca Raton, FL 33487-2742

© 2013 by European Union and Matthew Hansen,
CRC Press is an imprint of Taylor & Francis Group, an Informa business

No claim to original U.S. Government works

International Standard Book Number: 978-1-1380-7447-7 (Paperback)
International Standard Book Number: 978-1-4665-5201-2 (Hardback)

Library of Congress Cataloging-in-Publication Data

Global forest monitoring from earth observation / edited by Frederic Achard and
  Matthew C. Hansen.
    p. cm.
  Includes bibliographical references and index.
  ISBN 978-1-4665-5201-2
  1. Forests and forestry--Remote sensing. 2. Forest monitoring. I. Achard, Frédéric. II.
Hansen, Matthew C.

  SD387.R4G56 2012
  333.75--dc23   2012018562

Visit the Taylor & Francis Web site at
http://www.taylorandfrancis.com

and the CRC Press Web site at
http://www.crcpress.com

Contents

Preface

Forest resources are crucial in the context of sustainable development and climate change mitigation. Dynamic information on the location and evolution of forest resources are needed to properly define, implement, and evaluate strategies related to multilateral environmental agreements such as the UN Framework Convention on Climate Change (UNFCCC) and the Convention on Biological Diversity. For the global change scientific community and the UNFCCC process, it is important to tackle the technical issues surrounding the ability to produce accurate and consistent estimates of greenhouse gas emissions and removals from forest area changes worldwide and at the country level.

The following compilation of chapters constitutes a review of why and how researchers currently use remotely sensed data to study forest cover extent and loss over large areas. Remotely sensed data are most valuable where other information, for example, forest inventory data, are not available, or for analyses of large areas for which such data cannot be easily acquired. The ability of a satellite sensor to synoptically measure the land surface from national to global scales provides researchers, governments, civil society, and private industry with an invaluable perspective on the spatial and temporal dynamics of forest cover changes. The reasons for quantifying forest extent and change rates are many. In addition to commercial exploitation and local livelihoods, forests provide key ecosystem services including climate regulation, carbon sequestration, watershed protection, and biodiversity conservation, to name a few. Many of our land use planning decisions are made without full understanding of the value of these services, or of the rate at which they are being lost in the pursuit of more immediate economic gains through direct forest exploitation. Our collection of papers begins with an introduction on the roles of forests in the provision of ecosystem services and the need for monitoring their change over time (Chapters 1 and 2).

We follow this introduction with an overview on the use of Earth observation datasets in support of forest monitoring (Chapters 3 through 5). General methodological differences, including wall-to-wall mapping and sampling approaches, as well as data availability, are discussed. For large-area monitoring applications, the need for systematically acquired low or no cost data cannot be overstated. To date, data policy has been the primary impediment to large-area monitoring, as national to global scale forest monitoring requires large volumes of consistently acquired and processed imagery. Without this, there is no prospect for tracking the changes to this key Earth system resource.

The main section of the book covers forest monitoring using optical data sets (Chapters 6 through 14). Optical datasets, such as Landsat, constitute the longest record of the Earth surface. Our experience of using them in mapping and monitoring forest cover is greater than that of other datasets due to the relatively rich record of optical imagery compared to actively acquired data sets such as radar imagery. The contributions to this section range from indicator mapping at coarse spatial resolution to sample-based assessments and wall-to-wall mapping at medium spatial resolution. The studies presented span scales, environments, and themes. For example, forest degradation, as opposed to stand-replacement disturbance, is analyzed in two chapters. Forest degradation is an important variable regarding biomass, emissions, and ecological integrity, as well as being a technically challenging theme to map. Chapters 6 through 14 also present a number of operational systems, from Brazil's PRODES and DETER products, to Australia's NCAS system. These chapters represent the maturity of methods as evidenced by their incorporation by governments into official environmental assessments. The fourth section covers the use of radar imagery in forest monitoring (Chapter 15). Radar data have a long history of experimental use and are presented here as a viable data source for global forest resource assessment.

We believe that this book is a point of departure for the future advancement of satellite-based monitoring of global forest resources. More and more observing systems are being launched, methods are quickly maturing, and the need for timely and accurate forest change information is increasing. If data policies are progressive, users of all kinds will soon have the opportunity to test and implement forest monitoring methods. Our collective understanding of forest change will improve dramatically. The information gained through these studies will be critical to informing policies that balance the various demands on our forest resources. The transparency provided by Earth observation data sets will, at a minimum, record how well we perform in this task.

We deeply thank Prof. Emilio Chuvieco from the University of Alcalá (Spain) who gave us the opportunity to publish this book and supported and encouraged us in its preparation. We also sincerely thank all the contributors who kindly agreed to take part in this publication and who together have produced a highly valuable book.

Frédéric Achard and Matthew C. Hansen

Editors

Dr. Frédéric Achard is a senior scientist at the Joint Research Centre (JRC), Ispra, Italy. He first worked in optical remote sensing at the Institute for the International Vegetation Map (CNRS/University) in Toulouse, France. Having joined the JRC in 1992, he started research over Southeast Asia in the framework of the TREES (TRopical Ecosystem Environment observations by Satellites) project. His current research interests include the development of Earth observation techniques for global and regional forest monitoring and the assessment of the implications of forest cover changes in the tropics and boreal Eurasia on the global carbon budget. Frédéric Achard received his PhD in tropical ecology and remote sensing from Toulouse University, Toulouse, France, in 1989. He has coauthored over 50 scientific peer-reviewed papers in leading scientific journals including *Nature, Science, International Journal of Remote Sensing, Forest Ecology and Management, Global Biogeochemical Cycles,* and *Remote Sensing of Environment.*

Dr. Matthew C. Hansen is a professor in the Department of Geographical Sciences at the University of Maryland, College Park, Maryland. He has a bachelor of electrical engineering degree from Auburn University, Auburn, Alabama. His graduate degrees include a master of engineering in civil engineering and a master of arts in geography from the University of North Carolina at Charlotte and a doctoral degree in geography from the University of Maryland, College Park, Maryland. His research specialization is in large-area land cover monitoring using multispectral, multitemporal, and multiresolution remotely sensed data sets. He is an associate member of the MODIS (Moderate Resolution Imaging Spectroradiometer) Land Science Team and a member of the GOFC-GOLD (Global Observations of Forest Cover and Land Dynamics) Implementation Working Group.

Contributors

Frédéric Achard
Institute for Environment
    and Sustainability
Joint Research Centre of
    the European Commission
Ispra, Italy

Olivier Arino
Earth Observation Directorate
European Space Agency
Frascati, Italy

Sergey Bartalev
Space Research Institute
Russian Academy of Sciences
Moscow, Russian Federation

Alan Belward
Institute for Environment
    and Sustainability
Joint Research Centre of
    the European Commission
Ispra, Italy

René Beuchle
Institute for Environment
    and Sustainability
Joint Research Centre of
    the European Commission
Ispra, Italy

Peter Caccetta
Mathematics, Statistics
    and Informatics
Commonwealth Scientific and
    Industrial Research Organisation
Floreat, Western Australia,
    Australia

Daniel Clewley
Institute of Geography and Earth
    Sciences
Aberystwyth University
Aberystwyth, United Kingdom

Rémi D'Annunzio
Forestry Department
Food and Agriculture Organization
    of the United Nations
Rome, Italy

Ruth DeFries
Department of Ecology, Evolution
    and Environmental Biology
Columbia University
New York, New York

Valdete Duarte
Remote Sensing Division
Brazilian Institute for Space
    Research
São Paulo, Brazil

Victor Efremov
Space Research Institute
Russian Academy of Sciences
Moscow, Russian Federation

Vyacheslav Egorov
Space Research Institute
Russian Academy of Sciences
Moscow, Russian Federation

Evgeny Flitman
Space Research Institute
Russian Academy of Sciences
Moscow, Russian Federation

Antonio Roberto Formaggio
Remote Sensing Division
Brazilian Institute for Space Research
São Paulo, Brazil

Suzanne Furby
Mathematics, Statistics
   and Informatics
Commonwealth Scientific
   and Industrial Research
   Organisation
Floreat, Western Australia, Australia

Matthew C. Hansen
Department of Geographical
   Sciences
University of Maryland
College Park, Maryland

Sean P. Healey
Rocky Mountain Research Station
U.S. Forest Service
Ogden, Utah

Dirk Hoekman
Department of Environmental
   Sciences, Earth System Science
   and Climate Change Group
Wageningen University
Wageningen, The Netherlands

Richard A. Houghton
Woods Hole Research Center
Falmouth, Massachusetts

Josef Kellndorfer
Woods Hole Research Center
Falmouth, Massachusetts

Pieter Kempeneers
Earth Observation Department
Flemish Institute for Technological
   Research
Mol, Belgium

Lars Laestadius
People and Ecosystems
   Department
World Resources Institute
Washington, DC

Erik Lindquist
Forestry Department
Food and Agriculture Organization
   of the United Nations
Rome, Italy

Evgeny Loupian
Space Research Institute
Russian Academy of Sciences
Moscow, Russian Federation

Thomas R. Loveland
Earth Resources Observation
   and Science Center
U.S. Geological Survey
Sioux Falls, South Dakota

Richard Lucas
Institute of Geography
   and Earth Sciences
Aberystwyth University
Aberystwyth, United Kingdom

Jeffrey G. Masek
NASA Goddard Space Flight Center
Greenbelt, Maryland

Daniel McInerney
Institute for Environment
   and Sustainability
Joint Research Centre of the
   European Commission
Ispra, Italy

Humberto Navarro de Mesquita, Jr.
National Forest Registry
Brazilian Forest Service
Brasilia, Brazil

Anssi Pekkarinen
Forestry Department
United Nations Food and
　Agriculture Organization
Rome, Italy

Peter Potapov
Department of Geographical
　Sciences
University of Maryland
College Park, Maryland

Gary Richards
Department of Climate Change
　and Energy Efficiency
Canberra, Australia
and
Fenner School of Environment
　and Society
Australian National University
Canberra, Australia

Ake Rosenqvist
solo Earth Observation
Tokyo, Japan

**Bernardo Friedrich Theodor
Rudorff**
Remote Sensing Division
Brazilian Institute for Space Research
São Paulo, Brazil

Jesús San-Miguel-Ayanz
Institute for Environment
　and Sustainability
Joint Research Centre of
　the European Commission
Ispra, Italy

João Roberto dos Santos
Remote Sensing Division
Brazilian Institute for Space
　Research
São Paulo, Brazil

Fernando Sedano
Institute for Environment
　and Sustainability
Joint Research Centre of
　the European Commission
Ispra, Italy

Lucia Seebach
Department of Forest and
　Landscape
University of Copenhagen
Copenhagen, Denmark

Yosio Edemir Shimabukuro
Remote Sensing Division
Brazilian Institute for Space
　Research
São Paulo, Brazil

Masanobu Shimada
Earth Observation Research
　Center
Japan Aerospace Exploration
　Agency
Tokyo, Japan

Carlos Souza, Jr.
Amazon Institute of People
　and the Environment
Belém, Pará, Brazil

Stephen V. Stehman
College of Environmental Science
　and Forestry
State University of New York
Syracuse, New York

Hans-Jürgen Stibig
Institute for Environment
　and Sustainability
Joint Research Centre
　of the European
　Commission
Ispra, Italy

Peter Strobl
Institute for Environment
  and Sustainability
Joint Research Centre of
  the European Commission
Ispra, Italy

Fedor Stytsenko
Space Research Institute
Russian Academy of Sciences
Moscow, Russian Federation

Svetlana Turubanova
Department of Geographical
  Sciences
University of Maryland
College Park, Maryland

Wayne Walker
Woods Hole Research Center
Falmouth, Massachusetts

Jeremy Wallace
Mathematics, Statistics
  and Informatics
Commonwealth Scientific and
  Industrial Research Organisation
Floreat, Western Australia,
  Australia

Robert Waterworth
Department of Climate Change
  and Energy Efficiency
Canberra, Australia

Xiaoliang Wu
Mathematics, Statistics
  and Informatics
Commonwealth Scientific
  and Industrial Research
  Organisation
Floreat, Western Australia,
  Australia

Alexey Yaroshenko
Greenpeace Russia
Moscow, Russian Federation

Ilona Zhuravleva
Greenpeace Russia
Moscow, Russian Federation

1

Why Forest Monitoring Matters for People and the Planet

Ruth DeFries

Columbia University

CONTENTS

1.1 Introduction

In children's tales, forests loom as dark and dangerous places holding mysterious and magical secrets. Hansel and Gretel ventured into the forbidden forest to encounter a child-eating witch. A vicious wolf tricked Little Red Riding Hood when she strayed into the forest. Forests are also places of enchantment, the home of Snow White's seven dwarfs, elves and nymphs, and the castle of the ill-fated prince in *Beauty and the Beast*. The stories revere forests for their magic and revile them for the perils that lurk within.

This dual view of forests persists until today. On the one hand, forests are roadblocks to progress that occupy space more productively used for agriculture. As slash and burn agriculture made its way northward from the Mediterranean coast through Europe, beginning about 4,000 years ago until the first centuries of the common era, forests were replaced by settled agriculture (Mazoyer and Roudart 2006). A similar story played out in North America in the last few centuries, with European expansion preceded by the

Native American's use of fire to manage forests (Williams 2006). Throughout the currently industrialized world, wholesale clearing of forests enabled agriculture to expand and economies to grow. A similar dynamic is currently underway in tropical regions, where economic growth often goes hand-in-hand with agricultural expansion into forested areas (DeFries et al. 2010). There is no doubt that clearing of forests for agriculture played a major role in the expansion of the human species into new areas, the growth in population from 5 million during the dawn of agriculture to over 7 billion today, and increasing prosperity (Mazoyer and Roudart 2006). In this sense, the fairy tale's view of forests as harmful places that are better off cleared resonates with the experience of human history.

The opposite side of the dual view reveres forests for the large range of beneficial services they provide for humanity. Tangible goods such as timber or recreation are apparent. Less apparent are intangible services such as climate regulation, biodiversity, and watershed protection. These regulating ecosystem services are only beginning to be quantified and understood (Millennium Ecosystem Assessment 2005). Without consideration of regulating services from forests, if the economic value of land use following clearing is greater than the economic value of standing forests, the decision to deforest is likely to ensue. This has been the calculus for millennia of forest clearing that has reduced over 40% of the world's forest cover (Figure 1.1).

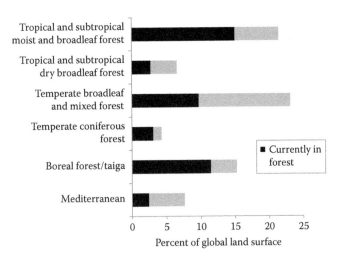

FIGURE 1.1
Approximate percent of the global land surface currently (ca. 1990) occupied by major forests types and the percent previously converted to agriculture. (Values for current percent from Wade, T., et al., *Conserv. Ecol., 7,* 7, 2003 and values for converted percent derived from Stokstad, E. *Science,* 308, 41, 2005, except for boreal forests which is from Table C2 in Scholes, R., et al. Summary: Ecosystems and their services around the year 2000. In Hassan, R., et al., eds. *Ecosystems and Human Well-Being: Current State and Trends,* vol 1. Washington, DC: Island Press, 2005, 2–23.)

Forest conversion varies greatly in different forest types in different parts of the world. Nearly 70% of Mediterranean forests and almost 60% of temperate deciduous and dry tropical forests have been converted to agriculture. Tropical moist broadleaf forest and boreal forests still have substantial areas of forest remaining.

Remaining forests and the services they provide are increasingly under pressure from both economic and biophysical forces. With increases in population, per capita consumption, and shifts to animal-based diets, demand for agricultural products is estimated to increase by at least 50% by 2050 (Godfray et al. 2010; Nelleman et al. 2009; Royal Society of London 2009). Increasing yield rather than expansion explains the bulk of the vast increase in agricultural production in the last century and is likely to continue to be the main factor in meeting future food demand (Mooney et al. 2005), but agricultural expansion is also likely to continue into the future. Tropical forest and woodlands are the only biomes with substantial area remaining for agricultural expansion. In the past few decades, over 80% of agricultural expansion in the tropics occurred into intact and disturbed forests (Gibbs et al. 2010). Rapid clearing of tropical forests in the last few decades has enabled escalating production of commodities such as oil palm, soy, and sugarcane in response to rising demand (Johnston and Holloway 2007). This pressure on tropical forests and woodlands, particularly in South America and Africa, will only continue in the future with competition of land for food production and biofuels.

Ecological and climatic factors in addition to economic forces are creating pressures on forests. In tropical forests, dry conditions combined with ignition sources create conditions conducive to fires (Chen et al. 2011; van der Werf et al. 2008). In temperate and boreal latitudes, anomalously dry years lead to large forest fires, such as the Russian fires of 2010 (Baltzer et al. 2010). Warmer conditions promote insect outbreaks, such as the pine beetle infestation of western North America, leading to loss of forest stands (Kurz et al. 2008).

These multiple economic, climatic, and ecological forces acting in different parts of the world reverberate to alter the services that forests perform, including habitats that forests provide for other species and the ability of forests to sequester carbon and regulate climate. As both knowledge of the role of forests in providing ecosystem services and the pressures on forests increase, the ability of communities, countries, and global-scale policy makers to monitor forests becomes paramount.

Forests in different parts of the world contribute differentially to ecosystem services, depending on the economic and ecological setting. For example, from an ecological point of view, boreal and peat forests regulate climate through their large stores of belowground carbon while tropical forests contain nearly all of their carbon aboveground. From a socioeconomic point of view, in dry tropical forests with relatively dense populations of poor, forest-dependent people, for example, forests contribute substantially to livelihood needs such as fuel wood and fodder for livestock (Miles et al. 2006). In temperate forests, the recreation value of forests for populations with disposable income for tourism or the need to protect watersheds for large urban centers becomes more important. This heterogeneity in services and pressures on forests create varying needs for monitoring in different parts of the world.

This introductory chapter describes a framework for assessing land use and ecological processes affecting forests and the implications for a range of ecosystem services. The chapter then addresses the evolving needs for forest monitoring in light of information needs to maintain these services.

1.2 Socioeconomic and Ecological Processes Affecting Forests: What Processes Need to Be Monitored?

Methods and approaches to monitor forest extent and condition depend on the processes of interest to the user of the information. These processes—for example, changes in productivity, deforestation, or increases in forest cover—vary greatly in different forest regions around the world and change over time depending on economic and ecological factors. These myriad processes acting on forests require considerable thought in designing monitoring efforts that are flexible and appropriate to the processes occurring in different forest regions.

1.2.1 Land Use Processes

The generalized schematic of land use transitions that accompany economic development provides a framework to view pressures on forests and implications for ecosystem services (DeFries et al. 2004; Mustard et al. 2004). The extent and condition of forests are intricately tied to land use change, as demand for timber, food, and other agricultural products creates pressures to use forests or clear them to make way for croplands and pasture. Pressure to use forested land, in turn, is connected to transitions that typically occur in the course of urbanization, development, and structural transformations in the economy from predominance of agrarian to industrial sectors. Land use typically follows a trajectory from presettlement wildlands with low population densities, to frontier clearing and subsistence agriculture with people reliant on local food production, to higher yield intensive agriculture to support urban populations. Although the details and speed of transitions vary greatly in different places and at different times in history, this general pattern describes the overall trajectory. Different places around the world can be viewed from a lens of their position within this stylized trajectory. On the one hand, the southern Brazilian state of Mato Grosso, for example,

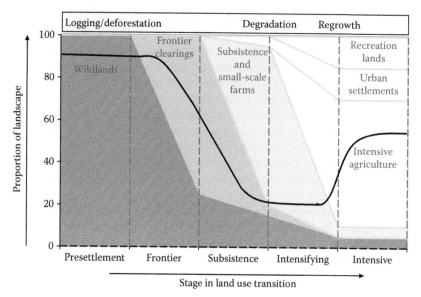

FIGURE 1.2
Generalized land use transition that accompanies economic development, urbanization, and shift from agrarian to industrial economies (DeFries et al. 2004; Mustard et al. 2004). Accompanying proportion of landscape in forest cover (dark line) first declines and then increases with the forest transition (Mather 1992; Rudel et al. 2005; Walker 1993). Proportions of landscape are hypothetical, do not represent actual data, and depict only general patterns that vary in different places. Processes shift from logging and deforestation to degradation and regrowth as regions progress through stages in land use and forest transitions.

is currently undergoing a very rapid transition from wildlands to intensive agriculture, with rapid frontier clearing that largely bypasses the step of subsistence agriculture. South Asia, on the other hand, moved through the frontier clearing of wildlands millennia ago, but much of the land remains in small-scale farming for subsistence and local markets (Figure 1.2).

In forested areas, land use transitions accompany a characteristic trajectory in forest extent and condition. In the early, wildland stage of the land use transition, forests cover extensive areas with low-intensity use for hunting, collection of foods and medicines, or shifting cultivation by low densities of indigenous peoples. With frontier clearing, logging of valuable tree species might occur followed by deforestation and an increasingly fragmented forest. As the transition moves into a period of subsistence agriculture, remaining forest patches are likely to be heavily used for fuel wood, fodder, and nontimber forest product collection. Forest degradation, currently extensive in dry tropical forests of Asia, is the main pressure on forests during a subsistence stage of a land use transition. With urbanization, economic growth, and agricultural intensification, the well-known "forest transition" of increasing forest cover has been observed in many countries (Mather 1992; Rudel et al. 2005; Walker 1993). Rudel et al. (2005) identify two pathways through which increasing forest cover occurs. One pathway is an increase in planted trees incentivized by a shortage of timber; such was the case in Europe. The other pathway is through abandonment of less productive agricultural land as economic growth brings small-scale farmers to urban areas and food production is transported from productive agricultural areas. Such was the case in New England, where forest cover rebounded in areas of abandoned agriculture.

Land use and forest transitions provide a framework to assess monitoring needs in light of the varying pressures on forests at different stages along the transition. Forest areas in distant wildlands are not likely to be undergoing rapid change, consequently requiring less frequent monitoring for human impacts. In frontier forests undergoing a transition from wildlands, deforestation and degradation from unsustainable logging are the activities requiring monitoring. Places in a mode of small-scale farming with local reliance on forest patches for livelihood needs are subject to degradation. Monitoring for deforestation in such locations is less relevant and degradation is more likely to be important. Finally, in the later stages of a land use transition, regrowth of forests becomes an important process, requiring a monitoring strategy to identify increases rather than decreases of tree cover.

As different places move through land use and forest transitions, the processes that require monitoring will shift. Monitoring efforts for deforestation might most usefully focus on frontier regions and monitoring for degradation in postfrontier remaining forest patches. Monitoring to identify regrowth is most relevant in those places undergoing agricultural abandonment. Methods vary to monitor these different processes, requiring flexibility in monitoring efforts as processes requiring monitoring change.

1.2.2 Ecological Processes

As with land use processes, ecological processes affecting forests vary in different places. The types of ecological processes that may be important for monitoring systems to identify include:

Biome shifts in response to climate change: Climate change is already leading to shifts in boundaries of forests biomes in high latitudes (Beck et al. 2011). In the tropics, a biome shift between savanna and forest has been hypothesized with a drier climate and increased fires (Hirota et al. 2010). As the process of biome shifts is heterogeneous and conflicting evidence arises from different places, a remote sensing approach is critical to enable observations over large areas. Shifts in forest boundaries have major consequences for carbon storage and biophysical feedbacks to climate through changes in albedo and evapotranspiration of the land surface. A long-term monitoring system that enables observations of changes in forest boundaries allows earth system models to incorporate dynamic interactions between vegetation and climate in the growing field of dynamic vegetation models (Gonzalez et al. 2010). The ability to monitor such changes over large areas at fine spatial resolution is becoming more feasible.

Changes in forest ecosystems in response to atmospheric chemistry: Enhanced forest productivity and biomass accumulation attributable to fertilization from elevated carbon dioxide concentrations is controversial but may explain increased productivity and biomass accumulation in tropical forests (Lewis et al. 2009). Nitrogen deposition is another forcing factor on forest productivity, with studies suggesting an effect on species composition and ecosystem function in temperate and northern Europe and North America (Bobbink et al. 2010). Long-term monitoring of productivity cannot attribute the cause of any observed changes, but is a critical piece to unraveling the responses of forests to changing atmospheric chemistry.

Fire: The ability to monitor active fires (Justice et al. 2002) and burned areas (Giglio et al. 2010) with remote sensing has developed rapidly. Many types of fires affect forests, including intentionally set deforestation fires, fires escaped from land management, and fires ignited by lightning. The extent to which these fires occur depends on multiple factors such as climate, fuel loads, and ignition sources. Fire is a particularly complex phenomenon that combines climatic, ecological, and human factors (Bowman et al. 2009).

A framework to identify monitoring needs through a lens of economic and ecological processes creates the need for multiple approaches that can vary through space and time. To date, global monitoring with remote sensing has focused predominantly on forest extent. As methods develop, robust global forest monitoring in the longer term should assess changes occurring in response to the full suite of processes affecting forests throughout the world.

1.3   Ecosystem Services from Forests

Monitoring systems aim to identify changes in the extent and condition of forests so that timely and effective policies can be put in place to avoid negative consequences for ecosystem services. Forests provide many ecosystem services that accrue benefits at proximal, downstream, and distal scales. Similar to the processes affecting forests discussed above, ecosystem services from forests and their beneficiaries vary across forest regions according to socioeconomic and ecological settings. Consequently, monitoring methods and approaches need to vary depending on the ecosystem services of concern. A monitoring system that aims to be applied to the implementation of

TABLE 1.1

Some Ecosystem Services Accruing to Beneficiaries at Different Spatial Scales from Forests in Varying Stages of Land Use Transitions

Location of Beneficiary	Forest Condition by Stage of Land Use Transition		
	Wildlands Prior to Frontier Clearing	*Forest Fragments Embedded in Small-Scale Agricultural Land*	*Regrowth with Agricultural Intensification*
Proximate	Livelihood needs and local regulating services (e.g., pollination) for low density of forest-dependent people	**Livelihood needs and local regulating services for high density of forest-dependent people**	
Downstream		Prevention of soil erosion, flood regulation, water purification	**Prevention of soil erosion, flood regulation, water purification**
Distal	**Carbon storage, biodiversity**	Biodiversity in forest fragments	Carbon sequestration, biodiversity in secondary forest

Note: Dominant ecosystem service of each stage based on author's judgment is in bold.

REDD (reducing emissions from deforestation and degradation), for example, requires observations of forest extent and biomass while a system aimed at biodiversity requires monitoring of habitat and forest structure. The following highlights the range of ecosystem services from forests at different scales (Table 1.1).

Proximal: Ecosystem services from forests play a particularly essential role for forest-dependent people throughout the global South (Agrawal et al. 2011). Natural capital from forests is a disproportionately large component for millions of poor households and communities relying directly on forests for livelihood needs. Services from forests include fuel wood, fodder for livestock, nontimber forest products to generate income, meat for protein, and medicinal plants. On the one hand, regulating services such as clean water, pollination, disease regulation, and pest control as well as spiritual and cultural importance of forests are more difficult to quantify but are important locally. On the other hand, forests and, particularly, protected areas harbor species that provide a disservice to local communities by crop raiding and livestock predation affecting local residents (White and Ward 2010) and spread of zoonotic diseases (Keesing et al. 2010).

Downstream: The watershed protection value of forests has garnered the most tractable implementation of payment for ecosystem service schemes. Forests buffer runoff to regulate floods and filter water to improve water quality.

Well-known examples of forest conservation for watershed protection include watersheds for the surface water supply of urban areas such as New York City and Quito, Ecuador (Postel and Thompson 2005). In addition to downstream users, another example of the role of forests at a regional scale is through energy balance and evapotranspiration, such as the Amazon basin where deforestation leads to decreases in basin-wide precipitation of climate and downwind transport of vapor (Davidson et al. 2012).

Distal: Global-scale services from forests accrue to beneficiaries living far away. Carbon storage to maintain carbon in vegetation rather than as a greenhouse gas in the atmosphere is a critical role for forests. Terrestrial vegetation and litter combined contain approximately the same amount of carbon as the atmosphere (850 and 780 Pg, respectively), with forests a particularly important reservoir for carbon (Houghton 2007). Tropical forests are exceptionally valuable for biodiversity in terms of species richness, family richness, and species endemism (Mace et al. 2005). Distal beneficiaries of biodiversity value the knowledge of existence as well as the functional role of biodiversity for disease regulation, resilience to disturbance, and other functions (Thompson et al. 2011).

In sum, forests provide a myriad of ecosystem services that vary in different forest regions. Aboveground carbon storage and biodiversity are particularly relevant in humid tropical forests. Local livelihood needs are relevant in dry tropical forests, and watershed protection is particularly relevant in forests upstream of urban centers reliant on surface water. Communities, national governments, and global policy makers place varying priorities on different ecosystem services. For example, local communities may place little value on carbon and biodiversity services that accrue to distal beneficiaries, while global policy makers may place little value on forest products and other livelihood needs for local communities. This mismatch in scales and differences in priorities about which ecosystem services are most important create tensions for designing monitoring systems.

The importance of different ecosystem services may vary through time as places move through land use transitions. Monitoring systems designed to address particular ecosystem services might require flexibility as priorities shift. For example, if carbon storage is the rationale for a monitoring program, the focus might be on frontier regions aimed at reducing deforestation and on late-stage transitions aimed at sequestering carbon through regrowth (Lambin and Meyfroidt 2011). If the rationale were rather on local livelihood needs for forest products, a monitoring system would focus on places in a subsistence stage of the land use transition to monitor degradation. For watershed protection, riparian forest cover would be of primary importance.

In reality, existing monitoring systems have not explicitly identified the rationale in terms of ecosystem services. Monitoring systems ideally would be relevant for multiple ecosystem services to make effective use of the investment. As monitoring systems are implemented in different countries throughout the developing world in different stages of land use transitions, explicit consideration of the ecosystem services of interest may be a useful undertaking.

1.4   Evolving Capabilities for Forest Monitoring

Forest monitoring to date (FAO 2010; Forest Survey of India 2005; INPE 2007) has mainly focused on the areal extent of forest cover and changes over time. Other variables of forest condition are increasingly becoming possible to monitor from satellites. Biomass, a key variable for carbon storage, has traditionally been collected through ground-based inventories. Recent abilities to assess biomass using remote sensing (Saatchi et al. 2006) are promising technological advances that are becoming more amenable to operational implementation. Monitoring degradation from logging with the spatial pattern characteristic of the Amazon has also advanced to be operational (Asner et al. 2006; Souza Jr. et al. 2005). These advances represent major progress for subnational, national, and global efforts to monitor forests and the ecosystem services they provide.

While these advances are major achievements, several aspects of forest condition are still in need of methodological development to address the full range of ecosystem services and socioeconomic and ecological processes affecting forests in different parts of the world. One such need is forest degradation related to local uses such as fuel wood collection and forest grazing, such as occurring extensively in Asian forests with high density of poor populations dependent on local ecosystem services. While monitoring of degradation characteristics of logging in the Amazon has advanced, monitoring of degradation from other local uses has not progressed to the same degree. Another aspect that has not been incorporated in monitoring is postclearing land use. The land use and management following deforestation, such as fertilizer use, agricultural activity, and crop type and diversity, has implications for ecosystem services and is required information to assess the impact of deforestation (Galford et al. 2010). While methods have advanced to assess postclearing land use in terms of pasture versus crop (Macedo et al. 2012), other aspects of land management require attention. Finally, the importance of lands outside forests for ecosystem services such as biodiversity, so-called land sharing, is evident, given the inability to protect enough lands to preserve all biodiversity. India's national monitoring efforts to assess trees outside forests (Forest Survey of India 2005) is a step toward addressing this need. Additional forest variables including vegetation structure and connectivity are integral yet unrealized aspects of monitoring to maintain ecosystem services.

1.5 Conclusion

Interest and investments in forest monitoring systems have risen sharply, mainly in anticipation of REDD. Monitoring systems at global, national, subnational, and community levels are all components of the interest in establishing monitoring systems. As these investments move forward, it is timely to consider the purposes of a monitoring system in terms of which land use-driven and ecological processes need to be captured and how the information can be used to track changes in ecosystem services.

Forests in different parts of the world are facing pressures from both economic and biophysical factors. For instance, tropical forests are under pressure from economic forces for agricultural expansion, while forests in high latitudes are moving northward due to climate change. Land use and forest transition frameworks provide a context to identify the processes affecting forests in varying paths along a development trajectory, with deforestation and degradation altering forests in early stages and regrowth in later stages with agricultural intensification and urbanization. From a biophysical point of view, ecological processes related to biome shifts from climate change, enhanced productivity from changing atmospheric chemistry, and fire are altering forest extent and biomass. Monitoring approaches vary depending on which processes are of interest. For example, a monitoring system to track human land use change would most effectively focus on frontier regions and less on wildlands. If the process of interest is productivity change, a comprehensive monitoring of biomass in wildlands is needed.

Approaches for monitoring systems also vary depending on which ecosystem services are of interest to the user. Forests provide a multitude of ecosystem services at a range of scales. Some services accrue benefits at proximal (e.g., forest products for local livelihoods), some downstream (e.g., watershed protection), and some at distal scales (e.g., carbon storage and biodiversity). Perspectives on which ecosystem services are most important depend on the user. Local communities are likely to place more importance on those ecosystem services of value to their needs while global policy makers are likely to place importance on global-scale, distal services.

Traditionally, forest monitoring and inventories have been designed around the commercial value of forests. With increasing emphasis on the value of forests for carbon storage, conservation of biodiversity, watershed protection, and a myriad of other ecosystem services, the focus for monitoring systems becomes more complex. Explicit consideration of the ecosystem services of interest and the methods required to monitor changes in those services require attention to design systems that are relevant for a country's circumstances. Advancements in technologies that enable monitoring of biomass, postclearing land use, forest structure, and other attributes are rapidly developing and offer a wide menu of possibilities for monitoring systems.

About the Contributor

Ruth DeFries is Denning Professor of Sustainable Development in the Department of Ecology, Evolution, and Environmental Biology at Columbia University, New York. Her research uses remote sensing and field data to analyze patterns of land use change and consequences for climate regulation, biodiversity, and other ecosystem services. A particular focus is tropical landscapes and the trade-offs and synergies between agricultural production and maintenance of ecosystem services and human well-being.

References

Agrawal A, Nepstad DC, and Chhatre A, Reducing emissions from deforestation and forest degradation. *Annual Review of Environment and Resources*, 36, 2011, 373–396.

Asner GP, Broadbent EN, Oliviera P, Keller M, Knapp DE, and Silva JNM, Condition and fate of logged forests in the Brazilian Amazon. *Proceedings of the National Academy of Sciences*, 103(34), 2006, 12947–12950.

Baltzer H, Tansey K, Kaduk J, George C, Gerard F, Gonzales MC, Sukhinin A, and Ponomarev E, Fire/climate interactions in Siberia. in Baltzer H, ed. *Environmental Change in Siberia: Earth Observation, Field Studies and Modelling*, vol. 40, *Advances in Global Change Research*. Dordrecht, The Netherlands: Springer, 2010, 21–36.

Beck P, Juday G, Alix C, Barber V, Winslow S, Sousa E, Heiser P, Herriges J, and Goetz S, Changes in forest productivity across Alaska consistent with biome shift. *Ecology Letters*, 14, 2011, 373–379.

Bobbink R, et al., Global assessment of nitrogen deposition effects on terrestrial plant diversity: A synthesis. *Ecological Applications*, 20(1), 2010, 30–59.

Bowman D, et al., Fire in the earth system. *Science*, 324, 2009, 481–484.

Chen Y, Randerson J, Morton D, DeFries R, Collatz GJ, Kasibhatla P, Giglio L, Jin Y, and Marlier M, Fire season severity in South America using sea surface temperature anomalies. *Science*, 334(6057), 2011, 787–791.

Davidson E, et al., The Amazon basin in transition. *Nature*, 481, 2012, 321–328.

DeFries R, Asner GP, and Houghton RA, Trade-offs in land-use decisions: Towards a framework for assessing multiple ecosystem responses to land use change. in DeFries R, Asner GP, and Houghton RA, eds. *Ecosystems and Land Use Change*, vol. 153. Washington, DC: American Geophysical Union, 2004, 1–12.

DeFries R, Rudel TK, Uriarte M, and Hansen M, Deforestation driven by urban population growth and agricultural trade in the twenty-first century. *Nature Geoscience*, 3, 2010, 178–181.

FAO, *Global Forest Resources Assessment 2010: Main Report*. Rome, Italy: Food and Agriculture Organization. 2010.

Forest Survey of India, *State of Forest Report 2005*. Dehradun, India: Ministry of Environment and Forests. 2005.

Galford GL, Melillo J, Kicklighter D, Cronin T, Cerri CEP, Mustard J, and Cerri CC, Estimating greenhouse gas emissions from land-cover and land-use change: Future scenarios of deforestation and agricultural management. *Proceedings of the National Academy of Sciences*, 107, 46, 2010.

Gibbs HK, Ruesch AS, Achard F, Clayton MK, Holmgren P, Ramankutty N, and Foley J, Tropical forests were the primary sources of new agricultural land in the 1980s and 1990s. *Proceedings of the National Academy of Sciences*, 107(38), 2010, 16732–16733.

Giglio L, Randerson JT, Van der Werf G, Kasibhatla P, Collatz GJ, Morton D, and DeFries R, Assessing variability and long-term trends in burned area by merging multiple satellite fire products. *Biogeosciences Discussion*, 7, 2010, 1171–1186.

Godfray H, Beddington J, Crute I, Haddad L, Lawrence D, Muir J, Pretty J, Robison S, Thomas S, and Toulmin C, Food security: The challenge of feeding 9 billion people. *Science*, 327, 2010, 812–818.

Gonzalez P, Neilson R, Lenihan J, and Drapek R, Global patterns in the vulnerability of ecosystems to vegetation shifts due to climate change. *Global Ecology and Biogeography*, 19, 2010, 755–768.

Hirota M, Nobre CA, Oyama M, and Bustamante M, The climatic sensitivity of the forest, savanna and forest-savanna transition in tropical South America. *New Phytologist*, 187(3), 2010, 707–719.

Houghton RA, Balancing the global carbon budget. *Annual Review of Earth and Planetary Sciences*, 35, 2007, 313–347.

INPE, PRODES DIGITAL. (December 30, 2006 http://www.obt.inpe.br/prodes/prodes_1988_2007.htm), 2007.

Johnston M and Holloway T, A global comparison of national biodiesel production potentials. *Environmental Science and Technology*, 41(23), 2007, 7967–7973.

Justice CO, Giglio L, Korontzi S, Owens J, Morisette J, Roy D, Descloitres J, Alleaume S, Petitcolin F, and Kaufman Y, The MODIS fire products. *Remote Sensing of Environment*, 83(1–2), 2002, 244–262.

Keesing F, et al., Impacts of biodiversity on the emergence on transmission of infectious disease. *Nature*, 468, 2010, 647–652.

Kurz WA, Dymond CC, Stinson G, Rampley GJ, Neilson ET, Carroll AL, Ebata T, and Safranyik L, Mountain pine beetle and forest carbon feedback to climate change. *Nature*, 452, 2008, 987–990.

Lambin E and Meyfroidt P, Global land use change, economic globalization, and the looming land scarcity. *Proceedings of the National Academy of Sciences*, 108(9), 2011, 3465–3472.

Lewis S, Lloyd J, Sitch S, Mitchard E, and Laurance W, Changing ecology of tropical forests: Evidence and drivers. *Annual Review of Ecology, Evolution, and Systematics*, 40, 2009, 529–549.

Mace G, Masundire H, and Baillie J, Biodiversity. in Hassan R, Scholes R, and Ash N, eds. *Millennium Ecosystem Assessment Working Group on Conditions and Trends*, vol 1. Washington, DC: Island Press, 2005.

Macedo M, DeFries R, Morton D, Stickler C, Galford G, and Shimabukuro Y, Decoupling of deforestation and soy production in the southern Amazon during the late 2000s. *Proceedings of the National Academy of Sciences*, 109(4), 2012, 1841–1846.

Mather AS, The forest transition. *Area*, 24, 1992, 367–379.

Mazoyer M and Roudart L, *A History of World Agriculture: From the Neolithic Age to the Current Crisis*. New York: Monthly Review Press, 2006.

Miles L, Newton AC, DeFries R, Ravilious C, May I, Blyth S, Kapos V, and Gordon J, A global overview of the conservation status of tropical dry forests. *Journal of Biogeography*, 33, 2006, 491–505.

Millennium Ecosystem Assessment, *Ecosystems and Human Well-Being: Synthesis*. Washington, DC: Island Press, 2005.

Mooney HA, Cropper A, and Reid WV, Confronting the human dilemma: How can ecosystems provide sustainable services to benefit society? *Nature*, 434, 2005, 561–562.

Mustard J, DeFries R, Fisher T, and Moran EF, Land use and land cover change pathways and impacts. in Gutman G, Janetos J, Justice CO, Moran EF, Mustard J, Rindfuss R, Skole DL, Turner BL, and Cochrane MA, eds. *Land Change Science: Observing, Monitoring, and Understanding Trajectories of Change on the Earth's Surface*. Dordrecht, The Netherlands: Springer, 2004.

Nelleman C, MacDevette M, Manders T, Eickhout B, Svihus B, Prins A, and Kaltenborn B, *The Environmental Food Crisis: The Environment's Role in Averting Future Food Crises*. GRID-Arendal: United Nations Environment Program. 2009.

Postel S and Thompson B, Watershed protection: Capturing the benefits of nature's water supply services. *Natural Resources Forum*, 29, 2005, 98–108.

Royal Society of London, *Reaping the Benefits: Science and the Sustainable Intensification of Global Agriculture*. London: Royal Society of London. 2009.

Rudel TK, Coomes OT, Moran EF, Achard F, Angelsen A, Xu J, and Lambin E, Forest transitions: Towards a global understanding of land use change. *Global Environmental Change*, 15, 2005, 23–31.

Saatchi S, Houghton RA, Alvala RC, Soares J, and Yu Y, Distribution of aboveground live biomass in the Amazon Basin. *Global Change Biology*, 13(4), 2006, 816–837.

Scholes R, Hassan R, and Ash N, Summary: Ecosystems and their services around the year 2000. in Hassan R, Scholes R, and Ash N, eds. *Ecosystems and Human Well-Being: Current State and Trends*, vol 1. Washington, DC: Island Press, 2005, 2–23.

Souza Jr. CM, Roberts D, and Cochrane MA, Combining spectral and spatial information to map canopy damages from selective logging and forest fires. *Remote Sensing of Environment*, 98, 2005, 329–343.

Stokstad E, Taking the pulse of the Earth's life-support systems. *Science*, 308, 2005, 41–43.

Thompson I, Okabe K, Tylianakis J, Kumar P, Brockerhoff E, Schellhorn N, Pattotta J, and Nasi R, Forest biodiversity and the delivery of ecosystem goods and services: Translating science into policy. *BioScience*, 61(12), 2011, 972–981.

van der Werf GR, et al., Climate regulation of fire emissions and deforestation in equatorial Asia. *Proceedings of the National Academy of Sciences*, 105(51), 2008, 20350–20355.

Wade T, Riitters K, Wickham J, and Jones KB, Distribution and causes of global forest fragmentation. *Conservation Ecology*, 7(2), 2003, 7.

Walker R, Deforestation and economic development. *Canadian Journal of Regional Science*, 16(3), 1993, 481–497.

White P and Ward A, Interdisciplinary approaches for the management of existing and emerging human-wildlife conflicts. *Wildlife Research*, 37(8), 2010, 623–629.

Williams M, *Deforesting the Earth: From Prehistory to Global Crisis*. Chicago, IL: The University of Chicago Press, 2006.

2

Role of Forests and Impact of Deforestation in the Global Carbon Cycle

Richard A. Houghton

Woods Hole Research Center

CONTENTS

2.1   Introduction

Forests are important in the global carbon cycle because they hold in their vegetation and soils about as much carbon as is held in the atmosphere (Table 2.1), and, with an annual GPP of 65 PgC yr^{-1} (Beer et al. 2010), forests circulate about 8% of the atmosphere's carbon each year through photosynthesis and respiration. These exchanges are part of the natural carbon cycle. More important from the perspective of climate change is the role that forests play in altering the concentration of atmospheric CO_2 over decades to centuries. This chapter discusses forests in that role. It begins with a brief review of the global carbon cycle and goes on to discuss, first, the global carbon sink measured in forest inventories, second, sources and sinks of carbon that result from direct human use of forests, and, third, possible

TABLE 2.1

Stocks and Flows of Carbon

Carbon Stocks (PgC)	(Forests)
Atmosphere	825
Land	2,000
Vegetation	500 (436)
Soil	1,500 (426)
Ocean	39,000
Surface	700
Deep	38,000
Fossil fuel reserves	10,000
Annual Flows (PgC yr^{-1})	
Atmosphere–oceans	90
Atmosphere–land	120 (65)
Net Annual Exchanges (PgC yr^{-1} Averaged over 2000–2009)	
Fossil fuels	7.7
Land use change	1.1 (1.0)
Atmospheric increase	4.1
Oceanic uptake	2.3
Residual terrestrial sink	2.4 (2.4)

reasons why the results from inventories and analyses of land use change do not agree. The chapter ends with a discussion of the processes affecting carbon storage on land that are and are not amenable to monitoring with satellites.

Note that forests affect climate through emissions of chemically and radiatively active gases other than CO_2, including other carbon compounds. Further, changes in forest area affect climate biogeophysically as well as biogeochemically through effects on albedo, surface roughness, and evapotranspiration (e.g., Pongratz et al. 2010). Non-CO_2 gases and biophysical effects are not considered here.

2.2   Global Carbon Cycle

The global carbon cycle is the exchange of carbon between the four major reservoirs: atmosphere, oceans, land, and fossil fuels. This chapter, and most of carbon cycle science, is concerned with anthropogenic carbon, that is, the amount of carbon emitted each year from combustion of fossil fuels and land use change and the sinks for that carbon in the atmosphere, oceans, and land. Forests play a major role in both the emissions of carbon from land use change and the sinks of carbon on land.

Figure 2.1 shows the annual sources and sinks of carbon in the major global reservoirs over the last century and a half. The most noticeable

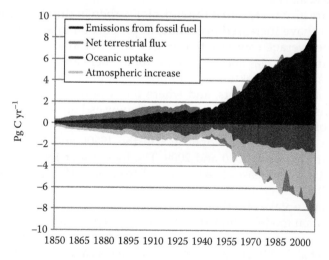

FIGURE 2.1
Annual sources (+) and sinks (–) in the global carbon budget. Note that the net terrestrial flux was consistently a net source before 1940, but has been a variable and growing sink in recent decades.

TABLE 2.2

Global Carbon Budget

	1980s	1990s	2000–2009
Fossil fuel emissions	5.5 ± 0.3	6.4 ± 0.4	7.7 ± 0.5
Land use change	1.5 ± 0.7	1.6 ± 0.7	1.1 ± 0.7
Atmospheric increase	−3.4 ± 0.1	−3.1 ± 0.2	−4.1 ± 0.1
Oceanic uptake	−2.0 ± 0.6	−2.2 ± 0.7	−2.3 ± 0.4
Residual terrestrial sink	−1.6 ± 1.0	−2.7 ± 1.0	−2.4 ± 1.0

Source: From Le Quéré, C., et al., *Nature GeoSci.*, 2, 831, 2009 and http://www.globalcarbonproject.org/carbonbudget/09/files/GCP2010_CarbonBudget2009.pdf.

Notes: Units are PgC yr^{-1}. Positive values indicate sources of carbon to the atmosphere; negative values indicate sinks, or removals from the atmosphere.

feature of the history is the increasing rate at which carbon has been emitted from combustion of fossil fuels (including cement production and gas flaring). In recent decades, the emissions have grown from 5.5 PgC yr^{-1} averaged for the 1980s to 6.4 PgC yr^{-1} for the 1990s to 7.7 PgC yr^{-1} over the period 2000–2009 (Table 2.2). After a slump in 2009 from the global financial crisis, fossil fuel emissions were above 9 PgC in 2010 (Peters et al. 2012). The annual emissions from fossil fuels are calculated from reports from the United National Energy Statistics. The error is thought to be ±6% (Le Quéré et al. 2009).

The figure also reveals that the sinks for carbon in the atmosphere, land, and oceans have increased over time, in proportion to annual emissions. In 1958 the average concentration of CO_2 in air at Mauna Loa was about 315 ppm; in 2010 it was about 390 ppm. Today there are nearly 200 stations, worldwide, where weekly flask samples of air are collected, analyzed for CO_2 and other constituents, and where the resulting data are integrated into a consistent global data set (http://www.esrl.noaa.gov/gmd/ccgg/). The rate of increase in concentrations averaged about 1 ppm yr^{-1} in the 1950s and 1960s, about 1.5 ppm yr^{-1} in the 1980s and 1990s, and about 1.9 ppm yr^{-1} between 2000 and 2009. The increase of 1.9 ppm CO_2 yr^{-1} is equivalent to an increase of ~4 PgC yr^{-1}. The error is 0.04 PgC yr^{-1} (Canadell et al. 2007).

The annual uptake of carbon by the world's oceans is based on ocean general circulation models coupled to ocean biogeochemistry models (Le Quéré et al. 2009), corrected to agree with the observed uptake rates over 1990–2000 (Canadell et al. 2007). The error in the modeled oceanic sink is thought to be 0.4 PgC yr^{-1}.

There are no direct measurements of terrestrial sources or sinks globally. Instead, the annual net exchange of carbon between land and the atmosphere is calculated by the difference between the annual release of carbon from fossil fuels and the annual accumulations in the atmosphere and oceans. The total emissions must balance the total sinks. The net terrestrial flux of carbon was a small source before 1940 and a sink after. That sink is variable year to year and appears to have grown in recent decades. It averaged 1.3 PgC yr^{-1} between 2000 and 2009. The role of forests in the historic source of carbon and the more recent sink is the topic of this chapter.

2.3 Forest Inventories

A recent paper by Pan et al. (2011) summarized the results of measurements obtained through forest inventories. Countries in temperate zone and boreal regions have systematic forest inventories that periodically measure the volumes of timber. Biomass and carbon densities can be calculated from these measurements of wood volume. The inventories often include measurement of belowground carbon stocks and coarse woody debris on the forest floor, and estimates are also made of the storage of carbon in wood products and land fills. Because nearly all forests are sampled in these inventories, the change in carbon storage from one inventory to another represents the total change in forest carbon, including wood products—a net sink in temperate and boreal forests of 1.22 PgC yr^{-1} averaged over the period 2000–2007 (Table 2.3).

This measured sink is a net sink composed of both releases of carbon from fire, storms, disease, and logging and uptake of carbon in growing forests. It is worth noting that the sampling used to obtain these estimates is arguably better for measuring changes in wood volume in existing forests than it is for measuring changes in forest area. A satellite-based approach might provide more accurate estimates of changes in forest area.

The net sink for the world's temperate zone and boreal forests does not mean that all such forests were sinks. Canadian forests, for example, were a small source over 1990–2007, and European forests were a net source over 2000–2007, according to analyses of forest inventories (Pan et al. 2011). Furthermore, studies based on analyses of satellite data suggest that forest area has been declining, for example, in the eastern United States (Drummond and Loveland 2010; Jeon et al. 2011).

Systematic inventories of forests are rare in tropical countries. However, small permanent plots (generally ~1 ha) have been inventoried for years in the unmanaged, or intact, forests of Amazonia (Phillips et al. 2004, 2008) and Africa (Lewis et al. 2009). These inventories show an average net accumulation of 0.84 MgC/ha yr^{-1} in biomass. The total area of tropical forests in 2010 was 1949 million ha (FAO 2010), but the area of intact

TABLE 2.3

Average Annual Net Source (+) or Sink (–) for Carbon Based on (A) Forest Inventories and (B) LULCC

	1980s	1990s	2000–2007
A. Forest Inventories			
Temperate and boreal forests (a)	–1.17 ± 0.11	–1.28 ± 0.12	–1.22 ± 0.11
Intact tropical forests (b)	–1.33 ± 0.35	–1.02 ± 0.47	–1.19 ± 0.41
Total	–2.50	–2.30	–2.41
B. LULCC			
Temperate and boreal forests			
Gross uptake (c)	–1.38	–1.48	–1.56
Gross emissions	1.51	1.53	1.52
Net flux	0.13	0.05	–0.04
Tropical forests			
Gross uptake (d)	–1.57 ± 0.50	–1.72 ± 0.54	–1.64 ± 0.52
Gross emissions	3.03 ± 0.49	2.82 ± 0.45	2.94 ± 0.47
Net tropical LULCC flux (e)	1.46 ± 0.70	1.10 ± 0.70	1.30 ± 0.70
Net flux for tropical forest (b + e)	0.13	0.08	0.11
Net global forest sink (a + b + e)	–1.04 ± 0.79	–1.20 ± 0.85	–1.11 ± 0.82
Gross global forest sink[a] (a + b + d)	–4.07	–4.02	–4.05
Gross global forest sink[b] (a + b + c + d)	–5.45	–5.50	–5.61

Source: From Pan, Y., et al., *Science* 333, 988, 2011.
[a] As reported by Pan et al. (2011).
[b] With gross uptake in temperate and boreal forests (c) included.

forests, for which this average accumulation applies, was smaller. At least 557 million ha of forest are estimated by Houghton (2010, unpublished data; global results reported in Friedlingstein et al. 2010) to be managed, that is, recovering from wood harvest or in the fallow portion of shifting cultivation. Subtracting this area of managed forests from the total area of tropical forests yields the area of intact forests (1,392 million ha) and thus a net carbon sink in these unmanaged tropical forests of 1.19 PgC yr^{-1} (Table 2.3).

The carbon sink in the world's inventoried forests was 2.4 PgC yr^{-1} (1.22 in temperate zone and boreal forests and 1.19 PgC yr^{-1} in the unmanaged forests of the tropics). In contrast, the net terrestrial sink calculated from the global carbon balance (Section 2.2) was 1.3 PgC yr^{-1} in the same period (1990–2007). The difference implies a source of 1.1 PgC yr^{-1} either in ecosystems other than forests or in the managed forests of the tropics not included in the inventories. The source/sink for managed tropical forests is determined from an analysis of land use change, described below (Houghton 2010, unpublished data).

2.4 Land Use Change (Direct Anthropogenic Effects)

Managed lands, or those lands directly affected by land use and land cover change (LULCC), can lead to either sources or sinks of carbon, and many analyses of LULCC have attempted to estimate those sources and sinks. "Land use" refers to management within a land cover type. For example, the harvest of wood does not change the designation of the land as forest although the land may be temporarily treeless. "Land cover change," in contrast, refers to the conversion of one cover type to another, for example, the conversion of forest to cropland. Note that "deforestation" as used in this chapter refers to the conversion of forest to another land cover. Logging, even clear-cut logging, is not deforestation unless it is followed by a land use without forest cover, for example, cropland.

Ideally, LULCC would be defined broadly to include not only human-induced changes in land cover, but all forms of land management (e.g., techniques of harvesting). The reason for this broad ideal is that the net flux of carbon attributable to management is that portion of a terrestrial carbon flux that might qualify for credits and debits under a post-Kyoto agreement. However, it is perhaps impossible to separate management effects from natural and indirect effects (e.g., CO_2 fertilization, N deposition, or the effects of climate change). Furthermore, the ideal requires more data, at higher spatial and temporal resolutions, than have been practical (or possible) to assemble at the global level. Thus, most analyses of the effects of LULCC on carbon storage have focused on the dominant (or documentable) forms of management and, to a large extent, ignored others.

Recent estimates of the flux of carbon from LULCC are shown in Figure 2.2. Most of these emissions in recent decades have been from the tropics, while the net annual flux of carbon from regions outside the tropics has been nearly zero (Houghton 2010, unpublished data). This near neutrality does not indicate a lack of activity outside the tropics. Rather, the sources of carbon from wood harvest are largely balanced by the sinks in regrowing forests harvested in previous years. Annual gross emissions and rates of uptake from LULCC are nearly as great in temperate and boreal regions as they are in the tropics (Richter and Houghton 2011). Rates of wood harvest, for example, are nearly the same in both regions. The main difference between the two regions is that forests are being lost in the tropics, while forest area has been expanding in Europe, China, and the United States.

The global net flux of carbon from LULCC based on these estimates is approximately 1.0 PgC yr^{-1} for the last three decades and 1.1 PgC yr^{-1} for the years 2000–2009 (Houghton 2010, unpublished data). Forests accounted for 90%–95% of this net source, and the global carbon budget is essentially balanced: the emissions from LULCC in the tropics (1.3 PgC yr^{-1}) are more than offset by a sink in the forests of all regions (2.4 PgC yr^{-1}) as determined from forest inventories (see more details in Section 2.4.3).

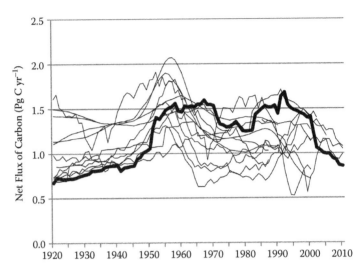

FIGURE 2.2
Recent estimates of the net emissions of carbon from land use and land cover change (LULCC). Houghton's estimate (2010, unpublished data), which is used as an example throughout this chapter, is highlighted.

The discussion below focuses on identifying the reasons underlying differences among the many estimates in Figure 2.2. Differences are grouped into two major categories: (1) data for rates of LULCC and carbon density and (2) the types of LULCC processes included in the analyses.

2.4.1 Data Used to Define Changes in Forest Area and Carbon Density

All of the approaches for calculating the emissions of carbon from LULCC consider the areas affected (e.g., deforested or reforested) and the emission coefficients (carbon lost or gained per hectare following a change in land management). The approaches differ, first, in the data used to define changes in the areas of croplands and pastures; and, second, in the way carbon stocks and changes in carbon stocks are estimated (some are modeled; others are specified from observations).

A significant difference among approaches is the spatial resolution of the analysis. The nonspatial approach of bookkeeping models (e.g., Houghton 2010, unpublished data) cannot represent the spatial heterogeneity of biomes, and thus the emissions calculated with mean carbon densities for large regions may be biased. In contrast, spatially explicit information on changes in forest area, especially when combined with spatially explicit estimates of biomass density, should provide more accurate estimates of the carbon emissions from LULCC. Compared with nationally aggregated estimates of change used in bookkeeping models, spatially explicit data reduce uncertainties by identifying where and which forests types have undergone change.

As biomass density can vary substantially within a country and across forest types, satellite data provide a clear benefit. The spatial colocation of deforestation with carbon density will greatly improve the precision of carbon emissions, including the sources and sinks from ecosystems not directly affected by land use or land cover change (Houghton and Goetz 2008).

Note that although process-based models are spatially explicit (Pongratz et al. 2009; Shevliakova et al. 2009), the historical data for simulating land cover change rarely are. Maps, at varying resolutions, exist for many parts of the world, but only during the satellite era (Landsat began in 1972) are spatial data on land cover change available, in theory. In fact, there are many holes in the coverage of the earth's surface until 1999 when the first global acquisition strategy for moderate spatial resolution data was undertaken with the Landsat Enhanced Thematic Mapper Plus Sensor (Arvidson et al. 2001). The long-term acquisition plan of Landsat ETM+ data ensures annual global acquisitions of the land surface. However, cloud cover and phenological variability limit the ability to provide annual global updates of forest extent and change. The only other satellite system to provide global coverage of the land surface is the ALOS PALSAR instrument, which also includes an annual acquisition strategy for the global land surface (Rosenqvist et al. 2007).

Remote sensing–based information on recent land cover change has been combined with regional statistics, such as from FAO, to reconstruct spatially explicit land cover reconstructions covering more than the satellite era (Ramankutty and Foley 1999; Goldewijk 2001; Pongratz et al. 2008). Historical changes in LULCC are important for today's sources and sinks of carbon because the emissions of carbon from deforestation are not instantaneous. Woody debris generated at the time of disturbance may take decades to decompose. Similarly, the uptake of carbon by secondary forests continues for decades and centuries after these forests begin to grow. In the absence of spatial data on biomass density, the long-term history of LULCC is necessary to simulate changes in biomass density resulting from management. The biomass density of forests cleared for agriculture today depends, in large part, on how long those forests have had to recover from previous harvests. On the other hand, if spatial estimates of biomass density are obtained directly, documentation of disturbance history may no longer be required.

2.4.2   Other Differences among Estimates of Carbon Emissions from Land Use Change

Besides differences in data used to estimate deforestation rates and carbon density, the variability in flux estimates also results from the types of land use included. All of the analyses in Figure 2.2 included deforestation, either by using satellite data on forest cover or by inferring changes in forest area by combining data on the expansion and abandonment of agricultural area (cropland and pasture) with information on the distribution of natural vegetation.

Forest degradation: Some of the estimates in Figure 2.2 also included forest management, wood harvest, or other management practices that change the carbon density within forests. The reduction in biomass density within forests as a result of land use is defined here as degradation. Logging in Amazonia, for example, added 15%–19% to the emissions of carbon from deforestation alone (Huang and Asner 2010). For all the tropics, harvests of wood and shifting cultivation, together, added 28% to the net emissions calculated on the basis of land cover change alone (Houghton 2010, unpublished data). Globally, these rotational uses of land added 32%–35% more to the net emissions from deforestation (Shevliakova et al. 2009). Thus, those analyses that have included wood harvest and shifting cultivation yield higher, and presumably more comprehensive, estimates of the net emissions from LULCC.

Indirect anthropogenic effects: While bookkeeping models use rates of growth and decay that are fixed for different types of ecosystems, process-based models simulate the processes of growth and decay as a function of climate variability and trends in atmospheric composition. Because effects are partly compensating (e.g., deforestation under increasing CO_2 leads to higher emissions because CO_2 fertilization increases carbon stocks, but regrowth is also stronger as CO_2 fertilization has a more pronounced impact on regrowing than on mature forest), a CO_2 fertilization effect is not likely to be a major factor in accounting for differences among emission estimates. In one study, the combined effect of changes in climate and atmospheric composition increased LULCC emissions by about 8% over the industrial era (Pongratz et al. 2009). There are doubtlessly other interactions as well between environmental changes and management. These interactions make attribution difficult; that is, are the sources and sinks the result of management or the indirect effect of environmental change?

There is another (indirect) effect of deforestation. As forests are lost, the sink capacity on land is diminished. This effect has been called the "net land use amplifier effect" (Gitz and Ciais 2003) and the "loss of additional sink capacity" (Pongratz et al. 2009). In models, the strength of this effect depends on the atmospheric CO_2 concentration. These indirect effects account for a portion of the variability among emission estimates.

2.4.3   Sources and Sinks of Carbon from Land Use Change

The sources and sinks of carbon from LULCC are significant in the global carbon budget (Table 2.2). Globally, the annual emissions of carbon from LULCC were larger than the emissions from fossil fuels until ~mid twentieth century. Since ~1945, the emissions from fossil fuels have increased dramatically, while the emissions from land use have remained nearly constant at 1–1.5 PgC yr^{-1}. Thus the contribution of LULCC to anthropogenic carbon emissions has varied from about 33% of total emissions over the last

150 years (Houghton 1999) to about 12% in 2008 (van der Werf et al. 2009). The declining fraction is largely the result of the accelerated rise in fossil fuel emissions.

It is important to note that these emissions from LULCC are net emissions. They include both sources and sinks of carbon from land use—sources when forests are converted to croplands or pastures and sinks when forests regrow following harvest or following abandonment of croplands or pastures. In fact, the gross sources and sinks from land use and recovery are two to three times greater than the net source (Richter and Houghton 2011). The error associated with the net flux of carbon from LULCC is thought to be ±0.7 PgC yr^{-1} (Le Quéré et al. 2009).

It should be clear that the net flux of carbon from LULCC is not equivalent to the "emissions of carbon from deforestation," although the terms are used interchangeably in the literature. The former includes other forms of management besides deforestation, for example, degradation of forests. Further, the net flux of carbon from LULCC includes sources and sinks of carbon from nonforests. Cultivation of prairie soils, for example, results in a loss of soil carbon unrelated to forests. Over the last 150 years, forests accounted for between 84% and 96% of the annual net flux from LULCC. The fraction has varied through history; in recent decades forests have accounted for 90%–95%.

2.4.3.1   Land Use Change in Tropical Forests

Recall that managed forests were not included in the forest inventories of the tropics (Section 2.3). The net carbon balance for managed forests was determined by simulating LULCC, specifically deforestation for crops, pasture, and shifting cultivation; reforestation following abandonment of these land uses; and harvest of wood products (Houghton 2010, unpublished data). LULCC in the tropics is estimated to have caused a net source of 1.3 (±0.7) PgC yr^{-1} over the period 1990–2007. The gross emissions were 2.9 PgC yr^{-1} (from deforestation and harvests); gross uptake in secondary forests averaged 1.6 ± 0.5 PgC yr^{-1} (Table 2.3).

2.4.3.2   Land Use Change in Boreal and Temperate Zone Forests

The forest inventories of boreal and temperate zone forests included both managed and unmanaged forests and thus provide enough information to determine the net effect of forests in the carbon cycle. This inventory-based estimate of flux is very different from the flux determined from analysis of LULCC. The net sink obtained from forest inventories was 1.22 PgC yr^{-1} over the period 2000–2007 (Table 2.3). In contrast, the net sink obtained from LULCC was nearly zero (a net sink of 0.04 PgC yr^{-1}), with gross emissions of 1.52 PgC yr^{-1} and a gross sink of 1.56 PgC yr^{-1} (Houghton 2010, unpublished data). The major reason for the large difference in the two estimates of the sink, aside from errors, is believed to be that forests are accumulating carbon in response to environmental changes, and these environmental responses are not included in Houghton's (2010, unpublished data) bookkeeping model (see Section 2.5.1.2).

2.4.3.3 Global Summary of LULCC

The world's forests were a net sink of 1.1 PgC yr^{-1} over the period 2000–2007 (Pan et al. 2011) (Table 2.3). This net sink includes a source of 1.3 PgC yr^{-1} from deforestation and harvests (LULCC) and a sink of 2.4 PgC yr^{-1} measured in forest inventories. These estimates yield a balanced global carbon budget. The net terrestrial sink (1.3 PgC for the period 1990–2009) is approximately equal to the net sink in forests (1.1 PgC yr^{-1}).

The gross uptake of carbon by the world's forests was estimated by Pan et al. (2011) to be 4.05 ± 0.67 PgC yr^{-1} (2.41 in intact forests and 1.64 in managed forests in the tropics). But this estimate of a gross uptake is an underestimate because the sink of 1.22 PgC yr^{-1} in temperate zone and boreal forests is a net sink, not a gross sink. Adding the gross uptake in these forests, obtained from LULCC (Houghton 2010, unpublished data), yields a gross uptake of 5.61 PgC yr^{-1} (4.05 + 1.56) for the world's forests.

2.5 Global Carbon Cycle Revisited: Residual Terrestrial Sink

The source of carbon from LULCC explains a part of the net terrestrial carbon flux and, thereby, helps define a different residual terrestrial flux (Figure 2.3). Figure 2.3 is similar to Figure 2.1 except the net terrestrial flux of Figure 2.1 has been broken into a net flux from land use change (always a net source historically) and a terrestrial residual flux. The residual flux is calculated by difference, just as the net terrestrial flux was calculated by difference in Figure 2.1. It is noteworthy that the net terrestrial flux and the LULCC flux were approximately equal before ~1925. Before this date the LULCC flux *was* the net terrestrial flux. Only in recent decades has there been another terrestrial sink unexplained by LULCC. It should be recognized that terrestrial carbon models calculate an annual carbon sink consistent with the sink calculated by difference (Le Quéré et al. 2009). Differences among estimates for the future, however, suggest that those models are not reliable enough to predict the future terrestrial carbon sink/source (Cramer et al. 2001; Friedlingstein et al. 2006).

In sum, forests account (1) for 90%–95% of the net emissions from LULCC and (2) for nearly all the residual terrestrial sink (Pan et al. 2011). Thus, forests are important, both as a source of carbon to the atmosphere from human activity and as a sink for carbon through natural processes not entirely understood. Obviously, forest management can be used, and is, to

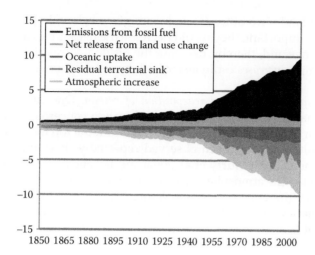

FIGURE 2.3
Annual sources (+) and sinks (−) in the global carbon budget. The terrestrial flux is partitioned into a flux from land use change and a residual terrestrial sink.

accumulate carbon on land (the gross sink from LULCC, globally, is about 3 PgC yr^{-1}) (Richter and Houghton 2011), but the emissions from deforestation have dominated the effects of management to date.

2.5.1   What Explains the Residual Terrestrial Sink?

The residual terrestrial sink incorporates all of the errors from the other terms in the global carbon budget and has an error on the order of 1 PgC yr^{-1}. The analysis of data from forest inventories suggests a net sink of 2.4 PgC yr^{-1} over the period 1990–2007 that was presumably driven by some combination of processes, some already considered in analyses of land use change and others not considered. The sections below consider potential carbon sinks driven by processes not yet included in analyses of land use change.

Aside from cumulative errors, the residual terrestrial sink may be attributed to two types of explanations: (1) omissions of management practices from analyses of LULCC and (2) factors other than management that affect terrestrial carbon storage.

2.5.1.1   *Management Effects Not Included in Analyses of Land Use Change*

Before discussing aspects of management that may account for the residual terrestrial sink, it is important to recall that the residual flux does *not* include the sinks of carbon in forests regrowing as a result of direct activity (logging, abandonment, etc.). These sinks are part of the global carbon source from LULCC.

Management activities not included in analysis of land use change (e.g., use of fertilizers in forest management) may have increased the storage of carbon on land. Two other examples are given below. To the extent these processes are important, they would decrease the net source calculated from land use change and, thereby, decrease the residual terrestrial sink, as well. A third example *increases* estimates of both terrestrial fluxes.

Aquatic transport: Erosion and redeposition of carbon: One uncertainty with respect to changes in soil carbon with cultivation concerns the fate of carbon lost from soil. A 25%–30% loss of carbon from the top meter in the years following cultivation has been observed repeatedly (Post and Kwon 2000; Guo and Gifford 2002; Murty et al. 2002) and is generally assumed to have been released to the atmosphere. However, some of it may have been moved laterally to a different location (erosion). Much of the transported carbon may be released to the atmosphere through subsequent decomposition, either during transport or once incorporated in sediment. If so, the loss of carbon was counted in analyses of land use change. However, if the organic carbon settles in anaerobic environments and decomposition is inhibited, the carbon will be sequestered, at least temporarily.

The carbon discharged to the oceans is only a fraction of the carbon entering rivers from terrestrial ecosystems by way of soil respiration, leaching, chemical weathering, and physical erosion. Although most of the carbon is released to the atmosphere in transport, as much as 0.6 PgC may be buried in the sediments of floodplains, lakes, reservoirs, and wetlands (Berhe et al. 2007; Tranvik et al. 2009; Aufdenkampe et al. 2011). If the sink includes some of the observed loss of carbon from the top meter of soil, then the emissions of carbon to the atmosphere from land use change have been overestimated. The estimated sink from erosion/deposition is large, responsive to both land use change and changes in climate, and ought to be considered in the global carbon balance. Furthermore, this buried carbon is important as a potential source of methane. Freshwater ecosystems release an estimated 0.1 PgC yr^{-1} as methane. The carbon emissions are small, but the radiative emissions are enough to account for 25% of the estimated terrestrial sink (Bastviken et al. 2011).

Woody encroachment: Another possible explanation for the residual sink is "woody encroachment." The expansion of trees and woody shrubs into herbaceous lands, although it cannot be attributed definitively to natural, indirect (climate, CO_2), or direct effects (fire suppression, grazing), is, nevertheless, happening in many regions. Scaling it up to a global estimate is problematical, however (Archer et al. 2001), in part because the areal extent of woody encroachment is unknown and difficult to measure. Also, the increase in vegetation carbon stocks observed with woody encroachment is in some cases offset by losses of soil carbon (Jackson et al. 2002). Finally, woody encroachment may be offset by its reverse process, woody elimination, an example of which is the fire-induced spread of cheatgrass (*Bromus tectorum*) into the native woody shrub lands of the Great Basin in the western United States (Bradley et al. 2006).

The net effect of woody encroachment and woody elimination is, thus, uncertain, not only with respect to net change in carbon storage, but also with respect to attribution. It may be an unintended effect of management, or it may be a response to indirect or natural effects of environmental change.

Emissions from draining and burning of peatlands: Not all of the processes left out of analyses of land use change would reduce the net carbon source if they were included. Some processes act to increase the emissions and increase the residual terrestrial sink as well. One such process is the draining and burning of tropical swamp forests for the establishment of oil palm plantations in Southeast Asia. This use of land is thought to add 0.3 PgC yr^{-1} to the net emissions of carbon from land use change (Hooijer et al. 2010). The elevated carbon emissions from these and other wetlands have not been included in global estimates of emissions from land cover change.

2.5.1.2   Indirect and Natural Effects (Processes Not Directly Related to Management)

Two other processes besides the direct effects of management (LULCC) account for changes in terrestrial carbon storage: indirect effects (rising concentrations of CO_2, deposition of reactive nitrogen, climate change) and natural effects, including changes in disturbance regimes (Marlon et al. 2008).

Effects of CO_2, N deposition, and climate change on carbon storage of forests (indirect anthropogenic effects): Three environmental factors are generally thought to explain increases in plant productivity and, thereby, carbon storage: CO_2 fertilization, nitrogen deposition, and changes in climate (Schimel 1995). Increased concentrations of CO_2 are thought to have caused increased biomass density in tropical forests (Lewis et al. 2004). Nitrogen deposition is believed to be especially important in the northern mid latitudes (Thomas et al. 2010). And changes in temperature and moisture are important, particularly through earlier and longer growing seasons. Competition among these factors to explain the residual terrestrial sink has existed for nearly as long as the sink has been recognized. The relative strengths are unknown.

Changes in disturbance regimes: Natural disturbance regimes (including recovery) may themselves change over decades or centuries, causing carbon to accumulate during some periods and to be lost during others (Marlon et al. 2008; Wang et al. 2010). A reduction in disturbances over the last decades may have shifted more forests to a phase of recovery with attendant sinks. It must be noted, however, that in many regions the effects of climate change (droughts and fires) appear to have caused additional carbon to be lost rather than accumulated (Gillett et al. 2004; Westerling et al. 2006; Kurz et al. 2008). Apparently the increased releases of carbon from fires, storms, diseases, and logging are offset by regrowth or enhanced growth elsewhere.

2.5.2   Sources and Sinks of Carbon in the Net Residual Terrestrial Sink

Like the net source of carbon from LULCC, the residual terrestrial sink is also a *net* sink, including both sources and sinks of carbon. Its existence today does not imply that it will continue to grow, or that it will continue at all. Model experiments suggest that the drying effects of a warmer climate may cause dieback of tropical forests in Amazonia (Cox et al. 2000), a prediction looking more reasonable after two 100-year droughts occurred there in the last decade (Phillips et al. 2009; Lewis et al. 2011). In boreal forests, too, not only have fires increased in recent decades (Stocks et al. 2003; Kasischke and Turetsky 2006; Westerling et al. 2006), but the productivity of the forests, at first observed to have increased, has declined since ~1990 (Goetz et al. 2007), most likely in response to drought stress. And an unusually large fire in the Alaskan tundra (Mack et al. 2011) may foreshadow increased sources of carbon from those ecosystems too.

2.5.3   Is the Residual Terrestrial Sink Changing? Or Will It Change?

Remarkably, the *proportions* of anthropogenic carbon emissions (fossil fuel and land use change) taken up by the atmosphere, oceans, and land have changed little in the last 50 years. In other words, the annual accumulations of carbon on land and in the oceans have increased in proportion to emissions. Over the years 2000–2009, the annual emissions from fossil fuels and land use change accumulated in the atmosphere (~47%), the oceans (~26%), and land (~27%) (Table 2.2). There is little sign of any saturation of these sinks. Some scientists argue that the airborne fraction (the increase in the atmosphere divided by total emissions) has increased slightly, suggesting that the sinks may be beginning to saturate (Canadell et al. 2007; Le Quéré et al. 2009), but others argue that that increase cannot be observed against the year-to-year variability in the airborne fraction and the uncertainty of the land use flux (Knorr 2009).

There are other problems with interpreting the airborne fraction. Changes in the airborne fraction may be influenced by the nonlinear responses of oceanic uptake to changes in the rate of emissions (Gloor et al. 2010). The oceanic sink is not determined by a single carbon reservoir that mixes infinitely fast, as assumed in the linear analyses. Rather, variations in the "CO_2 sink rate," if calculated with a single-box model, will result from variations in the growth rate of the sources, with no change in the rate constants of ocean mixing. The land and ocean sinks may, indeed, be slowing, but demonstrating it through observations of the airborne fraction will be difficult.

2.6 Which Sources and Sinks of Carbon Are Observable from Space?

Data from satellites have been used successfully to measure changes in forest area, but it has been more difficult to determine from satellite data alone whether those changes are anthropogenic or not, and, if they are, whether they represent a land cover change (e.g., conversion of forest to cropland) or a land use (logging and subsequent recovery).

Aside from changes in forest area, however (and changes in area are the changes that involve the greatest changes in carbon), there are other issues that need attention. This chapter concludes with a discussion of three questions:

- Can changes in terrestrial carbon be measured from space?
- Can the net carbon balance of terrestrial ecosystems be more easily measured if sources and sinks are unevenly distributed?
- Can losses and gains of terrestrial carbon be attributed to direct management, as opposed to indirect environmental effects?

2.6.1 Can Changes in Terrestrial Carbon Be Measured from Space?

For aboveground woody biomass, although different methods have yielded wildly different estimates for large regions in the past (Houghton et al. 2001), new satellite-based methods look promising (Hall et al. 2011; Le Toan et al. 2011). Mapping change in biomass density over large regions is in its infancy, and testing maps over large areas remains a challenge, but instruments coming online will most likely enable measurements at higher and higher spatial resolutions. The new study by Baccini et al. (2011) represents a step in this direction.

In contrast to aboveground biomass, changes in belowground carbon stocks, woody debris, and wood products will have to be modeled, but the good news is that changes in aboveground biomass account for ~90% of the net carbon flux (2000–2009), while changes in soil carbon, wood products, and woody debris account for only 20%, 10%, and 0% of the net flux, respectively (Figure 2.4). The sum is more than 100% because during this interval carbon accumulated in wood products, while it was lost from biomass and soils.

Large, rapid changes in aboveground biomass are more easily observed than small, slow changes. This observation means that satellites are biased toward detecting deforestation while missing the slower rates of accumulation of biomass during growth.

The existence of delayed fluxes implies that methods for estimating flux must include data on historical land cover activities and associated information on the fate of cleared carbon. Such historical data are not included in all

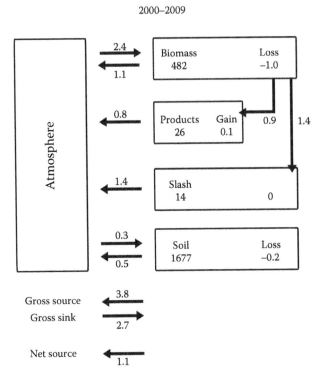

2000–2009

FIGURE 2.4
Average annual flows of carbon (PgC yr⁻¹) in the world's terrestrial ecosystems as a result of land use change over the period 2000–2009. The sum of exchanges with the atmosphere is equivalent to the sum of changes in the four pools (a flux of 1.1 PgC yr⁻¹ from land to atmosphere).

analyses, especially in those using remote sensing data where information is available only since the 1970s at best. How far back in time does one need to conduct analyses in order to estimate current emissions accurately, or, alternatively, how much are current emissions underestimated by ignoring legacy fluxes? Ramankutty et al. (2007) explored these questions using a sensitivity analysis of Amazonia. Their "control" study used historical land use information beginning in 1961 and calculated annual fluxes for the period 1961–2003. When they repeated the analysis ignoring historical land use prior to 1981, they underestimated the 1990–1999 emissions by 13%; when they repeated it ignoring data prior to 1991, they underestimated emissions by 62%. However, if more of the cleared carbon was burned and less decayed, the underestimated emissions were reduced to 4% and 21%, respectively.

In another analysis of deforestation and reforestation in Amazonia, Houghton et al. (2000) found that the annual emissions of carbon from accumulated wood products and slash were three to four times higher than the annual emissions from burning. The legacy from secondary forests was also large in this analysis, accounting for an annual sink as large as the annual source from burning. Sources and sinks of carbon from changes in aboveground biomass are amenable to measurement. Sources from accumulated wood products or downed woody debris will require historical information and modeling.

2.6.2 Can the Net Carbon Balance of Land Be More Easily Measured if Sources and Sinks Are Unevenly Distributed?

In the worst case, the net terrestrial sink is distributed evenly over the land surface and, thus, is so small per hectare that it would to be impossible to measure. On the other hand, many disturbances involve changes large enough to be observed remotely. Furthermore, the gross fluxes from disturbance and recovery are two to three times greater and thus more readily identified than the net source/sink (Richter and Houghton 2011). Several recent studies suggest that changes in forest biomass are more frequent than generally expected. More than half of the hectares of an old-growth tropical forest in Costa Rica, for example, showed (with airborne lidar) either losses or gains of carbon over 7 years (Dubayah et al. 2010), and a recent study with Landsat showed that small gaps associated with tree falls in Central Amazonia were numerous enough to account for an area equivalent to 40% of that region's annual deforestation (Negrón-Juárez et al. 2011).

These results on the one hand, raise the hope that change may be more common, and thus more readily detected and measured, than expected. That is, the net terrestrial sink is *not* distributed evenly over the land surface. On the other hand, the errors associated with the more easily measured sources and sinks may make estimation of a net global change no more accurate than it would be if the change were evenly distributed over the terrestrial surface. Furthermore, the recent examples of fine-scale changes in carbon density may be no more than "noise" in longer term trends or large-area averages. Changes might be better observed over large regions using coarse resolution imagery, sampled with high-resolution lidar, for a more accurate estimate of average change. If the goal is to understand individual trees in a stand, coarse resolution would, of course, not be appropriate.

2.6.3 Can Losses and Gains of Terrestrial Carbon Be Attributed to Direct Management, as Opposed to Indirect Environmental Effects?

Besides the policy reasons for distinguishing direct anthropogenic effects from environmental effects, the scientific reason for attribution is to better understand the current global carbon cycle and to better predict future changes. One goal is to understand the individual processes responsible for what is now referred to as the residual terrestrial flux. The global carbon budget has advanced from recognizing a single net terrestrial flux of carbon (Figure 2.1) to recognizing two terrestrial fluxes: an LULCC flux and a residual terrestrial flux (Figure 2.3). Both of these net fluxes can be further divided, for example, into gross fluxes or into different causal mechanisms. Changes driven by natural disturbances and recovery (structural changes) are clearly different from changes driven by enhanced or retarded growth rates (metabolic changes). Some will lend themselves to observation from space; others will remain in the residual category until models are good enough or data are specific enough to enable additional distinctions.

About the Contributor

Richard A. Houghton is a senior scientist at the Woods Hole Research Center in Falmouth, Massachusetts. The Center is an independent, nonprofit institute focused on environmental science, education, and policy. Houghton has studied the interactions of terrestrial ecosystems with the global carbon cycle and climate change for more than 30 years, in particular documenting changes in land use and determining the sources and sinks of carbon attributable to land management. He has participated in the IPCC Assessments of Climate Change and the U.S. Climate Change Science Program's *First State of the Carbon Cycle Report (SOCCR): The North American Carbon Budget and Implications for the Global Carbon Cycle.* He received his PhD in ecology from the State University of New York at Stony Brook in 1979 and has worked as a research scientist at Brookhaven National Laboratory in New York and the Marine Biological Laboratory in Woods Hole, Massachusetts. He has been at the Woods Hole Research Center since 1987, serving for 2 years (1993–1994) as a visiting senior scientist at the NASA headquarters in Washington, DC.

References

Archer, S., T.W. Boutton, and K.A. Hibbard. 2001. Trees in grasslands: biogeochemical consequences of woody plant expansion. In *Global Biogeochemical Cycles in the Climate System*, ed. E.-D. Schulze, S.P. Harrison, M. Heimann, E.A. Holland, J. Lloyd, I.C. Prentice, and D. Schimel, 115–137. Academic Press, San Diego, CA.

Arvidson, T., J. Gasch, and S.N. Goward. 2001. Landsat 7's long-term acquisition plan—an innovative approach to building a global imagery archive. *Remote Sensing of Environment* 78:13–26.

Aufdenkampe, A.K., E. Mayorga, P.A. Raymond, J.M. Melack, S.C. Doney, S.R. Alin, R.E. Aalto, and K. Yoo. 2011. Riverine coupling of biogeochemical cycles between land, oceans, and atmosphere. *Frontiers in Ecology and Environment* 9:53–60.

Baccini, A., S.J. Goetz, W. Walker, N.T. Laporte, M. Sun, D. Sulla-Menashe, J. Hackler, et al. 2011. New satellite-based estimates of tropical carbon stocks and CO_2 emissions from deforestation. *Nature Climate Change* 2:182–185.

Bastviken, D., L.J. Tranvik, J.A. Downing, P.M. Crill, and A. Enrich-Prast. 2011. Freshwater methane emissions offset the continental carbon sink. *Science* 331:50.

Beer, C., M. Reichstein, E. Tomelleri, P. Ciais, M. Jung, N. Carvalhais, C. Rödenbeck, et al. 2010. Terrestrial gross carbon dioxide uptake: Global distribution and covariation with climate. *Science* 329:834–838.

Berhe, A.A., J. Harte, J.W. Harden, and M.S. Torn. 2007. The significance of the erosion-induced terrestrial carbon sink. *BioScience* 57:337–347.

Bradley, B.A., R.A. Houghton, J.F. Mustard, and S.P. Hamburg. 2006. Invasive grass reduces aboveground carbon stocks in shrublands of the Western US. *Global Change Biology* 12:1815–1822.

Canadell, J.G., C. Le Quéré, M.R. Raupach, C.B. Field, E.T. Buitenhuis, P. Ciais, T.J. Conway, N.P. Gillett, R.A. Houghton, and G. Marland. 2007. Contributions to accelerating atmospheric CO_2 growth from economic activity, carbon intensity, and efficiency of natural sinks. *Proceedings of the National Academy of Sciences* 104:18866–18870.

Cox, P.M., R.A. Betts, C.D. Jones, S.A. Spall, and I.J. Totterdell. 2000. Acceleration of global warming due to carbon-cycle feedbacks in a coupled climate model. *Nature* 408:184–187.

Cramer, W., A. Bondeau, F.I. Woodward, I.C. Prentice, R.A. Betts, V. Brovkin, P.M. Cox, et al. 2001. Global response of terrestrial ecosystem structure and function to CO_2 and climate change: results from six dynamic global vegetation models. *Global Change Biology* 7:357–373.

Drummond, M.A. and T.R. Loveland. 2010. Land-use pressure and a transition to forest-cover loss in the eastern United States. *Bioscience* 60:286–298.

Dubayah, R.O., S.L. Sheldon, D.B. Clark, M.A. Hofton, J.B. Blair, G.C. Hurtt, and R.L. Chazdon. 2010. Estimation of tropical forest height and biomass dynamics using lidar remote sensing at La Selva, Costa Rica. *Journal of Geophysical Research* 115:G00E09, doi:10.1029/2009JG000933.

FAO. 2010. *Global Forest Resources Assessment 2010*. FAO Forestry paper 163, FAO, Rome.

Friedlingstein, P., P. Cox, R. Betts, L. Bopp, W. von Bloh, V. Brovkin, P. Cadule, et al. 2006. Climate-carbon cycle feedback analysis: results from the c4mip model intercomparison. *Journal of Climate* 19:3337–3353.

Gillett, N.P., A.J. Weaver, F.W. Zwiers, and M.D. Flannigan. 2004. Detecting the effect of climate change on Canadian forest fires. *Geophysical Research Letters* 31:L18211.

Gitz, V. and P. Ciais. 2003. Amplifying effects of land-use change on future atmospheric CO_2 levels. *Global Biogeochemical Cycles* 17:1024, doi:10.1029/2002GB001963.

Gloor, M., J.L. Sarmiento, and N. Gruber. 2010. What can be learned about carbon cycle climate feedbacks from the CO_2 airborne fraction? *Atmospheric Chemistry and Physics* 10:7739–7751.

Goetz, S.J., M.C. Mack, K.R. Gurney, J.T. Randerson, and R.A. Houghton. 2007. Ecosystem responses to recent climate change and fire disturbance at northern high latitudes: Observations and model results contrasting northern Eurasia and North America. *Environmental Research Letters* 2:045031, doi:10.1088/1748-9326/2/4/045031.

Goldewijk, K. 2001. Estimating global land use change over the past 300 years: The HYDE Database. *Global Biogeochemical Cycles* 15:417–433.

Guo, L.B. and R.M. Gifford. 2002. Soil carbon stocks and land use change: A meta analysis. *Global Change Biology* 8:345–360.

Hall, F.G., K. Bergen, J.B. Blair, R. Dubayah, R. Houghton, G. Hurtt, J. Kellndorfer, et al. 2011. Characterizing 3-D vegetation structure from space: Mission requirements. *Remote Sensing of Environment* 115:2753–2775.

Hooijer A., S. Page, J.G. Canadell, M. Silvius, J. Kwadijk, H. Wösten, and J. Jauhiainen. 2010. Current and future CO_2 emissions from drained peatlands in Southeast Asia. *Biogeosciences-Discussions* 7:1505–1514.

Houghton, R.A. 1999. The annual net flux of carbon to the atmosphere from changes in land use 1850–1990. *Tellus* 51B:298–313.

Houghton, R.A. and S.J. Goetz. 2008. New satellites help quantify carbon sources and sinks. *Eos* 89(43):417–418.

Houghton, R.A., K.T. Lawrence, J.L. Hackler, and S. Brown. 2001. The spatial distribution of forest biomass in the Brazilian Amazon: A comparison of estimates. *Global Change Biology* 7:731–746.

Houghton, R.A., D.L. Skole, C.A. Nobre, J.L. Hackler, K.T. Lawrence, and W.H. Chomentowski. 2000. Annual fluxes of carbon from deforestation and regrowth in the Brazilian Amazon. *Nature* 403:301–304.

Huang, M. and G.P. Asner. 2010. Long-term carbon loss and recovery following selective logging in Amazon forests. *Global Biogeochemical Cycles* 24:GB3028, doi:10.1029/2009GB003727.

Jackson, R.B., J.L. Banner, E.G. Jobbágy, W.T. Pockman, and D.H. Wall. 2002. Ecosystem carbon loss with woody plant invasion of grasslands. *Nature* 418:623–626.

Jeon, S.-B., P. Olofsson, C.E. Woodcock, F. Zhao, X. Yang, and R.A. Houghton. 2011. The effect of land use change on the terrestrial carbon budget of New England. PhD Dissertation, Boston University, Boston, MA.

Kasischke, E.S. and M.R. Turetsky. 2006. Recent changes in the fire regime across the North American boreal region—Spatial and temporal patterns of burning across Canada and Alaska. *Geophysical Research Letters* 33:L09703, doi:10.1029/2006GL025677.

Knorr, W. 2009. Is the airborne fraction of anthropogenic CO_2 emissions increasing? *Geophysical Research Letters* 36:L21710, doi:10.1029/2009GL040613.

Kurz, W.A., C.C. Dymond, G. Stinson, G.J. Rampley, E.T. Neilson, A.L. Carroll, T. Ebata, and L. Safranyik. 2008. Mountain pine beetle and forest carbon feedback to climate change. *Nature* 452:987–990.

Le Quéré, C., M.R. Raupach, J.G. Canadell, G. Marland, L. Bopp, P. Ciais, T.J. Conway, et al. 2009. Trends in the sources and sinks of carbon dioxide. *Nature GeoScience* 2:831–836.

Le Toan, T., S. Quegan, M.W.J. Davidson, H. Balzter, P. Paillou, K. Papathanassiou, S. Plummer, et al. 2011. The BIOMASS mission: Mapping global forest biomass to better understand the terrestrial carbon cycle. *Remote Sensing of Environment* 115:2850–2860.

Lewis, S.L., P.M. Brando, O.L. Phillips, G.M.F. van der Heijden, and D. Nepstad. 2011. The 2010 Amazon drought. *Science* 331:554.

Lewis, S.L., G. Lopez-Gonalez, B. Sonké, K. Affum-Baffo, and T.R. Baker. 2009. Increasing carbon storage in intact African tropical forests. *Nature* 477:1003–1006.

Lewis, S.L., O.L. Phillips, T.R. Baker, J. Lloyd, Y. Malhi, S. Almeida, N. Higuchi, et al. 2004. Concerted changes in tropical forest structure and dynamics: Evidence from 50 South American long-term plots. *Philosophical Transactions of the Royal Society of London, Series B* 359:421–436.

Mack, M.C., M.S. Bret-Harte, T.N. Hollingsworth, R.R. Jandt, E.A.G. Schuur, G.R. Shaver, and D.L. Verbyla. 2011. Carbon loss from an unprecedented Arctic tundra wildfire. *Nature* 475:489–492.

Marlon, J.R., P.J. Bartlein, C. Carcaillet, D.G. Gavin, S.P. Harrison, P.E. Higuera, F. Joos, M.J. Power, and I.C. Prentice. 2008. Climate and human influences on global biomass burning over the past two millennia. *Nature Geoscience* 1:697–701.

Murty, D., M.F. Kirschbaum, R.E. McMurtrie, and H. McGilvray. 2002. Does conversion of forest to agricultural land change soil carbon and nitrogen? A review of the literature. *Global Change Biology* 8:105–123.

Negrón-Juárez, R.I., J.Q. Chambers, D.M. Marra, G.H.P.M. Ribeiro, S.W. Rifai, N. Higuchi, and D. Roberts. 2011. Detection of subpixel treefall gaps with Landsat imagery in Central Amazon forests. *Remote Sensing of Environment* 115:3322–3328, doi:10.1016/j.rse.2011.07.015.

Pan, Y., R.A. Birdsey, J. Fang, R. Houghton, P.E. Kauppi, W.A. Kurz, O.L. Phillips, et al. 2011. A large and persistent carbon sink in the world's forests. *Science* 333:988–993.

Peters, G.P., G. Marland, C. Le Quéré, T. Boden, J.G. Canadell, and M.R. Raupach. 2012. Rapid growth in CO_2 emissions after the 2008–2009 global financial crisis. *Nature Climate Change* 2:2–4.

Phillips, O.L., L.E. Aragão, S.L. Lewis, J.B. Fisher, J. Lloyd, G. López-González, Y. Malhi, et al. 2009. Drought sensitivity of the Amazon rainforest. *Science* 323:1344–1347.

Phillips, O.L., T.R. Baker, L. Arroyo, N. Higuchi, T.J. Killeen, W.F. Laurance, S.L. Lewis, et al. 2004. Pattern and process in Amazon forest dynamics, 1976–2001. *Philosophical Transactions of the Royal Society, Series B* 359:381–407.

Phillips, O.L., S.L. Lewis, T.R. Baker, K.-J. Chao, and N. Higuchi. 2008. The changing Amazon forest. *Philosophical Transactions of the Royal Society, Series B* 363:1819–1828.

Pongratz, J., C. Reick, T. Raddatz, and M. Claussen. 2008. A reconstruction of global agricultural areas and land cover for the last millennium. *Global Biogeochemical Cycles* 22:GB3018, doi:10.1029/2007GB003153.

Pongratz, J., C.H. Reick, T. Raddatz, and M. Claussen. 2009. Effects of anthropogenic land cover change on the carbon cycle of the last millennium. *Global Biogeochemical Cycles* 23:GB4001, doi:10.1029/2009GB003488.

Pongratz, J., C.H. Reick, T. Raddatz, and M. Claussen. 2010. Biogeophysical versus biogeochemical climate response to historical anthropogenic land cover change. *Geophysical Research Letters* 37:L08702, doi:10.1029/2010GL043010.

Post, W.M. and K.C. Kwon. 2000. Soil carbon sequestration and land-use change: processes and potential. *Global Change Biology* 6:317–327.

Ramankutty, N. and J.A. Foley. 1999. Estimating historical changes in global land cover: Croplands from 1700 to 1992. *Global Biogeochemical Cycles* 13:997–1027.

Ramankutty, N., H.K. Gibbs, F. Achard, R. DeFries, J.A. Foley, and R.A. Houghton. 2007. Challenges to estimating carbon emissions from tropical deforestation. *Global Change Biology* 13:51–66.

Richter, D. deB. and R.A. Houghton. 2011. Gross CO_2 fluxes from land-use change: Implications for reducing global emissions and increasing sinks. *Carbon Management* 2:41–47.

Rosenqvist, A., M. Shimada, N. Ito, and M. Watanabe. 2007. ALOS PALSAR: A pathfinder mission for global-scale monitoring of the environment. *IEEE Transactions on Geoscience and Remote Sensing* 45:3307–3316.

Schimel, D.S. 1995. Terrestrial ecosystems and the carbon cycle. *Global Change Biology* 1:77–91.

Shevliakova, E., S.W. Pacala, S. Malyshev, G. Hurtt, P.C.D. Milly, J. Casperseon, L. Sentman, et al. 2009. Carbon cycling under 300 years of land use change: Importance of the secondary vegetation sink. *Global Biogeochemical Cycles* 23:GB2022, doi:10.1029/2007GB003176.

Stocks, B.J., J.A. Mason, J.B. Todd, E.M. Bosch, B.M. Wotton, B.D. Amiro, M.D. Flannigan, et al. 2003. Large forest fires in Canada, 1959–1997. *Journal of Geophysical Research* 108(D1):8149, doi:10.1029/2001JD000484.

Thomas, R.Q., C.D. Canham, K.C. Weathers, and C.L. Goodale. 2010. Increased tree carbon storage in response to nitrogen deposition in the US. *Nature Geoscience* 3:13–17.

Tranvik, L.J., J.A. Downing, J.B. Cotner, S.A. Loiselle, R.G. Striegl, T.J. Ballatore, P. Dillon, et al. 2009. Lakes and reservoirs as regulators of carbon cycling and climate. *Limnology and Oceanography* 54:2298–2314.

van der Werf, G.R., D.C. Morton, R.S. DeFries, J.G.J. Olivier, P.S. Kasibhatla, R.B. Jackson, G.J. Collatz, and J.T. Randerson. 2009. CO_2 emissions from forest loss. *Nature Geoscience* 2:737–738.

Wang, Z., J. Chappellaz, K. Park, and J.E. Mak. 2010. Large variations in southern hemisphere biomass burning during the last 650 years. *Science* 330:1663–1666.

Westerling, A.L., H.G. Hidalgo, D.R. Cayan, and T.W. Swetnam. 2006. Warming and earlier spring increase western U.S. forest wildfire activity. *Science* 313:940–943.

3

Use of Earth Observation Technology to Monitor Forests across the Globe

Frédéric Achard
Joint Research Centre of the European Commission

Matthew C. Hansen
University of Maryland

CONTENTS

3.1 Introduction

As presented in Chapters 1 and 2, forests provide crucial ecosystem services. In this respect, it is important to tackle the technical issues surrounding the ability to produce accurate maps and consistent estimates of forest type, location, area, condition, and changes in these factors at scales from global to local. Remotely sensed data from earth observing satellites are crucial to such efforts. Recent developments in regional and global monitoring of tropical forests from earth observation have profited immensely from changes made to the data policy and conditions of data access imposed by major providers of imagery from earth observing satellites—changes that have made access to suitably processed imagery far easier, far cheaper, and far more wide reaching in terms of both geographic coverage and time. On July 23, 1972, the United States launched Landsat 1. This civilian polar-orbiting imaging satellite carried a four-channel multispectral scanner (MSS), which provided images suitable for many forest mapping applications. Its successor is still flying on the quite remarkable Landsat 5. We thus have an unbroken record of observations stretching back over almost four decades.

Imaging sensors on earth observing satellites measure electromagnetic radiation (EMR) reflected or emitted from the Earth's surface and use these measurements as a source of information concerning our planet's physical, chemical, and biological systems. Satellites in geostationary orbit provide frequent images of a fixed view of one side of Earth (as often as every 15 minutes in the case of Europe's Meteosat second-generation instruments), while those in polar orbits, like Landsat, image the entire planet's surface every day or every couple of weeks or so, depending on the spatial characteristics of the sensor; images with detailed spatial measurements (1–30 m) are usually only available once or twice a month—for example, Landsat 5 and 7 (both still flying at the time of writing) image every 16 days at 30 m resolution, while coarser resolution imagery (e.g., the MODerate resolution Imaging Spectroradiometer [MODIS] sensor on Terra at 250 m or the SPOT satellites' VGT sensor at 1 km) is provided every day. Most satellite sensors record EMR beyond the sensitivity of the human eye-measurements in the near and shortwave infrared wavelengths, for example, help differentiate between vegetation types and condition; shortwave and thermal infrared wavelengths are essential for mapping and monitoring forest fires; and measurements in the microwave wavelengths (from imaging radar systems) can even "see" through clouds.

Because the information is captured digitally, computers can be used to process, store, analyze, and distribute the data; and because the information is an image captured at a particular time and place, it provides a permanent record of prevailing environmental conditions. As the same sensor on the same platform is gathering the images for all points on the planet's surface, these measurements are globally consistent and independent—important attributes where monitoring, reporting, and verification (MRV) linked to multilateral environmental agreements, such as the UN Framework Convention on Climate Change (UNFCCC) or the Convention on Biological Diversity, are concerned.

Earth observation from space has become more widely accepted and widely adopted as well as technologically more and more sophisticated. The latest systems launched, such as the Franco-Italian Pleiades system (the first of which was launched December 17, 2011), combine very high spatial resolution (70 cm) with a highly maneuverable platform, capable of providing an image of any point on the surface (cloud cover permitting) with a 24 h revisit period. Earth observation from space has also become more important due to the significant impact that modern human civilization is having on the Earth—over 7 billion people are putting relentless pressure on our planet, and the forests are certainly feeling this. Forty years ago, the United States was largely the only source of imagery—today there are more than 25 space-faring nations flying imaging systems. In 1972 Landsat 1 was the only civilian satellite capable of imaging Earth at a level of spatial detail appropriate for measuring any sort of quantitative changes in forests—today there are more than 40 satellites on orbit that can provide suitable imagery (or at least they could, *if* they had a suitable data acquisition, archiving, processing, access, and distribution policy). This chapter introduces the use of earth observation technology to monitor forests across the globe.

3.2 Scope of the Book

Monitoring forest areas on anything greater than local or regional scales would be a major challenge without the use of satellite imagery, in particular, for large and remote regions. Satellite remote sensing combined with a set of ground measurements for verification plays a key role in determining loss of forest cover. Technical capabilities and statistical tools have advanced since the early 1990s, and operational forest monitoring systems at the national level are now a feasible goal for most developing countries in the tropics (Achard et al. 2010). Improved global observations can support activities related to multilateral environmental agreements, such as the Reducing Emissions from Deforestation and Forest Degradation (REDD)-plus readiness mechanism of the UNFCCC. While the primary interest of countries in forest cover monitoring would occur at national or subnational levels, global or pan-tropical monitoring can contribute through (1) identifying critical areas of change, (2) helping to establish areas within countries that require detailed monitoring, and (3) ensuring consistency of national efforts. The main requirements of global monitoring systems are that they measure changes throughout all forested area, use consistent methodologies at repeated intervals, and verify results. Verification is usually a combination of finer resolution observations and/ or ground observations.

This chapter provides an overview of operational remote sensing approaches used to monitor forest cover over large areas. Many methods of satellite imagery analysis can produce adequate results from global to national scales. One of the key issues for forest cover monitoring is that satellite data need to be interpreted (digitally or visually) for forest cover change, i.e., focusing on the interdependent interpretation of multitemporal imagery to detect and characterize changes. Four general remote sensing–based approaches are currently used for capturing forest cover trends:

1. Statistical sampling designed to estimate deforestation from moderate spatial resolution imagery from optical sensors (typically 10–30 m resolution).

2. Global land cover mapping and identification of areas of rapid forest cover changes from coarse spatial resolution imagery from optical sensors (typically 250 m to 1 km resolution).

3. Nested approach with coarse and moderate spatial resolution imagery from optical sensors, i.e., analysis of wall-to-wall coverage from coarse-resolution data to identify locations of large deforestation fronts for further analysis with a sample of moderate spatial resolution data.

4. Analysis of wall-to-wall coverage from moderate spatial resolution imagery from optical or radar sensors.

The use of moderate-resolution satellite imagery for the historical assessment of deforestation has been boosted by changes to the policy that determines access and distribution of data from the U.S. Landsat archive. In the 1990s, the National Aeronautics and Space Administration (NASA) and the U.S. Geological Survey (USGS) developed a global dataset from the Landsat archives. Initially known as the GeoCover™ program, this became the Global Land Survey (GLS) and provided free and open access to selected scenes covering the whole surface of the planet making up the specific epochs (1990, 2000, 2005, and 2010) for the program. The GLS database is described in Chapter 4 together with the freely available complementary database of coarse-resolution MODIS imagery. In December 2008, the U.S. government revised its Landsat data policy and released the entire Landsat archive at no charge. Together the GLS and the U.S. open access data policy mean that anyone interested in global forest monitoring now has access to an archive of data spanning four decades and covering most points on the Earth's surface multiple times over this period. This powerful resource is now being used for statistical sampling on a global scale. The statistical sampling strategies for the use of moderate-resolution satellite imagery are described in Chapter 5. The technical details of the most prominent forest ecosystem monitoring approaches are provided in Chapters 6 through 14. Finally, Chapter 15 covers the use of synthetic aperture radar (SAR) technology and Chapter 16 gives some perspectives of future satellite remote sensing imagery and technology.

The content of the book is introduced briefly hereafter.

3.3   Use of Moderate Spatial Resolution Imagery

Nearly complete pan-tropical coverage from the Landsat satellites is now available at no cost from the Earth Resources Observation Systems (EROS) Data Center (EDC) of the USGS. A recent product, called the GLS, was derived by reprocessing GeoCover data, a selection of good quality, orthorectified, and geodetically accurate global land dataset of Landsat MSS, Landsat TM, and Landsat ETM+ satellite images with a global coverage, which was created by NASA for the epochs of the mid-1970s at 60 m × 60 m resolution and ca. 1990, ca. 2000, mid-2000s, and ca. 2010 at 28.5 m × 28.5 m resolution.

These GLS datasets play a key role in establishing historical deforestation rates, although in some parts of the tropics (e.g., Western Colombia, Central Africa, and Borneo) persistent cloud cover is a major challenge to using these data. For these regions, the GLS datasets can be complemented by remote sensing data from other satellite sensors with similar characteristics, in particular sensors in the optical domain with moderate spatial resolution (Table 3.1). The GLS datasets are described in full detail in Chapter 4.

3.4   Sampling Strategies for Forest Monitoring from Global to National Levels

An analysis that covers the full spatial extent of the forested areas with moderate spatial resolution imagery, termed "wall-to-wall" coverage, is ideal, but is certainly challenging over very large, heterogeneous areas and has commensurate constraints on resources for analysis. China's Institute for Global Change Studies at Tsinghua University and the National Geomatics Center of China have recently completed a first global wall-to-wall map at 30 m resolution, though this ground-breaking new map is still under validation. For digital analysis with moderate-resolution satellite images at pan-tropical or continental levels, sampling is, as of today, still the norm. Several approaches have been successfully applied by sampling within the total forest area so as to reduce costs of and time spent on analysis. A sampling procedure that adequately represents deforestation events can capture deforestation trends. Because deforestation events are not randomly distributed in space, particular attention is needed to ensure that the statistical design is adequately sampled within areas of potential deforestation (e.g., in proximity to roads or other access networks) using high-density systematic sampling when resources are available. The sampling strategies for forest monitoring from global to national levels are described in Chapter 5.

TABLE 3.1

Availability of Moderate Resolution (20 m × 20 m–50 m × 50 m) Optical Sensors

Nation	Satellite/Sensor	Resolution and Coverage	Feature
United States	Landsat 5 TM	30 m × 30 m 180 km × 180 km	This satellite offered images every 16 days to any satellite receiving station during its 27-year lifetime It stopped acquiring images in November 2011
United States	Landsat 7 ETM+	30 m × 30 m 180 km × 180 km	On May 31, 2003, the failure of the scan line corrector resulted in data gaps outside of the central portion of images (60 km wide)
United States/ Japan	Terra ASTER	15 m × 15 m 60 km × 60 km	Data are acquired on request and are not routinely collected for all areas
India	IRS-P6 LISS-III	23.5 m × 23.5 m 140 km × 140 km	Used by India for its forest assessments
China/Brazil	CBERS-2 HRCCD	20 m × 20 m 113 km swath	Experimental; Brazil uses on-demand images to bolster coverage
United Kingdom	UK-DMC	32 m × 32 m 160 km × 660 km	Commercial (DMCii); Brazil uses alongside Landsat data. Full coverage of sub-Saharan Africa acquired in 2010
France	SPOT-5 HRV	5 m × 5 m/ 20 m × 20 m 60 km × 60 km	Commercial; Indonesia and Thailand use alongside Landsat data
Spain/United Kingdom	Deimos-1 and UK-DMC2	22 m × 22 m 640 km swath	Commercial (DMCii); new version of UK-DMC; launched in July 2009
Japan	ALOS AVNIR-2	10 m × 10 m 70 km × 70 km	Launched in January 2006. Global systematic acquisition plan implemented 2007–2010. Stopped in April 2011

For the Forest Resources Assessment 2010 programme (FRA 2010), the Food and Agriculture Organisation of the UN (FAO) has extended its monitoring of forest cover changes at global to continental scales to complement national reporting. The remote sensing survey (RSS) of FRA 2010 has been extended to all lands. The survey aimed at estimating forest change for the periods 1990–2000–2005 based on a sample of moderate-resolution satellite imagery. The methodology used for this global survey is described in Chapter 7.

3.5 Identification of Hot Spots of Deforestation from Coarse-Resolution Satellite Imagery

Global land cover maps provide a static depiction of land cover and cannot be used to map changes in forest areas due to uncertainty levels that are higher than levels of area changes. However, land cover maps can serve as a baseline against which future change can be assessed and can help locate forest areas that need to be monitored for change.

Coarse spatial resolution (from 250 m × 250 m to 1 km × 1 km) satellite imagery is presently used for global land or forest cover mapping. In the late 1990s, global or pan-continental maps were produced at around 1 km × 1 km resolution from a single data source: the advanced very high-resolution radiometer, or AVHRR sensor (Table 3.2). From 2000 onward, new global land cover datasets were produced at similar resolution—1 km × 1 km— from advanced earth observation sensors (VEGETATION on board SPOT-4 and SPOT-5, and the MODIS, on board the Terra and Aqua platforms). These products, GLC-2000 (Bartholomé and Belward 2005) and MODIS global land cover product (Friedl et al. 2010), allowed for a spatial and thematic refinement of the previous global maps owing to the greater stability of

TABLE 3.2

Main Global Land Cover Maps Derived from Remote Sensing Data from 1 km × 1 km to 300 m × 300 m Spatial Resolution

Map Title	Domain	Sensor	Method
IGBP Discover	Global 1 km	NOAA-AVHRR	12 monthly vegetation indices from April 1992 to March 1993
University of Maryland (UMD)	Global 1 km	NOAA-AVHRR	41 multitemporal metrics from composites from April 1992 to March 1993
TREES	Tropics 1 km	NOAA-AVHRR	Mosaics of single date classifications (1992–1993)
FRA 2000	Global 1 km	NOAA-AVHRR	Updated from the IGBP dataset
MODIS Land Cover Product Collection 4	Global 1 km	TERRA MODIS	12 monthly composites from October 2000 to October 2001
Global Land Cover 2000 (GLC-2000)	Global 1 km	SPOT-VGT	Global 365 daily mosaics for the year 2000
VCF	Global 500 m	TERRA MODIS	Annually derived phenological metrics
MODIS Land Cover Product Collection 5	Global 500 m	TERRA MODIS	12 monthly composites plus annual metrics—version of year 2005 released in late 2008
GlobCover	Global 300 m	Envisat MERIS	6 bimonthly mosaics from mid-2005 to mid-2006

the platforms and spectral characteristics of the sensors. An international initiative was also carried out to harmonize existing and future land cover datasets at 1 km resolution to support operational observation of the Earth's land surface (Herold et al. 2006).

More recently, new global land cover datasets at finer spatial resolution (from 250 m × 250 m to 500 m × 500 m) were generated from TERRA-MODIS or ENVISAT-MERIS sensors. The two key products at this scale are the vegetation continuous field (VCF) product (Hansen et al. 2005) and the GlobCover map (Arino et al. 2008). The MODIS-derived VCF product depicts subpixel vegetation cover at a spatial resolution of 500 m × 500 m. The systematic geometric and radiometric processing of MODIS data has enabled the implementation of operational land cover characterization algorithms. Currently, 10 years (2000–2010) of global VCF tree cover are now available to researchers and are being incorporated into various forest cover and change analyses. The 2005 version of the MODIS global land cover product has been generated at 500 m × 500 m resolution, with substantial differences from previous versions arising from increased spatial resolution and changes in the classification algorithm (Friedl et al. 2010). The GlobCover initiative produced a global land cover map using the 300 m resolution mode from the MERIS sensor onboard the ENVISAT satellite. Data have been acquired from December 1, 2004, to June 30, 2006, and then during the full year 2009. A global land cover map was generated from these data from automatic classification tools using equal-reasoning areas. This product has complemented previous global products and other existing comparable continental products, with improvement in terms of spatial resolution. These global products can also be used as complementary forest maps (Figure 3.1) when they do not already exist at the national level, in particular, for ecosystem stratification to help in the estimation of forest biomass through spatial extrapolation methods.

Static forest cover maps are particularly useful as a stratification tool in developing sampling approaches for forest change estimation. For such purposes, reporting the accuracy of these products is essential through the use of agreed protocols. The overall accuracies of the GLC-2000, MODIS, and GlobCover global land cover products have been reported at 68%, 75%, and 73% respectively, though it is important to remember that these accuracy figures relate to all classes of land cover—the accuracy with which forest cover types are mapped are higher than these overall averages.

A first global map of the main deforestation fronts in the 1980s and 1990s has been produced in the early 2000s (Lepers et al. 2005). This map combines the knowledge of deforestation fronts in the humid tropics using expert knowledge, available deforestation maps, and a time-series analysis of tree cover based on NOAA AVHRR 8 km resolution data. In this exercise, the use of expert knowledge ensured that areas of major change not detected with the satellite-based approaches were not overlooked. More recently, a more

FIGURE 3.1
(See color insert.) Global forest cover map derived from the GlobCover Land Cover map at 300 m resolution. Forested areas appear in green. (From Arino, O. et al., *ESA Bull.*, 136, 24, 2008; The GlobCover Land Cover map is available from the European Space Agency website at http://ionia1.esrin. esa.int/.)

detailed quantification of gross forest cover loss at a global scale has been produced for the period 2000–2005 from MODIS imagery. MODIS-indicated change was used to guide sampling of Landsat image pairs in estimating forest extent and loss (Hansen et al. 2010). The MODIS forest cover loss mapping method is presented in Chapter 6.

The Brazilian PRODES monitoring system for the Brazilian Amazon also uses a hotspot approach to identify critical areas based on the previous year's monitoring. These critical areas are priorities for analysis in the following year. Other databases such as transportation networks, population changes in rural areas, and the locations of government resettlement program can be used to help identify areas where a more detailed analysis needs to be performed. Since May 2005, the Brazilian government also has been running the DETER (Detecção de Desmatamento em Tempo Real) system which serves as an alert in almost real time (every 15 days) for deforestation events larger than 25 ha. The system uses MODIS data and WFI data on board the CBERS-2 satellite (260 m × 250 m resolution) and a combination of linear mixture modeling and visual analysis. This approach is described in Chapter 8.

3.6 Nested Approach with Coarse- and Moderate-Resolution Data

Analysis of coarse-resolution data can identify locations of rapid and large deforestation fronts, though such data are unsuitable on their own to determine rates of deforestation based on changes in forest area. A nested approach in which wall-to-wall coarse-resolution data are analyzed to identify locations requiring further analysis with moderate-resolution data can reduce the need to analyze the entire forested area within a country. Coarse-resolution data have been available from the MODIS sensor for no cost since 2000 (see Chapter 4 for the description of this dataset). In some cases, it is possible to identify deforestation directly with coarse-resolution data. Clearings for large-scale mechanized agriculture are detectable with coarse-resolution data based on digital analysis. However, coarse spatial resolution data do not directly allow for accurately estimating forest area changes, given that most change occurs at subpixel scales. Small agricultural clearings or clearings for settlements require finer resolution data (<50 m × 50 m) to accurately detect clearings of 0.5–1 ha. A nested approach that takes advantage of both coarse spatial resolution satellite data and the large Landsat data archive to estimate humid tropical forest cover change is presented in Chapter 6. This method employs a fusion of coarse spatial resolution MODIS data and moderate spatial resolution Landsat data to estimate and map forest cover change as in the studies of Hansen et al. (2008; 2010).

Estimates of forest clearing are generated from the relatively fine-scale resolution Landsat and, through the use of the regression models, can be extended to the continuous MODIS data.

3.7 Analysis of Wall-to-Wall Coverage from Moderate Spatial Resolution Optical Imagery

A few large countries or regions, in particular India, the Congo Basin, Brazil, the European Union, the United States, Australia, and the Russian federation, have demonstrated for many years already that operational wall-to-wall systems over very large regions or countries can be established based on moderate-resolution satellite imagery.

The use of satellite remote sensing technology to assess the forest cover of the whole of India began in early 1980s. The first forest map of the country was produced in 1984 at 1:1 million scale by visual interpretation of Landsat data. The Forest Survey of India (FSI) has since been assessing the forest cover of the country on a 2-year cycle. Over the years, there have been improvements both in the remote sensing data and in the interpretation techniques. The 12th biennial cycle has been completed from digital interpretation of satellite data collected from October 2008 to March 2009 by the Indian satellite IRS P6 (sensor LISS III at 23.5 m × 23.5 m resolution) with a minimum mapping unit of 1 ha (FSI 2011). The entire assessment from the procurement of satellite data to the reporting, including image rectification, interpretation, ground truthing, and validation of the changes by the state/province forest department, takes almost 2 years. The interpretation involves a hybrid approach combining unsupervised classification in raster format and onscreen visual interpretation of classes. Accuracy assessment is carried out independently using randomly selected sample points verified on the ground (field inventory data) or with satellite data at 5.8 m × 5.8 m resolution and compared with interpretation results. In the last assessment, 4,291 validation points randomly led to an overall accuracy level of the assessment of 92%.

Data fusion approaches are also being employed to produce spatially exhaustive, or wall-to-wall, estimates and maps of forest cover clearing within the humid tropics. In the Congo Basin, MODIS and Landsat data are used to create time-series multi-spectral composites, forest area, and forest cover change maps of the entire basin at the Landsat scale for the years 2000, 2005, and 2010. MODIS data are used to radiometrically normalize Landsat data, which are then related to training sites using supervised classification algorithms. This approach, which is currently being applied pan-tropically, is presented in Chapter 8.

Brazil has been measuring deforestation rates in Brazilian Amazonia since the 1980s. The Brazilian National Space Agency (INPE) produces annual estimates of deforestation in the Legal Amazon using a comprehensive annual national monitoring program called PRODES. Spatially explicit results of the analysis of the satellite imagery are published every year (http://www.obt.inpe.br/prodes/). The PRODES project has been producing the annual rate of gross deforestation since 1988 using a minimum mapping (change detection) unit of 6.25 ha, with the release of estimates foreseen around the end of each year. This approach is presented in Chapter 9.

Selective logging and small-scale forest clearing in heterogeneous landscapes require data with moderate-to-fine spatial resolution, more complex computer algorithms capable of detecting less pronounced differences in spectral reflectance, and greater involvement of an interpreter for visual analysis and verification. Methods have been developed and applied for regional mapping of vegetation type and condition (forest cover, deforestation, degradation, regrowth) using Landsat imagery in annual time steps in the Amazon basin. A review of methods for the monitoring of forest degradation is made in Chapter 10.

Chapter 11 describes the development of two recently released high-resolution pan-European forest maps produced for the years 2000 and 2006. The underlying satellite and auxiliary datasets are presented with an overview of the methodology and the main processing steps that governed their production. Validation, as a most important aspect of applicability, receives special attention, and the outlook highlights some aspects, such as differences arising from "forest use" versus "forest cover" concepts, which are important for prospective users.

The United States relies on its national forest inventory for domestic and international reporting of forest change. The U.S. Forest Inventory and Analysis (FIA) program collects data on a set of over 300,000 plots across the United States. A range of attributes are collected in addition to stand volume, including stand age, species composition, and management practice. Plots are resampled on a 5- to 10-year cycle, depending on the state. While FIA is well suited for estimating national forest statistics, it is not designed to accurately capture local dynamics due to disturbance and other rare events. The desire for consistent, geospatial information on forest disturbance and conversion has invigorated the application of Landsat-type remote sensing technology for forest monitoring in the United States. Recent increases in computing power, coupled with the gradual opening of the Landsat archive for free distribution, have resulted in researchers undertaking increasingly ambitious programs in large-area forest dynamics monitoring. In Chapter 12, several of these efforts are described, focusing on national-scale work in the United States.

Australia has developed a system to account for carbon emissions and removals from the land sector, called the National Carbon Accounting System (NCAS). A key component of this system is to track areas of land use change. The NCAS Land Cover Change Program (NCAS-LCCP) produces fine-scale continental mapping and monitors the extent and change in vegetation cover using Landsat satellite imagery from 1972 to 2011 and continues on an annual update cycle, making it one of the most intensive land cover monitoring programs of its kind in the world. The approach is described in Chapter 13.

A forest fire monitoring information system (FIRMS) has been developed for the Russian territory by the Russian Academy of Sciences and is run by the Forest Fire Protection Service of the Federal Forest Agency since the year 2003. The system covers the entire territory of Russia and provides daily information on burned areas in support to fire management activities and fire impact assessments. Satellite remote sensing technology is the main source of data in the system, in particular data from Terra-MODIS and Landsat-TM/ETM+ sensors acquired since the year 2000. Three different burnt area products are generated: at 1 km resolution, at 250 m resolution, and at about 30 m resolution.

3.8   Forest Monitoring with Radar Imagery

Optical mid-resolution data have historically been the primary tool for forest monitoring. However, SAR provides opportunities for forest mapping and monitoring, not least because data can be acquired regardless of sun illumination and weather conditions, which is particularly relevant in the tropics where cloud cover, smoke and haze are prevalent. Through empirical relationships with SAR data or more complex algorithms based on polarimetry or interferometry, the three-dimensional structure of forests can be retrieved, particularly as transmitted microwaves of different frequency and polarization penetrate through and interact with components of the forest volume (e.g., leaves, branches, and/or trunks) and the underlying surface. Changes in vegetation cover and structure over time can also be detected and linked with the processes of deforestation, degradation, or regeneration. Despite the potential of SAR, users are still comparatively few because of the challenges in interpreting, processing, and analyzing radar data and until recently, the limited availability of consistent radar data at regional to global levels. SAR-operating space agencies are, however, beginning to acknowledge the data problem and, following the example of the global systematic acquisition strategy implemented for the Advanced Land Observing Satellite (ALOS) Phased Arrayed L-band SAR (PALSAR) through the Kyoto and Carbon (K&C) Initiative, are making efforts to ensure regular and systematic acquisitions over large regions as part of forthcoming satellite missions. Whilst SAR data are unlikely to fully replace optical sensors in forest monitoring activities, they provide a useful complementary, supplementary or additional resource for monitoring activities. A background to SAR and examples of its use for forest monitoring are provided in Chapter 15.

3.9 Use of Fine Spatial Resolution Imagery for Accuracy Assessment

Whether through wall-to-wall or sample-based approaches, key requirements lie in verification that the methods are reproducible, provide consistent results when applied at different times, and meet standards for assessment of accuracy. Ground reference data (or information derived from very fine spatial resolution imagery that can be considered as being surrogate to ground reference data) are generally recommended as the most appropriate data to assess the accuracy of forest cover change estimation, although their imperfections may introduce biases into estimators of change. Reporting the overall accuracy (i.e., not only the statistical accuracy usually called precision, but also the interpretation accuracy) is an essential component of a monitoring system. Interpretation accuracies of 80%–95% are achievable for monitoring changes in forest cover with moderate-resolution imagery when using only two classes: forest and nonforest. Interpretation accuracies can be assessed through *in situ* observations or analysis of very fine-resolution airborne or satellite data. While it is difficult to verify change from one time to another on the ground unless the same location is visited at two different time periods, a time series of fine- (to very fine) resolution data can be used to assess the accuracy of forest cover change maps.

A new challenge is to provide a consistent coverage of fine-resolution satellite imagery for global forest cover monitoring, i.e., at least a statistical sample or, more challenging, a wall-to-wall coverage. Current plans for the Landsat Data Continuity Mission, the launch of which is scheduled for early 2013, and the European Sentinel-2, scheduled for mid-2014, will both adopt global data acquisition strategies and both (at least at the time of writing) will allow free and open access to their data. The finer resolution (from 1 m × 1 m up to 10 m × 10 m) can be expected to facilitate the derivation of more precise forest area estimates and canopy cover assessment and therefore more reliable statistical information on forest area changes, in particular, for estimating forest degradation and forest regrowth.

About the Contributors

Frédéric Achard is a senior scientist at the Joint Research Centre (JRC), Ispra, Italy. He first worked on optical remote sensing at the Institute for the International Vegetation Map (CNRS/University) in Toulouse, France. Having joined the JRC in 1992, he started his research activities over Southeast Asia in the framework of the TREES project. His current research interests include the development of earth observation techniques for global and regional forest monitoring, and the assessment of the implications of forest cover changes in the tropics and boreal Eurasia on the global carbon budget. Achard received his PhD in tropical ecology and remote sensing from Toulouse University, France, in 1989. He has coauthored over 48 scientific peer-reviewed papers in leading scientific journals including *Nature, Science, International Journal of Remote Sensing, Forest Ecology and Management, Global Biogeochemical Cycles*, and *Remote Sensing of Environment*.

Matthew C. Hansen is a remote sensing scientist with research specialization in large-area land cover and land use change mapping. Hansen's research is focused on developing improved algorithms, data inputs, and thematic outputs that enable the mapping of land cover change at regional, continental, and global scales. Such maps enable better informed approaches to natural resource management, including deforestation and biodiversity monitoring and can also be used by other scientists as inputs to carbon, climate, and hydrological modeling studies. Hansen is currently an associate team member of NASA's MODIS Land Science Team, responsible for the algorithmic development and product delivery of time-series maps of global forest cover, croplands, and other vegetation cover types. He also works on mapping deforestation within the Congo Basin as part of the Central Africa Regional Program for the Environment, a USAID-funded project and in Indonesia as part of the Indonesia–Australia Forest Carbon Partnership. Other current research includes improving global cropland monitoring capabilities for the Foreign Agriculture Service of the USDA. Hansen entered the field of remote sensing after serving with the Peace Corps in Zaire and has a PhD in geography from the University of Maryland, College Park, Maryland, an MA in geography, an MSE in civil engineering from the University of North Carolina at Charlotte, North Carolina, and a BEE in electrical engineering from Auburn University, Auburn, Alabama.

References

Achard, F. et al., Estimating tropical deforestation. *Carbon Management*, 1, 271, 2010.
Arino, O. et al., The most detailed portrait of Earth. *European Space Agency Bulletin*, 136, 24, 2008.
Bartholomé, E. and Belward, A.S., GLC2000: A new approach to global land cover mapping from earth observation data. *International Journal of Remote Sensing*, 26, 1959, 2005.

Friedl, M.A. et al., MODIS Collection 5 global land cover: Algorithm refinements and characterization of new datasets. *Remote Sensing of Environment*, 114, 168, 2010.

FSI, *India State of Forest Report 2011*. Forest Survey of India, Ministry of Environment and Forest, Dehra Dun, India, 2011. Available at http://www.fsi.nic.in/. Accessed June 29, 2012.

Hansen, M.C. et al., Estimation of tree cover using MODIS data at global, continental and regional/local scales. *International Journal of Remote Sensing*, 26, 4359, 2005.

Hansen, M.C. et al., Humid tropical forest clearing from 2000 to 2005 quantified by using multitemporal and multiresolution remotely sensed data. *Proceedings of the National Academy of Sciences of the United States of America*, 105, 9439, 2008.

Hansen, M.C. et al., Quantification of global gross forest cover loss. *Proceedings of the National Academy of Sciences of the United States of America*, 107, 8650, 2010.

Herold, M. et al., A joint initiative for harmonization and validation of land cover datasets. *IEEE Transactions on GeoScience and Remote Sensing*, 44, 1719, 2006.

Lepers, E. et al., A synthesis of information on rapid land-cover change for the period 1981–2000. *Bioscience*, 55, 115, 2005.

4

Global Data Availability from U.S. Satellites: Landsat and MODIS

Thomas R. Loveland
Earth Resources Observation and Science Center

Matthew C. Hansen
University of Maryland

CONTENTS

4.1 Introduction

All land remote sensing data from the U.S. government earth observation missions are available to anyone worldwide on a nondiscriminatory basis. U.S. missions are global in scope and emphasis and follow practices that ensure systematic data acquisition, archiving, and accessibility. This chapter focuses solely on data from two U.S. government earth observation missions commonly used for global land studies: the Moderate Resolution Imaging Spectroradiometer (MODIS) and Landsat sensors. Another U.S. mission used for earlier global investigations, the Advanced Very High-Resolution Radiometer (AVHRR) from National Oceanic and Atmospheric Administration (NOAA) polar orbiters, will not be addressed since the end of the AVHRR era is imminent. The follow-on to AVHRR, the Visible Infrared Imager Radiometer Suite (VIIRS) instrument, is a new earth observation data source launched in late 2011 that will build on the MODIS and AVHRR data processing and dissemination models (Justice et al. 2010).

Acquisition practices determine the amount and extent of global imagery available to users. For most U.S. earth observation programs, systematic global collection strategies ensure the availability of imagery over time and space. NASA, NOAA, and the U.S. Geological Survey (USGS) earth observation missions all systematically acquire global data. The NOAA's AVHRR and NASA's MODIS acquire complete global coverage on a daily basis, and the USGS Landsat mission uses the long-term acquisition plan (LTAP) to guide the collection of global seasonal coverage (Arvidson et al. 2006).

However, data can be available yet not practically accessible. If data query and access tools associated with archived data sets are inadequate, efficient access to data may be cumbersome and reduce data use. Perhaps more significant for global studies is data policy. U.S. earth observation policy has long had unrestricted access to imagery. NASA and NOAA have historically stressed free and open access to archives, while the USGS followed a "cost of filling user request" (COFUR) policy and charged per image fees. The cost of those fees has varied over the 40-year history of Landsat, with per scene charges for electronic data ranging from a low of $200 per scene to a high of $4400 per scene. For studies spanning long temporal periods and/or large geographic areas, the cost of Landsat data was too often prohibitive. For Landsat, the cost of scenes made global land mapping applications effectively prohibitive for most researchers and organizations. Recognizing this limitation, the USGS, with NASA support, changed the Landsat data policy in late 2008, and now all Landsat data are available at no cost to any user (Woodcock et al. 2008).

For an earth observation system to enable large-area land cover characterization and monitoring, it must meet certain data requirements. These requirements include (1) systematic global acquisitions, (2) available at low or no cost, (3) with easy access, and (4) featuring geometric and/or radiometric preprocessing. AVHRR data were the first such data sets processed to this standard, for example, the Pathfinder (James and Kalluri 1994) and global inventory monitoring and modeling studies (GIMMS) data sets (Los et al. 1994). The MODIS has advanced this concept through the use of a land science team to develop, implement, and iterate standard image products (Justice et al. 2010). Data from other coarse spatial resolution sensors such as SPOT VEGETATION also meet the criteria outlined above (Maisongrande et al. 2004). For moderate spatial resolution satellite data sets such as Landsat, progress in achieving a data policy and processing system that fulfills these requirements has been more problematic. Future advancement of the earth observation science community will largely depend on applying the experiences developed with coarse spatial resolution data sets to those at moderate spatial resolution. Recent developments with Landsat indicate a promising future for global moderate-resolution data set availability.

4.2 Changing Medium-Resolution Data Policies to Enable Global Studies

The first freely available global coverage of medium spatial resolution imagery was processed by Earth Satellite Corporation as the GeoCover data set (Tucker et al. 2004) and more recently augmented and reprocessed by NASA and the USGS as the global land survey (GLS) data set (Gutman et al. 2008). GeoCover data were first distributed by the Global Land Cover Facility at the University of Maryland (http://glcf.umd.edu/) and the USGS, and download volumes demonstrated the high interest in and demand for free moderate spatial resolution data over large areas. The GLS data sets currently consist of single-best growing season images for decadal and middecadal epochs (1990, 2000, 2005, 2010) and have been used in a host of large-area mapping projects (Hansen et al. 2010; Huang et al. 2008; Masek et al. 2008).

In the mid-2000s, Brazil's Instituto Nacional de Pesquisas Espaciais (INPE) furthered the medium-resolution free data revolution by announcing that all Brazilian Landsat-class imagery would be available at no cost. This was the first official government data policy to institute a no-cost provision of medium spatial resolution data. The USGS followed suit, and since then, other providers are moving to more open pricing models (e.g., the European Space Agency for Sentinel-2). The Committee on Earth Observations Satellites (CEOS) recently established a data democracy initiative that is working toward improving access to earth observations and expanding their use through no-cost access to data, improved data dissemination, provision of affordable software and other analysis tools, and capacity building.

The 2008 decision by the USGS to make U.S. held Landsat data available to anyone at no cost serves as an example of the impact of a free and open data policy (Loveland and Dwyer in press; Wulder et al. in press). Late that year, the USGS announced the end of the Landsat data purchase era and the beginning of "Web-enabled" access to the USGS Landsat archive. Web-enabling was a euphemism for making all data available at no cost over the Internet. In addition to making data available at no cost, the USGS also began providing Landsat data in an orthorectified format. As a result, users now receive application-ready imagery processed to a single format—Level 1 Terrain (L1T). These changes immediately improved the cost-effectiveness and efficiency of most Landsat applications. Additionally, the long-established and studied radiometric calibration of Landsat (Chander et al. 2009) ensures consistent spectral response across space and through time.

The response to the Landsat policy change has been significant. Prior to the policy change, annual Landsat data sales peaked in 2001 when approximately 23,000 products were sold. In the first full year that Landsat data were free, more than 1.1 million images were distributed, and the following year, the number of scenes more than doubled to 2.4 million images and continues to rise. Users in more than 180 countries download Landsat data annually. Also noteworthy is that the demand for data from the historical archive increased significantly in addition to the demand for newer data. Considering the Landsat 7 ETM+ collection, prior to the free-data era, users had accessed approximately 7% of the ETM+ archive. Now, more than 65% of the archive has been used.

The new data policy truly revolutionized the use of Landsat data for education, research, and applications, which therefore increased societal benefits of the 40-year Landsat archive. With the USGS decision in late-2008 to make Landsat data available at no cost to users, all major sources of land remote sensing data from U.S. government programs are also free. There are significant signs that other earth observation data providers are moving toward more open, no-cost data policies.

4.3   MODIS Data

Since before its launch, MODIS has had a land science team tasked with generating data sets that meet the requirements of global land monitoring (Justice et al. 1998, 2002). The MODIS land science team is funded by NASA to develop and maintain the science algorithms and processing software used to generate the MODIS land products and is responsible for coordinating, developing, and undertaking protocols to evaluate product performance, both on a systematic basis through quality assessment activities and on a periodic basis through validation campaigns (Masuoka et al. 2010). The MODIS land products are generated in a gridded format with standard geometric and radiometric corrections and per-pixel quality information (Masuoka et al. 2010; Roy et al. 2002; Vermote et al. 2002; Wolfe et al. 1998). The MODIS archive is systematically reprocessed as new and improved versions of core land processing algorithms are developed.

MODIS products, constituting a 13-year record, are available online at discipline-specific data centers within the NASA Earth Observing System Data and Information System (EOSDIS). Portals for searching and downloading MODIS land products can be accessed via the Land Processes Distributed Active Archive Center (LP DAAC) (https://lpdaac.usgs.gov/). The products are also available through science team–led portals. Looking forward, the experience and lessons learned from MODIS processing and delivery will be a model for global processing of moderate spatial resolution data.

4.4   Landsat Data

The USGS at the Earth Resources Observation and Science (EROS) Center manages the global Landsat archive. EROS has been the steward of the Landsat archive since the first Landsat was launched in July 1972. The EROS archive currently includes over 3 million images with approximately 300 new Landsat ETM+ scenes added to the archive every day. The LTAP described previously is ensuring that seasonal global coverage is systematically acquired and added to the Landsat archive. If Landsat 7 continues to acquire data until its fuel-based end-of-life in 2017, and when the Landsat Data Continuity Mission (LDCM) begins collecting its planned 400 daily global images in January 2013 (Irons et al. in press), 700 Landsat images per day will be added to the archive. This should improve the role of Landsat for global investigations.

The depth of historical global Landsat coverage varies over the 40-year history of the program due to both technical and policy factors. For example, the commercialization of Landsat in the 1980s and 1990s resulted in a reduction of global acquisitions, and the loss of Landsat 5 data relay capabilities restricted TM acquisitions to regions with direct reception ground stations. In addition, a significant portion of global Landsat coverage resides in archives controlled by Landsat International Cooperators (ICs). Approximately 5 million Landsat scenes are estimated to be in international archives maintained by the ICs, and perhaps as many as 3 million of these scenes are unique and not duplicated in the EROS Landsat archive. The IC Landsat collections add significant historical depth and breadth for global studies—if the global science and applications user community has access (Loveland and Irons 2007). The USGS is working closely with the ICs to consolidate as much of these historical holdings as possible into the EROS Landsat archive. Most ICs recognize the value of this initiative and are strong participants.

All new and archived USGS EROS Landsat data are available to anyone at no cost. In order to provide data for free, EROS simplified and automated Landsat product-generation capabilities and data specifications. Using the modular Landsat product-generation system (LPGS), when new Landsat 7 data are received and archived at EROS, an automated cloud cover assessment algorithm computes the percentage of cloud cover for each scene as an attribute for inventory metadata. Scenes that are acquired with less than 60% cloud cover are immediately processed to generate L1T products. The processed L1T data are temporarily available in a disk cache for immediate download for approximately 90 days before new additions cause the older images to "roll off" the disk. However, all 3 million images in the EROS archive, regardless of cloud cover, are available "on demand." In cases where the needed data are not immediately available, an on-demand processing request can be submitted and when the data have been processed, an e-mail is sent to the requestor with a universal resource locator from which to retrieve the data. The current processing capacity of LPGS is approximately 3,500 scenes per day, although as many as 9,000 scenes have been processed in a single day. The LPGS will continue to evolve and improve data processing and access as resources allow.

Landsat L1T data sets provide consistent, orthorectified, and calibrated Landsat scenes for users. All EROS Landsat data are calibrated to a common radiometric standard, instrument performance is constantly monitored, and scenes are orthorectified to a consistent global set of ground control points (Table 4.1).

Access to both processed and archived Landsat data is available primarily through the EarthExplorer and Global Visualization Viewer (GloVis) interfaces, both of which can be used to search and query the archive. In addition to USGS Landsat holdings, the series of Landsat satellites have also collected scenes for locations outside the United States that are not archived or distributed by the USGS EROS Center (see Figure 4.1 for a map of active Landsat ground stations). Landsat ICs also have unique archives containing data that are not duplicated in the EROS archive. Landsat scenes from the IC ground stations must be ordered directly from the specific station that acquired the data. Data prices, formats, and/or processing options may vary according

TABLE 4.1

Landsat L1T Product Specifications

Product type	Systematic or precision terrain correction pending availability of ground control points
Pixel size	30 m (TM, ETM+), 60 m (MSS)
Map projection	Universal transverse mercator
Datum	WGS84
Orientation	North-up
Resampling method	Cubic convolution
Output format	GeoTIFF
Geometric accuracy	~30 m RMSE (United States), ~50 m RMSE (Global)

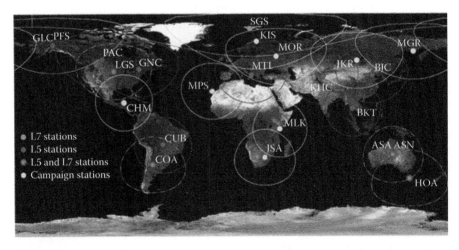

FIGURE 4.1
(See color insert.) Active Landsat ground stations. (More details are available at http://landsat.usgs.gov/about_ground_stations.php.)

to the data provider. A complete list of ground stations and Web addresses for accessing their Landsat collections is available at http://landsat.usgs.gov/about_ground_stations.php.

4.5 Accessing Data

There are a number of interfaces available for accessing MODIS and Landsat data. The GloVis is an intuitive, graphical-based tool for satellite and other image data products with access to several EROS data collections (http://glovis.usgs.gov). Through a graphical map display, any area of interest can be selected, and all available graphical images matching search criteria can immediately be viewed. For Landsat data, it is also possible to navigate to adjacent scene locations in order to identify additional compatible coverage. Controllable criteria include cloud cover limits, date limits, user-specified map layer displays, scene list maintenance, and access to metadata. An ordering interface allows the no-cost download of selected images.

EarthExplorer provides online search, graphical display, data download, and exports of metadata to support users with access to the broader collection of Earth science data sets within the EROS archive. It is a more complex and traditional query tool in comparison to GloVis. However, it offers a number of additional capabilities including:

- Map viewer for viewing overlay footprints and graphical overlays
- Data access tool to search and discover data
- Textual query capability
- Keyhole markup language (KML) export capability to interface with Google Earth
- Save or export queries, results, and map overlay for reuse
- User authentication service for access to specialized data sets and tools

A new tool named Reverb is now in operation and is planned as the "next generation Earth science discovery tool," providing a means for discovering, accessing, and invoking NASA data products and services (http://reverb.echo.nasa.gov). Searches can query by platform, instrument and sensor, or specific campaign and can be refined spatially, temporally, or by processing level and product type. Reverb is recommended for accessing MODIS data. There is considerable cross-fertilization between the various search systems. For example, Reverb can also serve as an interface to other archives, including those of Advanced Spaceborne Thermal Emission and Reflection Radiometer (ASTER) and AVHRR.

4.6 Conclusion

U.S. earth observation initiatives are now consistently committed and managed for use in global land studies. Especially critical are the use of systematic global acquisition strategies and nondiscriminatory, no-cost access to the acquired data. Continuation of these practices and the timely launch of follow-on missions are essential next steps in ensuring that current investments in global land studies are continued into the future. The launch of LDCM potentially extends the Landsat record for another 5–10 years (until 2018–2023), but after that no follow-on capability is currently authorized. On the other hand, the MODIS record is currently transitioning to the VIIRS era as this next generation of NOAA polar orbiters becomes operational. An operational moderate spatial resolution land monitoring program has been proposed, the National Land Imaging Program (Office of Science and Technology Policy 2007), but no substantive investment made to date for its implementation.

As moderate spatial resolution data policies and processing mimic those of coarser resolution data, new science capabilities will be enabled. The next few years are quite possibly going to be Landsat's "golden years," the time in which the Landsat program achieves its full potential for global studies. Free Landsat data, the consolidation of international holdings into the EROS archive, the expanded availability of these data in a consistently processed format, and new global coverage from Landsat 7 and the LDCM are enabling and improving the use of Landsat for global studies. Innovative improvements in Landsat data products and delivery systems, such as the Web-Enabled Landsat Data (WELD) system developed by Roy et al. (2010), will serve as catalysts for improved global use of Landsat. The integrated use of systematically acquired multiresolution, multitemporal, multispectral global data sets, such as MODIS and Landsat, will become a standard scientific practice.

About the Contributors

Thomas Loveland is a senior research scientist with the U.S. Geological Survey EROS Center in Sioux Falls, South Dakota. He has a BS and MS in geography from South Dakota State University, Brookings, South Dakota, and a PhD in geography from the University of California, Santa Barbara, California. Loveland currently coleads the USGS–NASA Landsat Science Team and is engaged in research in the use of remote sensing for land cover studies spanning local to global scales.

Matthew C. Hansen is a professor in the Department of Geographical Sciences at the University of Maryland, College Park, Maryland. He has a bachelor of electrical engineering from Auburn University, Auburn, Alabama. His graduate degrees include a master of engineering in civil engineering and a master of arts in geography from the University of North Carolina at Charlotte, North Carolina, and a doctoral degree in geography from the University of Maryland, College Park, Maryland. His research specialization is in large-area land cover monitoring using multispectral, multitemporal, and multiresolution remotely sensed data sets. He is an associate member of the MODIS Land Science Team and a member of the GOFC-GOLD Implementation Working Group.

References

Arvidson, T., et al., The Landsat-7 Long-Term Acquisition Plan: Development and validation. *Photogrammetric Engineering and Remote Sensing*, 72, 1137, 2006.

Chander, G., Markham, B.L., and Helder, D.L., Summary of current radiometric calibration coefficients for Landsat MSS, TM, and EO-1 ALI sensors. *Remote Sensing of Environment*, 113, 893, 2009.

Gutman, G., et al., Towards monitoring land-cover and land-use changes at a global scale: The global land survey 2005. *Photogrammetric Engineering and Remote Sensing*, 74, 6, 2008.

Hansen, M.C., Stehman, S.V., and Potapov, P.V., Quantification of global gross forest cover loss, *Proceedings of the National Academy of Sciences*, 107, 8650, 2010.

Huang, C., et al., Assessment of Paraguay's forest cover change using Landsat observations. *Remote Sensing of Environment*, 67, 1, 2008.

Irons, J.R., Dwyer, J.L., and Barsi, J.A., The next Landsat satellite: The Landsat Data Continuity Mission. *Remote Sensing of Environment*, in press (http://dx.doi.org/10.1016/j.rse.2011.08.026).

James, M.E. and Kalluri, S.N.V., The Pathfinder AVHRR land data set: An improved coarse resolution data set for terrestrial monitoring. *International Journal of Remote Sensing*, 15, 3347, 1994.

Justice, C.O., et al., The Moderate Resolution Imaging Spectroradiometer (MODIS): Land remote sensing for global change research. *IEEE Transactions on Geoscience and Remote Sensing*, 36, 1228, 1998.

Justice, C.O., et al., An overview of MODIS Land data processing and product status. *Remote Sensing of Environment*, 83, 3, 2002.

Justice C.O., et al., The evolution of U.S. moderate resolution optical land remote sensing from AVHRR to VIIRS. In B. Ramachandran, C.O. Justice, and M.J. Abrams (Eds.), *Land Remote Sensing and Global Environmental Change: NASA's EOS and the Science of ASTER and MODIS*. New York: Springer, 2010.

Los, S.O., Justice, C.O., and Tucker, C.J., A global 1° by 1° NDVI data set for climate studies derived from the GIMMS continental NDVI data. *International Journal of Remote Sensing*, 15, 3493, 1994.

Loveland, T.R. and Dwyer, J.L., Landsat—building a strong future. *Remote Sensing of Environment*, in press (http://dx.doi.org/10.1016/j.rse.2011.09.022).

Loveland, T.R. and Irons, J.R., Landsat Science Team meeting summary. *The Earth Observer*, 19, 10, 2007.

Maisongrande, P., Duchemin, B., and Dedieu, G., VEGETATION/SPOT: An operational mission for the Earth monitoring; presentation of new standard products. *International Journal of Remote Sensing*, 25, 9, 2004.

Masek, J., et al., North American forest disturbance mapped from a decadal Landsat record. *Remote Sensing of Environment*, 112, 2914, 2008.

Masuoka, E., et al., MODIS land data products: generation, quality assurance and validation. In B. Ramachandran, C. Justice, and M. Abrams (Eds.), *Land Remote Sensing and Global Environmental Change: NASA's EOS and the Science of ASTER and MODIS*. New York: Springer, 2010.

Office of Science and Technology Policy, *A Plan for a National Land Imaging Program*, Future of Land Imaging Interagency Working Group, August, 2007.

Roy, D.P., Borak, J.S., Devadiga, S., Wolfe, R.E., Zheng, M., and Descloitre, J., The MODIS land product quality assessment approach. *Remote Sensing of Environment*, 83, 62, 2002.

Roy, D.P., et al., Web-enabled Landsat Data (WELD): Landsat ETM+ composited mosaics of the conterminous United States. *Remote Sensing of Environment*, 114, 35, 2010.

Tucker, C.J., Grant, D.M., and Dykstra, J.D., NASA's global orthorectified Landsat data set. *Photogrammetric Engineering and Remote Sensing*, 70, 3131, 2004.

Vermote, E.F., El Saleous, N.Z., and Justice, C.O., Atmospheric correction of MODIS data in the visible to middle infrared: First results. *Remote Sensing of Environment*, 83, 97, 2002.

Wolfe, R.E., Roy, D.P., and Vermote, E.F., MODIS land data storage, gridding, and compositing methodology: Level 2 grid. *IEEE Transactions on Geoscience and Remote Sensing*, 36, 1324, 1998.

Woodcock, C., et al., Free access to Landsat imagery, *Science*, 320, 1011, 2008.

Wulder, M.A., et al., Opening the archive—how free data has enabled the science and monitoring promise of Landsat. *Remote Sensing of Environment*, in press (http://dx.doi.org/10.1016/j.rse.2012.01.010).

5

Sampling Strategies for Forest Monitoring
from Global to National Levels

Stephen V. Stehman

State University of New York

CONTENTS

5.1 Introduction

Remote sensing plays a key role in forest monitoring because it offers a cost-effective option for frequent observation of vast areas of forest. Forest attribute maps derived from remote sensing may be integrated with forest inventory data in a variety of ways within a forest monitoring framework (Corona 2010). The effective use of remote sensing to produce maps of forest attributes has been described and convincingly demonstrated elsewhere in this book. These maps serve the critical purpose of providing spatially explicit information for forest attributes. The focus of this chapter is not on monitoring forests by complete coverage mapping but on taking advantage of remote sensing via a sampling approach to forest monitoring. Whereas it is sometimes too costly and time consuming to obtain wall-to-wall coverage using the quality of imagery and processing desired for a particular forest monitoring objective, sampling provides the opportunity to apply measurement and observation protocols to a much smaller total area, and this may allow for the use of very high-resolution imagery or sophisticated classification methods that otherwise would not be practical for a complete coverage assessment. A sampling-based monitoring framework targets aggregate properties such as the total area of forest and the area of forest cover change. A traditional intensive ground-based forest inventory approach to forest monitoring is another option based on sampling. But in this chapter, remotely sensed data, defined as data from sensors placed on aircraft or space-based platforms, are assumed to be the basis for forest monitoring.

Forest monitoring can be applied to a variety of forest characteristics, for example forest cover and biomass. In this chapter, the focus will be on monitoring forest cover. The attention to forest cover allows for framing the monitoring objective as an area estimation problem, an objective commonly addressed in mapping applications using remotely sensed data (Gallego 2004). Area estimation can be approached in two ways. One approach is to compute area from a complete coverage map of the target region, for example, using a complete coverage map of deforestation to compute the area deforested. Mayaux et al. (2005, 374–375) review applications in which global land cover and forest mapping efforts are used as the basis for estimating the area of deforestation. The other approach is to estimate the area of deforestation from a sample. By requiring information on a smaller subarea of the full region, sampling offers advantages of significant cost reduction (e.g., fewer satellite images or fewer people to interpret aerial photographs) and better accuracy of the measurements of area. Mayaux et al. (1998) critique the limitations and practical advantages of the two approaches. A further advantage of remote sensing is that it offers an option for forest monitoring based on a consistent methodology that can allow for more direct regional comparisons, for example, of regional rates of forest change than is possible when methods used for monitoring vary by region. Hansen et al. (2010) and the FRA 2010 remote sensing survey (Ridder 2007; FAO 2009) are examples in which regional comparisons have been facilitated because regionally consistent sampling and analysis protocols have been applied to remote sensing assessments of forest change.

The area estimation objective highlights a distinction between two common uses of maps constructed from satellite imagery. The spatially explicit information of pattern and location conveyed by a map is critical to some applications, whereas in other applications, information aggregated over a specified region is sufficient. The latter applications address aggregate properties such as totals, means, or proportions, for example, area of forest cover, proportion of area of deforestation, or total biomass. These aggregate properties or population parameters can be estimated from a sample. When the objective is to estimate area, a statistical comparison between the mapping and sampling approaches can be framed in terms of accuracy and precision. Is the map sufficiently accurate to provide valid change estimates (i.e., bias attributable to classification error is negligible)? Is the sample-based estimate sufficiently precise to provide useful change estimates (i.e., sampling variability is small relative to the quantity being estimated)? Stehman (2005) provides guidance for evaluating the trade-off between precision (sampling variability) and accuracy (measurement or interpretation error) for estimating area.

Sample-based forest monitoring using remotely sensed data has been successfully implemented to provide estimates of forest cover and forest cover change over the tropics (e.g., Achard et al. 2002) and global forest biomes (e.g., Hansen et al. 2010). The global Forest Resources Assessment (FRA) remote sensing survey (FAO 2009) is another recent application of a sample-based forest monitoring activity. These successful operational monitoring efforts are the outcome of years of research and development probing the question of how large-area forest monitoring can be accomplished with the aid of remote sensing. The basic theory and methods underlying the sampling approach to forest monitoring are reviewed in this chapter. Although much progress has been made developing appropriate sampling methods, additional work is needed to further refine and understand the methods of current practice and to develop new methods for more cost-efficient and accurate forest monitoring using remotely sensed data. The prospects for sample-based forest monitoring in the future are discussed in the closing section of this chapter.

5.2 Fundamental Sampling Concepts and Methods

In this section, basic concepts and methods of sampling are defined to establish the context for sample-based forest monitoring. The approach described takes a finite population sampling perspective in which the region of interest (e.g., a country, a continent, or the forested biomes of earth) is partitioned into a set of N nonoverlapping elements or spatial units (e.g., 5 km × 5 km units) called the *universe*. For each element of the universe, one or more attributes or measurements may be obtained (e.g., area of forest cover or area of forest degradation for each unit). A *population* will refer to a collection of these measurements for all N units of the universe, and a *parameter* is defined as a number that describes an aggregate property of this population (e.g., total area of forest cover, or percent loss of forest cover). A *sample* is a subset of the N elements of the universe, and a sample therefore consists of one or more such elements.

Although landscapes are truly continuous, the finite population perspective usually provides a close approximation to reality. For example, if the objective is to obtain the total area of forest for a region, dividing the area into 5 km × 5 km units and summing the forest area over all N such units in the region will yield the same total area as a measurement of area from the unpartitioned (full) region. Some forest characteristics may be less amenable to a sampling approach; for example, certain landscape pattern metrics such as contiguity of patches or landscape diversity may not be estimated well via a sampling approach (Hassett et al. 2012). But for estimating area and change in area, the finite population sampling perspective provides a frequently used, familiar approach that is simple, practical, flexible, and effective.

A *sampling strategy* consists of three major components: the sampling design, response design, and analysis. The sampling design is the protocol by which a subset of the universe (i.e., the sample) is selected. For example, the subset could be 100 sampling units where each sampling unit is 5 km × 5 km. The response design is the protocol for obtaining the measurements of each sampling unit. For example, the response design for the objective of monitoring area of forest cover would be the protocol implemented to measure the area of forest cover of each unit sampled. The protocol may include specification of the imagery to use, the classification method applied to the imagery, and the definition of forest. The analysis protocol includes the formulas used to estimate parameters of interest and the standard errors associated with these estimates.

5.2.1   Basic Sampling Designs

Once the region to be monitored has been partitioned into N spatial units or elements that constitute the universe, a variety of sampling designs may be considered to select the sample. Choosing a sampling design requires three main decisions: (1) Will stratification be used? (2) Will the sampling unit be a cluster? (3) Will the primary selection protocol be simple random, systematic, or something else? The answers to these three questions will determine the sampling design. Examples of sampling designs created by different combinations of these decisions exist in applications to forest monitoring using remotely sensed data (Section 5.3). Considerations influencing each of these decisions are briefly reviewed.

Stratification is the process of grouping the N elements of the universe into strata such that each element belongs to one and only one stratum. Stratification is generally used for two purposes. If the objectives specify reporting forest characteristics by region (e.g., by continent, country, or provinces within a country), strata may be defined by these reporting regions. Typically, the sampling design is then developed with the goal of allocating the sample such that each stratum has a sufficient sample size to achieve acceptable standard errors for estimates of that stratum. Stratification thus can be used to avoid the problem that a reporting region that occupies a relatively small proportion of the full area monitored will have too few sample units to obtain precise estimates for that region.

Another use of stratification is to define strata to minimize the standard error of an estimate. The optimization is attained by defining strata such that strata means differ from one another and elements within a stratum have similar responses. For example, if the objective is to estimate forest cover loss, the strata could be advantageously defined by the amount of forest cover loss, and strata representing no loss, low loss, moderate loss, and high loss may be defined based on the available information of forest cover loss for each of the N elements. Stratifying for the purpose of improving precision requires that ancillary data related to the response of interest are available. For example, Hansen et al. (2010) used complete coverage, MODIS-derived forest cover loss as ancillary data to define strata related to Landsat-derived gross forest cover loss, where Landsat-derived loss was the target measurement for the assessment.

A cluster is a group of elements of the universe that is sampled as a single entity. For example, the basic element of the universe may be defined as a 1 km × 1 km unit, and a 10 km × 10 km group of 100 such units could be defined as a cluster. A cluster sampling protocol would then be applied to the 10 km × 10 km cluster units, but the data would be collected at the support of the 1 km × 1 km units within a cluster. In the terminology of cluster sampling, the 10 km × 10 km unit is labeled a primary sampling unit (PSU) and the 1 km × 1 km unit is called a secondary sampling unit (SSU).

Cluster sampling may be implemented as either one-stage or two-stage sampling (additional stages are possible but the discussion here will be limited to two stages). The first stage of sampling is always a selection of PSUs. For one-stage cluster sampling, all SSUs within each sampled PSU are observed so only one stage of sampling is used. One-stage cluster sampling is thus very similar to defining an element of the universe based on the PSU. For example, the 10 km × 10 km units (PSUs) could be considered the elements of the universe because the 1 km × 1 km units are always selected in groups of 100 defined by the PSU. The only difference between a sample of 10 km × 10 km units and a one-stage cluster sample of 1 km × 1 km units grouped into sets (PSUs) of 100 is that for the cluster sample, the data would be recorded for each 1 km × 1 km unit within the PSU, whereas this measurement on each 1 km × 1 km unit would likely not be retained if the 10 km × 10 km unit is defined as the element of the universe.

In two-stage cluster sampling, a sample of SSUs is selected within each sampled first-stage PSU. Two-stage cluster sampling is motivated by the recognition that typically units spatially proximate to each other will have relatively similar values, and this spatial correlation of the sample observations will tend to inflate the standard errors of estimates from cluster sampling relative to a more spatially dispersed sample of the same size. So instead of sampling all SSUs within a sampled PSU, a sample of SSUs is selected and the cost and time savings achieved by the lower effort per PSU can be allocated to increase the number of PSUs sampled.

The choice of whether to use clusters is typically driven by cost. When the primary data are obtained from remote sensing, the cost of the imagery and the time required to obtain and process the imagery are key considerations.

For example, if RapidEye imagery is used, the size of the PSU may be defined to be a portion of a RapidEye image so that the number of RapidEye images that must be purchased is limited. Cluster sampling allows control over the spatial distribution of the sample because of the spatial grouping of elements into a fixed number of sampled clusters.

Whether clusters or strata are present, it is necessary to specify a protocol for selecting the elements of the sample. For *simple random* selection of a sample size of n sampling units, the sample is selected such that all possible sets of n units have the same probability of being selected. For example, if the universe is first partitioned into strata and simple random selection is implemented in each stratum, the design is called stratified random sampling. For cluster sampling, the simple random selection protocol could be used to select a first-stage sample of PSUs, or applied within sampled PSUs to select a second-stage sample of SSUs. For a *systematic* selection protocol, a random starting element or location is selected, and the remaining sample elements are selected based on their location in a list of all N elements of the universe or based on their spatial location relative to the random starting location. Systematic selection can also be applied in combination with strata and clusters. For example, if strata are present, the elements sampled within a stratum can be selected via the systematic protocol. Similarly, both stages of two-stage cluster sampling could be implemented via a systematic selection protocol. Some considerations influencing the choice of selection protocol are discussed in Section 5.5.

5.2.2 Inclusion Probabilities and Probability Sampling

A useful general perspective of sampling design is obtained by focusing on *inclusion probabilities*. An inclusion probability is defined as the probability that a particular element of the universe is included in the sample. That is, prior to selecting the actual sample, for a given element of the universe, what is the probability of that element being included in the sample selected? Inclusion probabilities thus inform about the process of sample selection. For simple random sampling of n elements from a universe of N elements, the inclusion probability is n/N for each element. For systematic sampling from a list of N elements, if the sampling interval is K (i.e., select every Kth element after a random selection of the first sample element), the inclusion probability is $1/K$ for each element (see Overton and Stehman 1995 for additional examples).

Inclusion probabilities play an important role in defining a *probability sample*. Specifically, a probability sample is defined by two conditions: (1) the inclusion probabilities for all elements in the sample must be known and (2) the inclusion probabilities for all elements of the universe must be greater than zero. The rationale for these conditions is explained in Overton and Stehman (1995). For this chapter, it suffices to recognize that a probability sampling protocol conveys a degree of statistical rigor to the sample-based estimates and inference. For the basic sampling designs typically used in practice (e.g., simple random sampling, systematic sampling, stratified random sampling, and one-stage and two-stage cluster sampling with either simple random or systematic sampling for each stage), the inclusion probabilities are known and these designs meet the conditions of probability sampling (Särndal et al. 1992). If the sampling design does not follow a standard selection protocol, it is necessary to establish that the protocol meets the conditions defining a probability sample. Some practical, but ad hoc selection protocols may create very challenging problems for defining inclusion probabilities, and for very complex selection protocols the inclusion probabilities may be intractable.

5.2.3 Inference

The process of generalizing from the sample data to describe characteristics of the full population is called inference. Clearly, an understanding of inference is necessary when a sampling approach to forest monitoring is used. The two approaches to inference most frequently used in finite population sampling are design- and model-based inference. The two approaches differ primarily in how uncertainty or variability is represented as determined by the definition of the "variable" in each approach.

In design-based inference, the observations obtained for each element of the population are treated as fixed constants and therefore the response or observation is not considered a variable. The uncertainty in design-based inference is attributable to the randomization determining which elements of the universe are selected for observation. It is variation of the estimate from sample to sample that is the uncertainty of interest in design-based inference, and consequently the sampling design is of paramount importance. Specifically, for a given universe and sampling design, the *sample space* is defined as the set of all possible samples that could be selected by that particular design. For each possible sample from a given population, the estimate of the parameter of interest would differ for different samples. For example, suppose the target parameter is the area of deforestation over a 5-year period. A systematic sample of 10 km × 10 km units is selected by randomly locating a grid, with each grid point separated by 250 km. If the sample is repeated by a second random placement of the grid, the estimate of deforestation is likely to change. In design-based inference, it is the variability of an estimate over all possible samples comprising the sample space that characterizes uncertainty. Because the sampling design determines the sample space, the name "design-based" inference is naturally applied.

For model-based inference, the response observed for each element of the population is viewed as a variable, and inference is conditional on the sample obtained. For example, the values of a finite population $y_1, y_2, \ldots, y_N$ are viewed as realizations of the random variables $Y_1, Y_2, \ldots, Y_N$. The goal is to estimate some function of all the y's in the population, $h(y_1, y_2, \ldots, y_N)$, for example, the mean or total (Valliant et al. 2000, 2). After the sample of n elements has been obtained, estimating $h(y_1, y_2, \ldots, y_N)$ entails predicting a function of the unobserved Y's. A model is used for this purpose. The model typically incorporates an auxiliary variable (denoted x) that is related to Y. The model would then include a specification of how the variable Y is related to x, this relationship being represented by the model M. For example, the model M could be a simple linear relationship between the expected value of Y and x, $E_M(Y_i) = \beta x_i$ $(i = 1, 2, \ldots, N)$, with the covariance between the variables Y_i and Y_j specified as $\text{cov}_M(Y_i, Y_j) = \sigma^2 x_i$ if $i = j$ and $\text{cov}_M(Y_i, Y_j) = 0$ if $i \neq j$ (Valliant et al. 2000, 4). The model and observed sample data are the basis for predicting the unobserved Y's, so the probability model specified plays a key role in model-based inference. An example applying model-based inference is provided at the end of Section 5.4.

The choice of inference framework impacts sampling design decisions. Design-based inference is predicated on the sampling design being a probability sampling design. Therefore, if design-based inference will be used, only probability sampling designs should be considered. Conversely, model-based inference does not require a probability sample. The model specified for model-based inference may take into account the fact that the sample was obtained via cluster sampling or stratified sampling, but this would represent a model specification choice and not a required dependence of the inference on the sample. However, advocates of model-based inference often cite the potential advantage that randomization provides to avoid accusations that a sample was subjectively chosen to achieve certain outcomes. Model-based inference can be conducted with a probability sample, but design-based inference cannot be conducted unless a probability sampling design has been implemented.

5.2.4 Estimation

Once the sample has been selected and the data obtained, a variety of estimators may be available to estimate a parameter of interest. For probability sampling designs and design-based inference, a general unbiased estimator of a population total is the Horvitz–Thompson estimator. Suppose the observation on element u of the sample is denoted y_u and the inclusion probability for element u is denoted π_u. If Y is the population total (i.e., the sum of y_u over all N elements of the population), the Horvitz–Thompson estimator of Y is

$$\hat{Y} = \sum_{u \in s} y_u / \pi_u \tag{5.1}$$

where the summation is over the elements of the sample. For example, if y_u is the area of deforestation for element u and Y is the total area of deforestation for the region, then Y can be estimated from a probability sample using the Horvitz–Thompson estimator. For the basic sampling designs typically used in practice, the Horvitz–Thompson estimator simplifies to a special case formula. For example, for a simple random sample of n elements, the estimator simplifies to $\hat{Y} = N\bar{y}$, where $\bar{y}$ is the sample mean of the response y_u, and for stratified random sampling of n_h elements from the N_h available in stratum h (H strata total), the Horvitz–Thompson estimator simplifies to

$$\hat{Y} = \sum_{h=1}^{H} N_h \bar{y}_h \qquad (5.2)$$

where $\bar{y}_h$ is the sample mean in stratum h.

In most applications, it is possible to obtain an auxiliary variable x_u that is associated with the response of interest, y_u. Such an auxiliary variable may be used to advantage to reduce the standard error of the parameter estimate. A widely applicable estimator for this purpose is the generalized regression estimator (GRE) (see Särndal et al. (1992, 225) for full details of this estimator). More familiar simple estimators such as the ratio and regression estimators applied to simple random sampling are special cases of this general form. Because the GRE encompasses a variety of models of the relationship between the response y and one or more auxiliary variables, the GRE is almost always better (i.e., more precise) than the generalized difference estimator (Särndal et al. 1992, section 6.3). The GRE belongs to the class of "model-assisted estimators" (Särndal et al. 1992, 227). These estimators employ a model to information in one or more auxiliary variables to improve precision of estimates, but the estimators are not dependent on the validity of the model, and inference is still design based.

5.2.5   Desirable Design Criteria

Choosing a sampling design for forest monitoring using remote sensing should be guided by the monitoring objectives and by desirable design criteria specified for a particular application. A list of potential desirable criteria follows, but the prioritization of these criteria will be different depending on the specific application.

1. *The sampling protocol satisfies the requirements of a probability sampling design.* As previously stated, this criterion is essential to support design-based inference, but is optional for model-based inference.

2. *The sampling design is easy to implement.* Simplicity of design can be a major virtue. It is critical that the design is implemented correctly, so a simple protocol is advantageous in this regard. Also, a simple design is simpler to analyze, as, for example, when using a model-assisted estimator to improve precision (Section 5.2.4).

3. *The design is cost-effective.* The rationale for this criterion is obvious because a design goal should be to obtain adequately precise estimates (i.e., acceptably small standard errors) for the lowest cost possible. Of course, what constitutes "adequate precision" will be application dependent.

4. *The sample is spatially well distributed* (i.e., spatially balanced). If the sample units are spatially dispersed throughout the target region, the sample has intuitive appeal and often results in smaller standard errors.

5. *The standard errors of estimates resulting from the design are small.* In design-based inference, this would mean that estimates of the target parameter from different samples would be relatively similar.

6. *An unbiased or nearly unbiased estimator of variance is available.* This criterion specifies that standard errors quantifying the uncertainty of the estimates can be provided without undue reliance on approximations other than those related to the need for a large sample size to justify the variance approximation. This criterion becomes particularly relevant when considering the use of systematic sampling because a variance approximation will need to be used as an unbiased estimator of variance is not available for systematic sampling.

7. *A change in sample size can be accommodated before the full sample has been selected.* This criterion is valuable because the final cost of completing the sample data collection is often difficult to predict, so it may be necessary to reduce the sample from the initial target size, or in rare cases it may be possible to increase the sample size. Budgets also sometimes change, and the sample size may need to be reduced or increased accordingly.

8. *The design is transparent and familiar to users of the information.* This criterion may be particularly relevant if nonscientists will be using the monitoring results to inform policy decisions. Transparency may include information of actual plot locations or specific details of how randomization is incorporated into the selection protocol.

5.3 Applications of Sampling to Estimate Forest Cover Change from Remotely Sensed Data

Published studies demonstrating the application of a sampling approach to forest monitoring based on remote sensing are reviewed. The review focuses on two broad categories: actual applications in which forest monitoring based on remotely sensed data has been implemented and evaluative studies in which different sampling design and estimation strategies have been compared. The application studies are discussed first, followed by the design evaluation studies (Section 5.4). The applications are presented in chronological order.

The United Nations Food and Agriculture Organization's (FAO) FRA in 1990 is a landmark application of a sampling approach employing satellite imagery to derive estimates of forest change. The FRA 1990 design used 117 Landsat scenes as the sampling units (FAO 1996). The design was stratified based on three major geographic regions (Africa, Latin America, and Asia) and 10 subregions among the three major regions. The sample size allocated to these regions was based on the expected area of deforestation, as predicted for each subnational unit based on prevalence of forest, human population size, and per capita income. An additional level of stratification (FAO 1996, 8) was based on forest cover in Asia and Latin America (>70%, 40%–70%, and 10%–40%, where cover was derived from country-specific inventories) and on dominant forest types in Africa (forest, woodland, or tree savanna for the three strata). Thus both purposes of stratification were accommodated in this design: stratification for regional reporting and stratification for minimizing standard errors of estimates. Within each sampled Landsat scene, a subsample of points was obtained using a 2 km × 2 km grid. The land cover class was interpreted from Landsat imagery at each sample point of the dot grid to obtain area estimates for each frame or PSU. To assess change in forest cover, the sampling unit was defined as "the overlap area of a pair of multi-date Landsat scenes" (FAO 1996, 7). The FRA 2000 assessment employed the same sample as the FRA 1990, with an additional time period included to estimate change from 1990 to 2000. This design employs a combination of design elements discussed in Section 5.2.1. The sampling design may be labeled as a two-stage cluster sample, with stratified random sampling used at the first stage to select a sample of Landsat scenes (PSUs) and systematic sampling used at the second stage to select points (SSUs).

The TREES II design (Richards et al. 2000) was implemented for estimating deforestation in the humid tropical forests for the time period 1990–1997. This design employed full and quarter Landsat scenes as the sampling units, with $n = 104$ sampled out of a possible $N = 740$ units. The sampling design had five strata based on percent forest cover and percent deforestation within each of the 740 units (Richards et al. 2000, 1480). Gallego's (2005, 370) retrospective assessment of the TREES II design concluded that it was statistically sound but overly complicated. As a simplification of the TREES II design, Gallego (2005) proposed employing stratification to partition variability of change (i.e., low and high variation) and selecting sample locations from a systematic grid. Similar to the TREES II design, the proposed modification is still strongly linked to using Landsat scenes as the basis for defining the sampling unit. The study region would first be partitioned by a tessellation based on Landsat scenes that accounted for scene overlap. The sample units created by this partitioning are unequal in size (area), and Gallego (2005) suggested implementing a design where the units are sampled with probability proportional to their area.

Mayaux et al. (2005) provide a retrospective critique of both the FRA 1990 and TREES II designs. They suggest that stratified sampling based on forest distribution and fragmentation, as determined from coarse-resolution satellite imagery, should be considered (Mayaux et al. 2005, 382). Knowledge of deforestation hot spots should also be used, possibly via stratification, to improve precision. Mayaux et al. (2005) proposed a design for future FRA global assessments, suggesting a large systematic sample of 10 km × 10 km blocks located at the intersections of 1° lines of latitude and longitude. This sample would consist of approximately 10,000 sample units. Such a design represents a shift from the strong dependence on Landsat images of the TREES II and FRA 1990, but as described in Mayaux et al. (2005), it would not incorporate stratification based on the anticipated degree of deforestation.

Hansen et al. (2008) selected a stratified random sample of 18.5 km × 18.5 km units to estimate gross forest cover loss during 2000–2005 in the humid tropical forest biome. The strata were determined based on MODIS-derived forest cover loss for each of the N units, and the estimated gross forest cover loss was quantified using Landsat imagery. A similar stratified design was implemented in the boreal and temperate forest biomes (Potapov et al. 2008) and the dry tropical forest biome (Hansen et al. 2010). The use of a common stratified sampling design and Landsat-derived gross forest cover loss for all four forested biomes is an example of how application of a consistent methodology can facilitate comparisons of rates of change at a global scale (Hansen et al. 2010). Hansen et al. (2008, 2010) employed a regression estimator (Section 5.2.4) to estimate gross forest cover loss, and the reported standard errors from this model-assisted strategy were generally small.

The FRA 2010 remote sensing survey is another example in which the consistency of methodology leads to global comparisons of forest change uncompromised by confounding differences in methods of measuring forest change. The FRA 2010 remote sensing survey is a systematic sample with the sample units (10 km × 10 km blocks) centered at the intersections of 1° lines of latitude and longitude (Ridder 2007; FAO 2009). Duveiller et al. (2008) report results from an intensified FRA sample to estimate forest cover change in Central Africa between 1990 and 2000. The sample grid points were located at every 0.5° intersection of latitude and longitude, yielding a fourfold increase in sample size over the 1° intersection grid. A total of 571 sample blocks (10 km × 10 km) were selected, although cloud cover prevented analysis of some sample blocks. The estimates of forest change had reasonably low standard errors, demonstrating the operational success of the methodology (Duveiller et al. 2008, table 2).

Levy and Milne (2004) review sample-based studies for estimating afforestation and deforestation in Great Britain. The National Countryside Monitoring Scheme (NCMS) of Scottish Natural heritage is a sample of 487 1 km × 1 km plots, with change interpreted from aerial photographs taken in the 1940s and 1980s. The countryside survey is based on 381 plots, also 1 km × 1 km, distributed throughout Great Britain. The countryside survey incorporates stratification based on "underlying environmental characteristics such as climate, geology and physiology" (Fuller et al. 1998, 103).

Leckie et al. (2002) describe a study to report deforestation and its carbon consequences for Canada. The sampling design is linked to the ongoing Canadian National Forest Inventory sample of 2 km × 2 km photoplots centered at points on a 20 km × 20 km grid. Stratification by expected deforestation level is incorporated in the sampling design. In the high deforestation strata, the sampling grid is intensified to 10 km × 10 km to increase the sample size. Interpretation of Landsat imagery is proposed to obtain the deforestation data.

Dymond et al. (2008) employed a stratified sampling design to estimate change in forest area between 1990 and 2002 for a portion of the South Island of New Zealand. The six strata defined were nonforest no change, two-forest no change strata (one for which a spectral difference was noted, the other for which no spectral difference was observed), a forest to nonforest change stratum, a nonforest to forest change stratum, and a "big clumps" stratum that could include to forest or from forest change, with these changes occurring in clumps of 5 ha or more. This "big clumps" stratum was expected to contain most of the change that could be identified from Landsat imagery, so this stratum was exhaustively sampled (censused). For the other five strata, sample points were randomly selected within each stratum. Dymond et al. (2008) found that this stratified design was much more efficient than simple random sampling.

To summarize these application studies, a variety of sampling designs have proven to be effective for monitoring forest change from remotely sensed data. Many of the basic design options described in Section 5.2.1 have been implemented in practice. Most studies employed a spatial sampling unit, with the FRA 1990 design and Dymond et al. (2008) being exceptions for which point sampling was implemented (the FRA 1990 did use a spatial sampling unit at the first stage of the two-stage cluster design). The early use of Landsat scenes or quarter scenes as the sampling units has generally been replaced in favor of smaller spatial units. Stratification is present in the majority of the designs implemented, with the FRA 2010 remote sensing survey being the most notable application not using stratification. Two-stage sampling in which the PSU is subsampled was implemented in the FRA 1990 design, but was not present in any other design included in this review. Systematic sampling is used at some stage of the sampling design in the FRA 1990, FRA 2010, TREES II, and the Canadian inventory (Leckie et al. 2002). Simple random selection, usually within strata, was used in the applications of Hansen et al. (2008, 2010), Dymond et al. (2008), and the surveys of Great Britain (Levy and Milne 2004).

5.4   Studies Evaluating Sampling Design Options

As the first noteworthy effort to employ sampling of remotely sensed data
to monitor forests, the FRA 1990 remote sensing survey triggered a series
of studies evaluating the effectiveness of sampling for forest monitoring
using remotely sensed data. An early and influential study by Tucker and
Townshend (2000) expressed concern that the FRA 1990 sampling approach
would not yield sufficiently precise estimates of deforestation unless the
sample size was extremely large. Tucker and Townshend's (2000) conclusions
were based on an investigation of deforestation for country-specific estimation for Bolivia, Colombia, and Peru. The populations evaluated were based
on complete coverage deforestation for these countries. Each country was
partitioned by Landsat scenes (41, 61, and 45 for Bolivia, Colombia, and Peru,
respectively), and the variability of sample-based estimates for simple random sampling of these scenes was evaluated. Tucker and Townshend (2000)
found that a large proportion of the available scenes had to be sampled to
obtain precise estimates of deforestation. Sanchez-Azofeifa et al. (1997) also
noticed that high variances of deforestation estimates could occur when the
sampling unit was a satellite scene. Sanchez-Azofeifa et al. (1997) examined a
population of 202 Landsat scenes from the Brazilian Amazon for which complete coverage change information was available. They demonstrated that a
stratified design with strata defined by "persistence" improved the precision
of the sample estimates relative to simple random sampling, where Sanchez-
Azofeifa et al. (1997, 183) defined persistence in terms of "scenes presenting
some degree of deforestation on time Ti will present more but no less deforestation between time Ti and time Ti+1 of total deforestation." Czaplewski
(2003) presented evidence to indicate that the problems encountered by these
studies were diminished when sampling was applied to larger regions, such
as continental or global estimates of deforestation.

   These early studies initiated a healthy debate of central issues of the sampling approach including the choice of sampling unit and the trade-offs
between cost and variability of sampling more but smaller sampling units.
These initial studies focused on Landsat scenes as the sampling unit, but relatively quickly (e.g., Tomppo et al. 2002; Stehman et al. 2003) it became apparent that using such a large sampling unit was a major contributor to the poor
performance of the sampling approach observed by Tucker and Townshend
(2000) and Sanchez-Azofeifa et al. (1997). Tucker and Townshend's (2000)
Bolivia population of $N = 41$ Landsat-based sampling units included one
unit that comprised 40% of the total deforestation of the region, and four
scenes accounted for 70% of the total deforestation of Bolivia. Tucker and
Townshend's (2000) analysis of the Bolivia population is noteworthy because
it identified that one or a few units with very high deforestation may occur
and have substantial impact on the standard error of the sample-based estimate of change. Outliers and their effect on the precision of estimated change

is an issue to be taken seriously. The shift to using sampling units smaller than Landsat scenes diminishes the impact of such outliers on the precision of the area estimates.

Tomppo et al. (2002) continued the evaluation of potential designs for continental and global forest assessments such as the FRA. Their results were based on a meticulously constructed hypothetical population of deforestation. Two sizes of sampling units were evaluated: a 150 km × 150 km sampling unit (corresponding approximately to the area of a Landsat image) and a 10 km × 10 km sampling unit. Stratification was implemented geographically using 10 FRA ecological zones to control the distribution of the sample among zones, and an additional level of stratification was defined using the Dalenius–Hodges rule (Cochran 1977) to determine strata boundaries based on the continuous variable Advanced Very High Resolution Radiometer (AVHRR) change. The sample was then allocated equally to five strata created within each geographic stratum. Tomppo et al. (2002) found that the 10 km × 10 km unit was more effective than the 150 km × 150 km unit when the stratified sampling design was implemented. Further, stratification by AVHRR change improved the standard errors of the estimates.

The planned use of systematic sampling for the FRA 2010 remote sensing survey prompted several studies investigating this design. As noted earlier, the FRA 2010 sampling design is a systematic sample of 10 km × 10 km blocks located at the intersections of the 1° lines of latitude and longitude. Steininger et al. (2009) evaluated the estimates that would be obtained from the FRA 2010 design if that design were to be applied to digital maps of deforestation for six regions (the five countries of Bolivia, Colombia, Ecuador, Peru, and Venezuela and the Brazilian Amazon) and the area represented by all six regions combined. This study also included a comparison of different size sampling units ranging from 5 km × 5 km to 50 km × 50 km and investigation of various grid densities (0.25° intersections of latitude and longitude up to 2° intersections). Steininger et al. (2009) concluded that the FRA design is clearly acceptable at the continental level, but country-specific estimates may be problematic. For a fixed sample size, a larger sample unit is obviously better, but Steininger et al. (2009) present results that provide insight into the trade-offs between smaller standard errors but increasing cost as the area of the sampling units increases.

Eva et al. (2010) conducted a study analogous to that of Steininger et al. (2009) to evaluate the performance of the FRA 2010 design estimates when applied to French Guiana (1990–2006 change) and the Brazilian Legal Amazon (BLA) (2002–2003 change). Again complete coverage deforestation information derived from Landsat imagery was the basis for evaluating the sample-based estimates. The sampling unit was 20 km × 20 km, and the sample size was $n = 330$ for the BLA. The estimated standard error of 0.10 million ha (based on nine replicate samples of the 1° intersections of latitude and longitude) obtained for the BLA is miniscule relative to the estimate of 2.81 million ha of deforested area. For French Guiana, the systematic sample was intensified to 0.25° grid intersections (a 16-fold increase over the standard FRA grid spacing of 1° intersections), resulting in a sample size of 108 sample units (approximately 12% of the total area), and the size of the sampling unit was reduced to 10 km × 10 km. For this intensified sample, the estimated standard error was about 6.8% of the estimated area of deforestation. The design of the Eva et al. (2010) study did not include comparison of systematic sampling to simple random sampling, but it can be expected that the systematic design improved upon the standard errors that would have been obtained from simple random sampling.

Broich et al. (2009) investigated the relative precision of systematic, stratified random, and simple random sampling using a population of Landsat-derived 2000–2005 deforestation for the BLA. The strata were based on MODIS-derived change for the 18.5 km × 18.5 km units partitioning the study region. The systematic sampling design was modeled after the FRA 2010 design of sampling at 1° intersections of latitude and longitude and an intensified version of that design with sampling units at 0.5° intersections. Broich et al. (2009, table 3 and table 4) found that both systematic and stratified sampling were improvements over simple random sampling, and both were operationally very effective for estimating deforestation based on the standard errors relative to the annual rate of deforestation for the study area (population) that was 0.55% (percent of area). The 1° systematic sample (325 sample units) yielded a standard error of 0.05%, the stratified random sample (150 sample units) yielded a standard error of 0.03%, and the 0.5° systematic sample (1,310 sample units) yielded a standard error of 0.02%. For this particular study, the stratified design was more effective than systematic sampling, the advantage being attributable to the effectiveness of the MODIS-based stratification. Further investigation would be needed to confirm the utility of a similar approach to stratification for other locations and different time periods.

Stehman et al. (2011) used the same population of deforestation for the BLA investigated by Broich et al. (2009) to demonstrate the utility of stratified random sampling for adapting a global forest monitoring design to achieve regional reporting objectives. The stratified sampling design employed by Hansen et al. (2008) for the humid tropical forests could be augmented using the same stratified design to address the objective of estimating deforestation by states within the BLA. The ability to augment a stratified continental or global sample parallels the use of an intensified systematic sample (Eva et al. 2010) to produce country- or region-specific estimates for the FRA 2010 design. The analyses also permitted comparing the standard errors for simple random, systematic, and stratified random sampling for the states within the BLA. When compared on the basis of equal sample size, both systematic and stratified random sampling were better than simple random sampling, and for most states, stratified random sampling had a smaller standard error than systematic sampling (Stehman et al. 2011, table 5). Similar to the precautions expressed for interpreting the Broich et al.'s (2009) results, the strong advantage gained by the MODIS-based stratification in the BLA would not necessarily extend to other geographic locations or time periods.

These evaluative studies have progressed from the precautionary findings revealed by Tucker and Townshend (2000) to strong confirmation that the sampling approach can yield estimates with relatively small standard errors. However, the sampling design must be chosen based on recognizing some of the potential pitfalls, the foremost of which is that very large sampling units (e.g., Landsat scenes) should be avoided. The evaluative studies support the results of the actual applications (Section 5.3) of sample-based estimates of forest change in that the small standard errors observed in practice are substantiated by empirical investigation of the sampling designs applied to known populations of deforestation.

The majority of the research examining different sampling design options has focused on the basic sampling designs outlined in Section 5.2.1 (systematic, stratified random, and cluster sampling). Several designs outside this traditional realm have been considered. Magnussen et al. (2005) evaluated adaptive cluster sampling (ACS), a sampling design that is advocated as efficient and practical for rare but spatially clustered phenomena, exactly a scenario often envisioned for forest cover change. Magnussen et al.'s (2005) general recommendation was that "ACS remains attractive when the average cost of adaptively adding a PU [population unit] to the initial sample is low relative to the average cost of sampling a PU at random." This condition would not be met when working with a satellite scene as the PU. If the PU is smaller than a Landsat scene, for example, when using a 10 km × 10 km unit, the condition described may be satisfied because if the adaptive procedure calls for additional PUs (the 10 km × 10 km units) within a scene in which other PUs have been interpreted, this would be less costly than obtaining a new PU in a different Landsat scene. Magnussen et al. (2005) expressed several additional reservations regarding the use of ACS, noting that practical experience with ACS is still limited and that design effects (i.e., precision improvements) and costs can be highly variable. They further noted that it is likely that a rule for terminating the adaptive selection process would be needed to avoid cost overruns (i.e., to avoid uncontrolled progression to selecting new sample units from the adaptive steps of the protocol), thus adding complexity to the design, and that the effect of population structure on ACS is so complex that it is difficult to predict success of ACS for a given application. ACS is more complex to implement and analyze, so the advantages gained must be sufficient to overcome this burden of greater complexity.

When stratified sampling is used to increase the sample size of sampling units with anticipated high forest cover change, the design is an example of an unequal probability sampling design. That is, the inclusion probabilities for units in different strata are different. The extension of unequal probability sampling to a design for which the inclusion probabilities are proportional to an auxiliary variable x (denoted as πpx designs) is another option to consider.

Giree (2011) implemented a πpx design in a study of gross forest cover loss in Malaysia, where x was the area of change derived from AVHRR for 1990–2000. The rationale for implementing a πpx design instead of a stratified design was related to the options for estimation (Section 5.2.4). A special case of the general regression estimator applicable to a stratified random design is the separate regression estimator, and this estimator requires a sample size of 25–30 per stratum to ensure that the estimator is not biased. Because the sample size for the entire Malaysia study was a modest $n = 25$ units (each 18.5 km × 18.5 km), a stratified design combined with the separate regression estimator would have been a risky proposition. The πpx design allowed the option to use the auxiliary variable x to increase the sample size of higher change units, and the general regression estimator could still be applied to the sample of 25 units without concern for bias attributable to a small sample size. For the πpx design implemented and using the Horvitz–Thompson estimator (Equation 5.1), Giree (2011) estimated the annual gross forest cover loss for Malaysia during 1990–2000 to be 0.43 million ha per year with a standard error of 0.04 million ha per year. Thus despite the small sample size, the πpx design yielded a reasonably small standard error relative to the estimated rate of deforestation.

The sample obtained by Giree (2011) is useful to illustrate the application of model-based inference. Suppose that Y_i is the area of deforestation for 1990–2000 obtained from Landsat and x_i is the area of deforestation obtained from AVHRR on unit i (where each unit is 18.5 km × 18.5 km). The AVHRR value (x_i) is available for all $N = 958$ units comprising Malaysia (i.e., the entire population), but the Landsat deforestation is available for only the $n = 25$ sample blocks selected by the πpx design described in the preceding paragraph. Following Valliant et al. (2000, section 5.5.1), suppose that the model relating Y_i to x_i is a quadratic model of the form

$$Y_i = \beta_0 + \beta_1 x_i + \beta_2 x_i^2 + e_i,$$

where e_i is distributed with mean 0 and variance $v_i = x_i^2 \sigma^2$. The predicted value for unit (block) i, $i = 1, \ldots, N$, is

$$\hat{Y}_i = \hat{\beta}_0 + \hat{\beta}_1 x_i + \hat{\beta}_2 x_i^2$$

where the estimates of the β's are obtained by least squares. If s denotes the elements selected for the sample and r denotes the remaining (not sampled) elements in the population, the model-based estimator for the population total (based on the model specified above) is

$$T = \sum_s Y_i + \sum_r \hat{Y}_i$$

The estimator T does not take into consideration that the sampling design was πpx and instead is entirely dependent on the specified model. The "prediction theory" basis of the estimator is also apparent because the second term of T is a sum of the predicted values of Y_i for the elements of the population that were not observed in the sample. For the Giree (2011) sample data for 1990–2000 deforestation in Malaysia, the model-based estimator is 0.35 million ha per year (slightly below the 0.43 million ha per year for the design-based estimate). The standard error for the model-based estimate was 0.07 million ha per year (based on the specified model and equation 5.1.6, p. 130 of Valliant et al. 2000). Although it is tempting to compare the standard errors of the design-based and model-based estimators, the two approaches to inference employ very different definitions of variability, and it does not seem relevant to compare variances that constitute very different representations of uncertainty. In practice, the analysis using a model-based estimator should include evaluation of competing models and an assessment of the goodness of fit of the data to model assumptions. These details are omitted for reasons of brevity.

5.5 Discussion of Sampling Applications and Evaluative Studies

Several general tendencies emerge from this review of applications and evaluative studies of forest monitoring sampling designs for remotely sensed data. The degree to which the sampling design is tailored to the spatial characteristics of the satellite imagery ranges from a strong dependence in which Landsat scenes or quarter scenes are used as the sampling units (Richards et al. 2000; Tucker and Townshend 2000; Czaplewski 2003; Gallego 2005) to virtually no dependence on the imagery for defining sampling units (Leckie et al. 2002; Levy and Milne 2004; Mayaux et al. 2005; Hansen et al. 2008, 2010). Gallego (2005) notes that choosing the size of the sampling unit to correspond to the specific imagery to be used to interpret forest cover or change is justified when working with sensors with approximately fixed image frames (e.g., Landsat TM), but otherwise becomes more complicated. In a long-term monitoring program, or in cases where several sources of imagery might be used, the advantages of choosing the sampling unit linked to a single imaging framework are diminished.

For studies covering continental or global change, an initial stratification by biomes, ecoregions, or other large areas is typically implemented, although the FRA remote sensing survey is a notable exception. Geographic strata are typically meaningful regions for reporting results, and they also serve to aggregate relatively homogeneous forest types together, which may be advantageous for better precision of continental or global estimates of change. In most of these studies targeting the objective of estimating the area of forest change, stratification based on a proxy or surrogate for true change must be used. The goal is to create strata in which change is relatively uniform within each stratum, thus creating smaller within-stratum variances. Stratification also allows for increasing the sample size in the higher variability strata.

Many of the desirable design criteria specified in Section 5.2.5 are prominent in the sampling designs implemented in practice for forest monitoring using remotely sensed data. All of the sampling designs reviewed in this chapter satisfy the conditions defining a probability sampling design. This noteworthy feature suggests that the importance of rigorous design-based inference combined with a probability sampling design has been recognized at the design planning stage. Most of the applications reviewed met the second desirable design criterion of being simple to implement. The two most commonly used sampling designs, systematic (e.g., the FRA 2010 design) and stratified random (e.g., Dymond et al. 2008; Hansen et al. 2008, 2010), are straightforward to implement. The two examples of more complex sampling designs presented in this chapter were ACS, investigated by Magnussen et al. (2005), and sampling with probability proportional to an auxiliary variable x, where x could be a measure of forest cover loss from coarser resolution imagery (Giree 2011) or x could simply be the area (size) of each element in the partition of the universe (Gallego 2005). A majority of the designs reviewed included some capacity for distributing the sample spatially (criterion 4), either by implementing a systematic selection protocol or by incorporating geographic stratification. The sampling designs implemented in practice (Section 5.3) produced standard errors that were small enough that the estimates would likely be viewed as credible for most uses of the estimates (criterion 5).

An unbiased estimator of variance is not available for systematic sampling, and the estimated variance is then based on an approximation (desirable design criterion 6). A simple approximation is to use a variance estimator appropriate for simple random sampling, and this approximation is typically a biased overestimate of the variance for the systematic design. Such an overestimate of variance is often acceptable because it is conservative (i.e., it does not under-report the uncertainty of the estimate), but a conservative estimate also will not reflect the true precision of the estimate. Thus it may be that systematic sampling has produced a very precise estimate, but the estimated standard error, being a conservative overestimate, will not reflect that precision. Stratified random sampling does permit an unbiased estimator of variance.

Most sampling designs can be implemented in a manner that will allow for changing the sample size "in progress" (criterion 7). Simple random and stratified random protocols are particularly easy to truncate to reduce the target sample size or extend to increase the target sample size while still maintaining the fundamental features of the design (Stehman et al. 2011). Intensifying a systematic sample is straightforward simply by changing the grid density (e.g., decreasing the distance between grid points by half increases the sample size fourfold). Less severe changes in sample size will require breaking up the strict grid structure. For example, to add 10 new sample units, the original grid spacing could be halved and 10 units selected at random from the introduced new grid points. To reduce the sample size from the initial grid, sample units could be randomly deleted, although this assumes that the existing sample up to the point of sample termination had been selected in a random order. Both of these sample size modifications of a systematic grid will produce a final sample that does not adhere exactly to the initial full grid structure and will therefore diminish some of the advantages of the systematic sample.

The last desirable design criterion, "transparency," is difficult to assess because it depends on individual experience with sampling methods and theory. The designs implemented in practice for forest monitoring (Section 5.3) are probability sampling designs, which conveys a strong element of transparency to the process if one is familiar with the theory of design-based inference and estimation. Systematic sampling is intuitively appealing and therefore transparent to nonscientists because of the uniform spatial distribution of the sample across a region and because of the obvious explanation for why sample points are located where they are. A probability sample based on simple random selection may be misconstrued by laypersons as having been subjectively selected to focus on specific locations to bias the results in a particular fashion. Similarly, intensifying the sampling effort within some strata may be misunderstood by laypersons as an effort to increase the sample size within areas of high deforestation, thus "obviously" biasing the estimates in the minds of those not aware of the weighted estimation approaches required with unequal probability sampling designs (see Equations 5.1 and 5.2). It is an interesting question of how individual perceptions (e.g., various levels of understanding of sampling theory and practice) should influence the decision-making process when considering different sampling design options for a given application.

5.6 Sampling for Forest Monitoring Using Remotely Sensed Data: A Look Ahead

Despite past operational successes of remote sensing–based forest monitoring using a sampling approach, much room for improvement exists to develop more accurate, more precise, and more cost-effective methods. One of the biggest concerns with forest monitoring by remote sensing is measurement error—are the remote sensing measurements of forest attributes such as cover or deforestation sufficiently accurate? Measurement error can be viewed as having two components: bias and variability. Measurement bias refers to a consistent over- or under-representation of the true value of the response, and measurement variability refers to the differences in the observed response over multiple replications of the measurement process (Särndal et al. 1992). For example, if the area of deforestation for a 10 km × 10 km unit is obtained by a human interpreter working with satellite imagery or aerial photographs, we can envision replicated realizations of this measurement by different interpreters. If the average result of these repeated observations of deforestation differs from the true value of the unit, measurement bias is present. If the repeated observations vary from interpreter to interpreter, measurement variability is present. It is straightforward to quantify measurement variability by having different interpreters examine the same sampling unit, but it is less obvious how to quantify measurement bias.

A fundamental premise of the sampling approach to forest monitoring is that the best available protocols for obtaining the target forest measurements are being used. The assessment of measurement bias would require that a more accurate measurement protocol existed, and that it would be possible to estimate measurement bias based on what would likely be a relatively small sample (i.e., if a larger sample size using the more accurate protocol were available, this measurement protocol would be the basis of the monitoring estimates). For example, if Landsat is the best-quality imagery that can be affordably used in a sample-based monitoring program, then it would be possible to spot check the Landsat interpretations using very high-resolution imagery and a more detailed (i.e., more accurate) interpretation protocol, and this would provide a way to assess measurement bias. Specific sampling designs to incorporate the assessment of measurement error have not received much attention.

Another challenging question is how to construct the sampling design for long-term forest monitoring based on remotely sensed data. A number of factors play into this decision. Over time, it is possible that improved methods (e.g., better imagery, more accurate classification methods) will be developed for measuring the forest characteristics of interest. The sampling design should be able to incorporate these improved options. For example, if new sources of imagery prove to be better, the sampling design must be able to accommodate a potential change in the footprint of different imagery. A good illustration of this problem is the early emphasis on using Landsat scenes as sampling units. Even if these large sample units had proven to be effective for use with Landsat, it is likely that smaller sampling units would now be more desirable for the very high-resolution imaging options that subsequently have become available.

A number of challenging questions remain to be resolved regarding the three primary decisions that determine a sampling design (Section 5.2.1). Consider the cluster sampling decision first. The primary advantage of cluster sampling is the savings in time and cost of working with a sample that is spatially constrained in the sense that the sample may be controlled to fall within a fixed number of clusters or PSUs. When working with a specific source of imagery, cluster sampling allows for controlling the number of images that must be processed (e.g., a Landsat or an IKONOS image). Gallego (2012) demonstrated that sampling a relatively small number of SSUs within each PSU is adequate from the standpoint of statistical precision, and little advantage is gained by one-stage cluster sampling. The qualitative nature of Gallego's (2012) result is not surprising, but the quantitative revelation that such a small number of SSUs would generally be adequate is eye opening. Gallego's (2012) result suggests that two-stage cluster sampling should be given serious consideration. One-stage cluster sampling may still be a good design option for other reasons (e.g., when landscape pattern and other landscape context information is desirable), but two-stage sampling is clearly a strong option when estimating area is the primary objective.

Although stratification has been demonstrated to be effective for estimating area (Tomppo et al. 2002; Broich et al. 2009; Stehman et al. 2011), the precautions noted about the portability of these results to other regions for which forest change dynamics may be different should be heeded. In a long-term forest monitoring setting, the benefit of stratification would almost surely diminish over time. However, it may still be worthwhile to include stratification simply because estimating a relatively rare event such as change with acceptably small standard errors may be difficult otherwise. If the monitoring is retrospective (e.g., estimating forest change from 1980 to 2010), then even though multiple time periods of change may be of interest (e.g., every 5-year period), it may still be possible to develop an effective stratification based on change throughout the full monitoring period. Because archival imagery and other information exist pertaining to changes that have taken place, it is possible to stratify by change based on auxiliary information. In the design of a forward-looking (prospective) monitoring program, the ability to choose an effective stratification may become more tenuous. In the prospective setting, the strata must be defined by expected change if the sample data must be obtained in real time (i.e., when it is not feasible to use archival imagery).

For long-term monitoring with periodic reporting (e.g., 5-year time periods), the question of permanent sample plots versus allowing the sample locations to change over time is another important consideration. For example, if estimates are desired for each 5-year period over a 30-year total period of monitoring, sample locations will need to be paired (i.e., the initial and end date) for any given 5-year period to estimate gross change. But the decision of whether to use permanent plots for the entire 30-year monitoring window will depend on the situation. For example, in a region of rapid cycling from forest clearing to regrowth to clearing, the 30-year time series from permanent plots may prove invaluable. Conversely, in a less dynamic region in which at most one change will occur in the 30-year period, it may be advantageous to focus more on the individual 5-year estimates. This may lead to implementing a stratification that is advantageous for each 5-year estimate, but not necessarily a stratification useful for any other time period, and consequently a new set of paired plots would be selected for each 5-year period. In a prospective monitoring program, particularly one that may have regulatory ramifications, it would be preferable to have the sample locations "hidden" from the parties involved so that forest management of the sample locations is not different from forest management of the general population. However, not revealing sample locations would seem to conflict with the desirable design criterion of transparency. Consequently, permanent plot locations for prospective regulatory monitoring could be problematic. If new sample locations are selected for each reporting interval, these problems with permanent plots would be avoided. Sampling design decisions will be strongly influenced by practical considerations. Additionally, studies investigating the precision of permanent sample locations versus more flexible sample arrangements should be conducted for various scenarios of forest change.

Two-phase sampling is often an effective design for general-purpose monitoring (see Fattorini et al. 2004 for a specific example application) and has a relatively long history of use for forest inventory. In two-phase sampling, a large first-phase sample is selected, and one or more auxiliary variables are measured for each unit sampled. A second-phase sample is then selected, typically from the first-phase sample units, and the target measurements are obtained for the smaller second-phase sample. In contrast to two-stage cluster sampling in which the sampling units are different sizes for the two stages, it will be assumed that the sampling units are defined similarly at both phases for two-phase sampling. The auxiliary information from the larger first-phase sample may be used in two ways. One option is to use the auxiliary variables in a model-assisted estimator. The other option is to use the auxiliary information to stratify the first-phase sample units and to then select a stratified sample at the second phase. Two-phase sampling for stratification is a practical option when it is not feasible to stratify all N elements of the universe.

5.7 Conclusions

The complete coverage mapping and sampling-based approaches should coexist in a forest monitoring program as both approaches address important and sometimes different objectives. The full coverage, spatially explicit information provided by maps is an invaluable resource. But typically there will be higher quality information than what was used to construct the map, and this higher quality information becomes affordable and practically manageable for only a sample of the full region. Thus a sample in which higher quality imagery and more accurate measurement protocols can be applied becomes the basis of an estimate for aggregate properties of the forest characteristics to be monitored. The sample-based approach to monitoring forest cover and change in forest cover has been proven to be operationally effective in a number of studies. Efforts to refine these methods to produce more accurate and precise estimates of forest characteristics should continue to take advantage of new developments of higher quality imagery and better classification methods.

About the Contributor

Stephen V. Stehman has been employed at State University of New York–College of Environmental Science and Forestry, New York (SUNY ESF), since 1989. He teaches courses in sampling methods and design of experiments and provides statistical consulting service for faculty and students. Stehman was introduced to sampling designs for environmental monitoring while working for Scott Overton during the design phase of the United States Environmental Protection Agency's Environmental Monitoring and Assessment Program (EMAP). Stehman also conducts research on sampling design and analysis methods for assessing accuracy of land cover and land cover change maps.

References

Achard, F. et al., Determination of deforestation rates of the world's humid tropical forests. *Science*, 297, 999, 2002.

Broich, M. et al., A comparison of sampling designs for estimating deforestation from Landsat imagery: A case study of the Brazilian Legal Amazon. *Remote Sensing of Environment*, 113, 2448, 2009.

Cochran, W.G., *Sampling Techniques* (3rd edn.). Wiley, New York, 1977.

Corona, P., Integration of forest mapping and inventory to support forest management. *iForest*, 3, 59, 2010.

Czaplewski, R.L., Can a sample of Landsat sensor scenes reliably estimate the global extent of tropical deforestation? *International Journal of Remote Sensing*, 24, 1409, 2003.

Duveiller, G. et al., Deforestation in Central Africa: Estimates at regional, national and landscape levels by advanced processing of systematically-distributed Landsat extracts. *Remote Sensing of Environment*, 112, 1969, 2008.

Dymond, J.R. et al., Estimating area of forest change by random sampling of change strata mapped using satellite imagery. *Forest Science*, 54, 475, 2008.

Eva, H. et al., Monitoring forest areas from continental to territorial levels using a sample of medium spatial resolution satellite imagery. *ISPRS Journal of Photogrammetry and Remote Sensing*, 65, 191, 2010.

FAO, *Forest Resources Assessment 1990: Survey of Tropical Forest Cover and Study of Change Processes.* FAO Forestry Paper 130, FAO, Rome, 1996.

FAO, *The 2010 Global Forest Resources Assessment Remote Sensing Survey: An Outline of Objectives, Data, Methods, and Approach.* FAO Forest Resources Assessment Working Paper 155, FAO, Rome, 2009.

Fattorini, L., Marcheselli, M., and Pisani, C., Two-phase estimation of coverages with second-phase corrections. *Environmetrics,* 15, 357, 2004.

Fuller, R.M., Wyatt, B.K., and Barr, C.J., Countryside survey from ground and space: Different perspectives, complementary results. *Journal of Environmental Management,* 54, 101, 1998.

Gallego, F.J., Remote sensing and land cover area estimation. *International Journal of Remote Sensing,* 25, 3019, 2004.

Gallego, F.J., Stratified sampling of satellite images with a systematic grid of points. *ISPRS Journal of Photogrammetry & Remote Sensing,* 59, 369, 2005.

Gallego, F.J., The efficiency of sampling very high resolution images for area estimation in the European Union. *International Journal of Remote Sensing,* 33, 1868, 2012.

Giree, N., Quantifying gross forest cover loss in Malaysia between 1990 and 2005, Unpublished MS Thesis, South Dakota State University, Brookings, SD, 2011.

Hansen, M.C., Stehman, S.V., and Potapov, P.V., Quantification of global gross forest cover loss. *Proceedings of the National Academy of Sciences,* 107, 8650, 2010.

Hansen, M.C. et al., Humid tropical forest clearing from 2000 to 2005 quantified using multi-temporal and multi-resolution remotely sensed data. *Proceedings of the National Academy of Sciences,* 105, 9439, 2008.

Hassett, E.M., Stehman, S.V., and Wickham, J.D., Estimating landscape pattern metrics from a sample of land cover. *Landscape Ecology,* 27, 133, 2012.

Leckie, D.G., Gillis, M.D., and Wulder, M.A., Deforestation estimation for Canada under the Kyoto Protocol: A design study. *Canadian Journal of Remote Sensing,* 28, 672, 2002.

Levy, P.E. and Milne, R., Estimation of deforestation rates in Great Britain. *Forestry,* 77, 9, 2004.

Magnussen, S. et al., Adaptive cluster sampling for estimation of deforestation rates. *European Journal of Forest Research,* 124, 207, 2005.

Mayaux, P., Achard, F., and Malingreau, J.-P., Global tropical forest area measurements derived from coarse resolution satellite imagery: A comparison with other approaches. *Environmental Conservation,* 25, 37, 1998.

Mayaux, P. et al., Tropical forest cover change in the 1990s and options for future monitoring. *Philosophical Transactions of the Royal Society B,* 360, 373, 2005.

Overton, W.S. and Stehman, S.V., The Horvitz-Thompson theorem as a unifying perspective for probability sampling: With examples from natural resource sampling. *American Statistician,* 49, 261, 1995.

Potapov, P. et al., Combining MODIS and Landsat imagery to estimate and map boreal forest cover loss. *Remote Sensing of Environment,* 112, 3708, 2008.

Richards, T., Gallego, J., and Achard, F. Sampling for forest cover change assessment at the pan-tropical scale. *International Journal of Remote Sensing,* 21, 1473, 2000.

Ridder, R.M., *Global Forest Resources Assessment 2010. Options and Recommendations for a Global Remote Sensing Survey of Forests.* FAO Forest Resources Assessment Working Paper 141, FAO, Rome, 2007.

Sanchez-Azofeifa, G.A., Skole, D.L., and Chomentowski, W., Sampling global deforestation databases: The role of persistence. *Mitigation and Adaptation Strategies for Global Change,* 2, 177, 1997.

Särndal, C.-E., Swensson, B., and Wretman, J., *Model-Assisted Survey Sampling*. Springer-Verlag, New York, 1992.

Stehman, S.V., Comparing estimators of gross change derived from complete coverage mapping versus statistical sampling of remotely sensed data. *Remote Sensing of Environment*, 96, 466, 2005.

Stehman, S.V., Sohl, T.L., and Loveland, T.R., Statistical sampling to characterize land-cover change in the U.S. Geological Survey land-cover trends project. *Remote Sensing of Environment*, 86, 517, 2003.

Stehman, S.V. et al., Adapting a global stratified random sample for regional estimation of forest cover change derived from satellite imagery. *Remote Sensing of Environment*, 115, 650, 2011.

Steininger, M.K., Godoy, F., and Harper, G., Effects of systematic sampling on satellite estimates of deforestation rates. *Environmental Research Letters*, 4, 034015, 2009.

Tomppo, E., Czaplewski, R., and Mäkisara, K., *The Role of Remote Sensing in Global Forest Assessments*. FAO Forest Resources Assessment Working Paper 61, FAO, Rome, 2002.

Tucker, C.J. and Townshend, J.R.G., Strategies for monitoring tropical deforestation using satellite data. *International Journal of Remote Sensing*, 21, 1461, 2000.

Valliant, R., Dorfman, A.H., and Royall, R.M., *Finite Population Sampling and Inference: A Prediction Approach*. Wiley, New York, 2000.

6

Use of Coarse-Resolution Imagery to Identify Hot Spots of Forest Loss at the Global Scale

Matthew C. Hansen, Peter Potapov, and Svetlana Turubanova
University of Maryland

CONTENTS

6.1 Introduction

6.1.1 MODIS

The MODIS (Moderate Resolution Imaging Spectroradiometer) sensor onboard NASA's Terra spacecraft has advanced large-area land monitoring during its 10-plus years of operation. Compared to heritage instruments such as the advanced very high-resolution radiometer (AVHRR) meteorological sensor, MODIS represented a significant gain in global land mapping and monitoring capabilities. First, the MODIS sensor has a finer instantaneous field of view compared to other global daily observing systems, including bands with 250, 500, and 1000 m spatial resolutions. Second, MODIS was built with seven bands specifically designed for land cover monitoring by avoiding wavelengths affected by atmospheric scattering and absorption. Third, the 250 m spatial resolution of the red and near-infrared bands was designed specifically to enable the monitoring of land cover change (Justice et al. 1998). Other sensors with global land monitoring capabilities, including SPOT VEGETATION and ENVISAT MERIS, with 1 km and 300 m spatial resolutions, respectively, have also been designed for land monitoring applications. However, MODIS retains the finest spatial resolution observational capability for this class of sensors. While a second MODIS sensor onboard NASA's Aqua spacecraft was launched in 2002, MODIS Terra data have been more widely used in land cover analyses and are the data used in the study presented here.

6.1.2 Global Forest Cover Mapping to Date

A viable solution to examining trends in forest cover change over large areas is to employ remotely sensed data. Satellite-based monitoring of forest clearing can be implemented consistently across large regions at a fraction of the cost of obtaining extensive ground inventory data. Forest inventories are typically unable to quantify forest dynamics at annual intervals due to the costs and logistical challenges of frequently revisiting plots. On the other hand, remotely sensed data enable the synoptic quantification of forest cover and change at regular intervals, providing information on where and how fast forest change is taking place at annual or finer time scales (INPE 2008). While numerous national-scale forest change products exist, global forest change characterizations are comparatively rare. Initial global forest mapping efforts focused on static map products of forest cover, typically as part of multiclass land cover classifications. The IGBP DISCover project (Loveland et al. 2000) used 1 km AVHRR data to produce a global land cover product that included forest leaf type and longevity classes, as did Hansen et al. (2000) with the University of Maryland (UMD) land cover map. Friedl et al. (2002) advanced these efforts in creating the standard MODIS land cover product (MOD12Q1), and Bartholomé et al. (2005) used SPOT VEGETATION data to produce the Global Landcover 2000 (GLC2000) product, both of which contained multiple forest type/density classes. Similarly, the Globcover initiative used 300 m ENVISAT MERIS data to produce a global multiforest class land cover map for 2005–2006 (Arino et al. 2007). Forests as a specific target have been mapped at the global scale as well. Global subpixel percent tree cover maps have been generated using AVHRR data (Hansen and DeFries 2004) and as a standard product using MODIS data, the vegetation continuous field (VCF) of percent tree cover (Hansen et al. 2003). Regarding global forest change, the 8 km AVHRR Pathfinder data set was used to estimate tree cover change from 1982 to 1999 from time-sequential percent tree cover maps (Hansen and DeFries 2004).

6.1.3 Global Forest Cover Loss Mapping Using MODIS

A more recent global forest cover change assessment employed MODIS data to quantify gross forest cover loss (Hansen et al. 2010). In this study, MODIS 500 m forest cover loss indicator maps were used to stratify biomes into homogeneous regions with respect to change (high, medium, and low forest cover loss strata). Within each stratum, samples of Landsat data were drawn and analyzed in order to estimate forest cover extent in 2000 and forest cover loss from 2000 to 2005. Stratum-specific regression estimators incorporating the MODIS-derived forest cover loss data as an auxiliary variable were applied to generate the final forest cover loss estimates. These results demonstrated the effectiveness of using the MODIS forest cover loss data to provide a spatially fine-grained stratification that offered an improvement over more generalized hot spot stratifications subjectively delineated to define low and high forest clearing strata (Achard et al. 2002).

The focus of this study is to extend this previous MODIS work and map indicated forest cover loss at 250 m spatial resolution over the 2000–2010 period. To do so, a turn-key algorithm is run on the 2000–2005 and 2005–2010 epochs. Previous work on multiyear forest cover change quantification using AVHRR data employed a recalibrated model for each year of analysis (Hansen and DeFries 2004). However, as MODIS data feature consistent radiometric calibration (Vermote et al. 2002), it is expected that the change signal being trained upon may be reliably and repeatedly captured over time. Our previous work with MODIS has employed turn-key models applied annually to identify change (Hansen et al. 2008; Potapov et al. 2008). For this study, we employ a fixed characterization algorithm for the 2000–2005 and 2005–2010 epochs. Calibration issues with MODIS have been studied, and a degradation of the near-infrared band quantified for MODIS Terra (Wang et al. 2012).

Given this fact, the use of turn-key approaches to repeatedly mapping land cover with the Terra instrument has come into question (Vermote E., personal communication). We present the following results more as a demonstration of global change mapping methods and not as a definitive long-term environmental change record. MODIS data are imaged nearly daily at the global scale, improving the probability of cloud-free acquisitions. This high-temporal acquisition frequency ensures a consistent and largely cloud-free image feature space at annual time scales. However, the moderate spatial resolution of MODIS is a limitation for area estimation of forest cover loss as much forest disturbance occurs at sub-MODIS pixel scales. The most appropriate use of MODIS for forest monitoring is as an alarm or hot spot indicator (INPE 2008; Hansen et al. 2010; Shimabukuro et al. 2012). Area estimation requires the integration of MODIS with a higher spatial resolution sensor, such as Landsat or another medium spatial resolution data source. MODIS-only products such as the ones presented in this study capture relative rates of forest cover loss across space and through time, with a considerable omission rate for small-scale forest disturbances.

The method presented here demonstrates a global assessment of forest cover loss using MODIS data from 2000 to 2010. For this study, forest clearing equals gross forest cover loss during the study period without quantification of contemporaneous gains in forest cover due to reforestation or afforestation. Forest cover loss is defined as a stand-replacement disturbance of a forest, where forest is defined as an assemblage of trees having a height of 5 m or greater and a canopy crown cover in excess of 25% at the MODIS pixel scale. The method could be implemented repeatedly for both forest cover loss and gain in establishing internally consistent biome-scale trends in both gross and net forest cover loss and gain.

6.2   Data

The 2000–2011 global Terra/MODIS 250 m data 16-day composite data set (MOD44C, collection 5) from the University of Maryland was used. This data set was originally created as an input to the vegetative continuous fields and vegetative cover conversion product and is described in Carroll et al. (2010). Four reflective bands—band 1/red (620–670 nm), band 2/near infrared (841–876 nm), band 6/shortwave infrared (1,628–1,652 nm), and band 7/shortwave infrared (2,105–2,155 nm), along with band 31/thermal (10,780–11,280 nm) and computed normalized difference vegetation index (NDVI)—were used.

Six-year MODIS metrics were derived for 2000 through 2005 and 2005 through 2011. Metrics have been shown to enable large-area mapping by generalizing the multispectral feature space, enabling signature extension over large areas (Reed et al. 1994; DeFries et al. 1995; Hansen et al. 2005). Each band was ranked individually and by temperature and NDVI. Ranked metrics calculated for all bands included 0, 10, 25, 50, 75, 90, and 100 percentiles. Averages between percentiles were also calculated. Annual metrics were generated and used as metrics and as inputs to a time-series regression calculation. Means of the three values corresponding to highest annual NDVI and band 31 brightness temperature were derived and used as the annual inputs and for the regression calculation.

An extensive Landsat-scale training data set was produced for calibrating the algorithm. National-scale products for Indonesia (Broich et al. 2011); the Democratic Republic of the Congo (Potapov et al. in press); European Russia (Potapov et al. 2011); Quebec, Canada; and Brazil, along with an additional 203 image pairs, were used as training data. The majority of the training data were from the 2000 to 2005 epoch. Only the Indonesia and Democratic Republic of the Congo data included 2005–2010 change data. The Landsat-scale forest cover loss maps were aggregated to the MODIS grid as percent forest cover loss. A total of over 23,000,000 pixels at MODIS scale were available as training data.

6.3   Algorithm

Decision trees are a type of distribution-free machine learning tool appropriate for use with remotely sensed data sets (Michaelson et al. 1994;

Hansen et al. 1996; Freidl and Brodley 1997). They are the primary algorithmic tool used in the standard MODIS land VCF products (Hansen et al. 2003). The VCF products depict the per pixel percent cover of basic vegetation traits, such as herbaceous and tree cover. As trees are distribution free, they allow for the improved representation of training data within the multispectral space. The relationship between the independent and dependent variables need not be monotonic or linear. This allows for more flexible subsetting of the multispectral image space not feasible with many other methods and is most appropriate for large-area studies that feature complicated multispectral signatures. In addition, the tree structure enables the interpretation of the explanatory nature of the independent variables.

Trees can accept either categorical data in performing classifications (classification trees) or continuous data in performing subpixel percent cover estimations (regression trees) (Breiman et al. 1984). For this study, we used the regression tree algorithm of the S-Plus statistical package (Clark and Pergibon 1992) to depict percent forest cover loss. Methods to avoid overfitting of tree models are available. One such approach entails performing multiple, independent runs of decision trees via sampling with replacement. This procedure is called bagging (Breiman 1996). A 10% sample of the training data was used to create each tree, which related the dependent percent forest cover loss variable to the set of MODIS-independent variables. Eleven trees were generated, and the median percent forest cover loss from all bagged trees was retained as the per pixel result. To reduce errors of commission, we thresholded the output product at 30% forest cover loss, converting each map to a yes/no forest cover loss estimate per 250 m MODIS pixel.

6.4 Results

Figure 6.1 shows a global-scale annual growing season metric derived from shortwave infrared, near-infrared, and red growing season imagery from 2000. The spectral feature space is largely cloud free, but persistent haze and partial cloud cover exist in the Andes Mountains of Colombia, northern Brazil, the central African coast along the Gulf of Guinea, and montane Borneo and New Guinea (the haze and residual cloud cover are not visible in the figure). The humid tropics are the only region where atmospheric effects are present in the MODIS metric feature space. Other potential limitations, such as seasonal forests and variable growing season length, are not readily apparent in the metric feature space.

Figures 6.2 and 6.3 provide an example of the derived metric feature space for an area of Mato Grosso, Brazil, and Quebec, Canada, respectively. For these subsets, blue represents year 2000 growing season band 7 shortwave infrared reflectance (mean of the band 7 values corresponding to the three

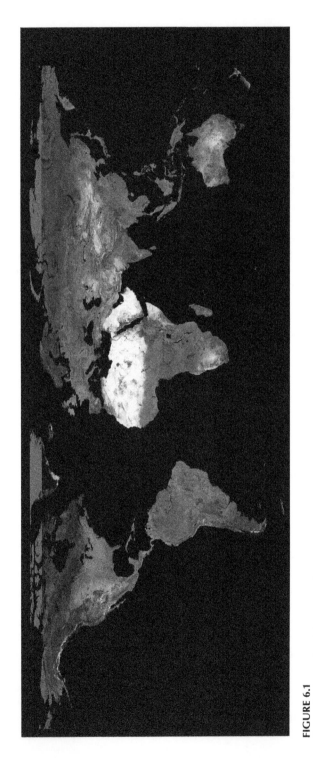

FIGURE 6.1
(See color insert.) MODIS annual growing season image composite of shortwave, near-infrared, and red band, enhanced to appear as true color.

FIGURE 6.2
(See color insert.) 400 km × 400 km subset centered on 12° 4' S, 55° 59' W in Mato Grosso, Brazil. False-color composite of MODIS band 7 growing season metrics—*blue*: 2000 mean band 7 shortwave infrared reflectance from the three greenest 16-day composite periods, *green*: difference in the 2000 and 2005 mean band 7 shortwave infrared reflectance from the three greenest 16-day composite periods, and *red*: difference in the 2005 and 2010 mean band 7 shortwave infrared reflectance from the three greenest 16-day composite periods.

FIGURE 6.3
(See color insert.) 400 km × 400 km subset centered on 51° 45' N, 72° 8' W in Quebec, Canada. False-color composite of MODIS band 7 growing season metrics—*blue*: 2000 mean band 7 shortwave infrared reflectance from the three greenest 16-day composite periods, *green*: difference in the 2000 and 2005 mean band 7 shortwave infrared reflectance from the three greenest 16-day composite periods, and *red*: difference in the 2005 and 2010 mean band 7 shortwave infrared reflectance from the three greenest 16-day composite periods.

greenest 16-day composite periods). Areas that are dark in this metric are typically forest (water has been masked out prior to analysis). Green represents the difference for this metric from 2000 to 2005 and red the difference from 2005 to 2010. Pixels that have high increases for this metric, and have an initial dark state (~<5% reflectance), are likely to represent forest disturbance. For the Brazil subset, a dramatic reduction in forest cover loss can be inferred from this false-color composite image. The proportion of 2000–2005 change dwarfs that from 2005 to 2010. For the Canada subset, a less dramatic reduction is observed, related to a predominantly fire-driven dynamic. The tree bagging algorithm formalized the labeling of all forest cover loss pixels.

The global total of MODIS hot spot pixels covered 500,000 km^2 from 2000 to 2005 and 360,000 km^2 from 2005 to 2010. The total MODIS-indicated forest cover loss represents 50% of the total area of gross forest cover loss from the MODIS/Landsat study of Hansen et al. (2010). In other words, the Landsat sample-based area estimate of gross forest cover loss equaled 1,011,000 km^2, while the MODIS hot spot mapped area equaled 500,000 km^2. The MODIS-indicated forest cover loss pixels were aggregated to the same sampling grid as the Hansen et al. study and compared. The following relation yielded an r^2 of 0.64 and a standard error of 1.73%:

$$MODIS/Landsat\ area = MODIS\text{-indicated change} \times 0.86 + 0.68$$

Areas from the Hansen et al. (2010) study were reported only for those regions or nations that had sufficient Landsat samples to provide a reasonable uncertainty estimate. These areas included the four major forested biomes (humid tropical, dry tropical, temperate, and boreal), all continents except Antarctica, and countries with over 1,000,000 km^2 of forest cover in 2000. The gross forest cover loss data from Hansen et al. (2010) are plotted against the MODIS-indicated change in Figure 6.4.

The degree of forest cover loss omission in the MODIS data is clear. As stated before, fully half of the global forest cover loss from the Hansen et al. (2010) study is not mapped with MODIS. Regardless, there is a strong overall relationship. Areas where small-scale disturbance predominates, such as Africa, feature the highest proportion of omitted, or cryptic, change. In Figure 6.4 the continent of Africa and the nation of the Democratic Republic of the Congo have the highest ratio of MODIS/Landsat area of forest loss to MODIS-indicated forest loss. This reflects the finer and more diffuse pattern of forest change in Africa where most clearing is performed in swidden agricultural settings too small for quantification using MODIS data. Areas with large agroindustrial clearing, such as Brazil, South America as a whole, and Indonesia, have the lowest omission rates.

The model was applied to the two study intervals, and a comparison of the amount of change hot spots was made. Figures 6.5 and 6.6 illustrate the

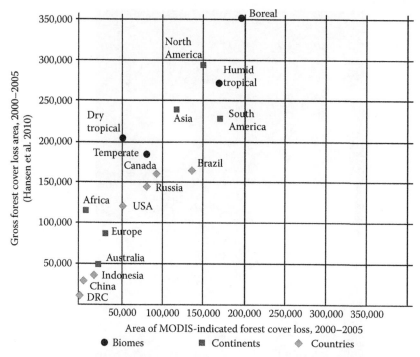

FIGURE 6.4
Plot of area of MODIS-indicated forest cover loss versus gross forest cover loss area for reported regions. (From Hansen, M.C., et al., *Proc. Natl. Acad. Sci.*, 107, 8650, 2010.)

global distribution of MODIS-indicated forest cover loss. The most obvious change in the patterns of forest cover loss is found in Brazil. As Shimabukuro et al. (2012) report, the Brazilian government has sought to reduce the clearing of Amazonian forests, efforts that have included the use of satellite data as an enforcement tool. The global results from Figures 6.5 and 6.6 confirm this reduction. Contrary to this trend is a marked increase in the clearing of the Chaco woodlands of Bolivia, Paraguay, and Argentina between the two periods. Africa is largely absent of large-scale change, with only the agroforestry of South Africa evident at this scale. For tropical Asia, Indonesia exhibits a rise in forest cover loss over the study period. Epochal variation at higher latitudes is less evident and largely due to variations in high latitude fire dynamics as well as storm damage. In general, forest cover losses due to fire appear greater in the 2000–2005 interval than in the 2005–2010 interval (see Alaska, Siberia, and Australia). Areas of active forestry practices feature prominently in both epochs.

Figures 6.7 through 6.9 show the change in MODIS-indicated forest cover loss over the study period. At the biome scale, significant reductions in forest cover loss within the humid tropical and boreal biomes are found.

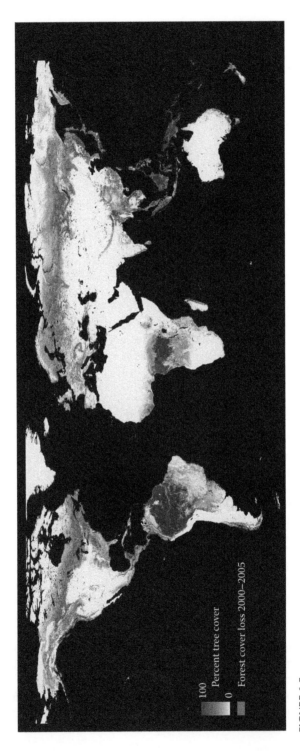

FIGURE 6.5
(See color insert.) MODIS percent tree cover 2000 and indicated forest cover loss from 2000 to 2005.

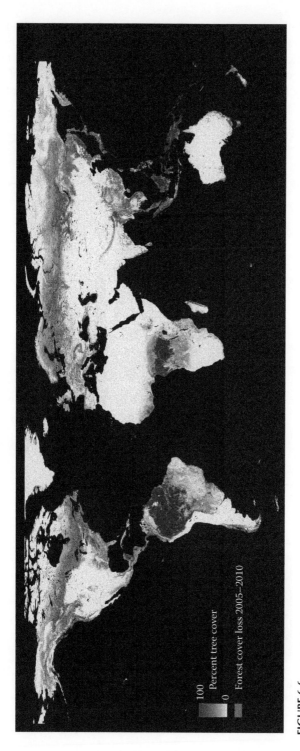

FIGURE 6.6
(See color insert.) MODIS percent tree cover 2000 and indicated forest cover loss from 2005 to 2010.

Brazil's reduced clearing drives the humid tropical change, while less forest cover loss due to fire drives the boreal forest change. At the continental scale, the same dynamics are evident, with Europe and Africa exhibiting little or no change in forest cover loss. For countries with greater than 1 Mha of year 2000 forest cover, only Indonesia exhibits a clear increase in forest cover loss.

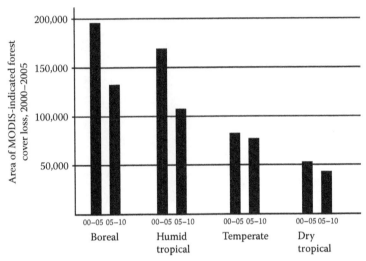

FIGURE 6.7
MODIS-indicated forest cover loss totals per forested biome for the 2000–2005 and 2005–2010 epochs.

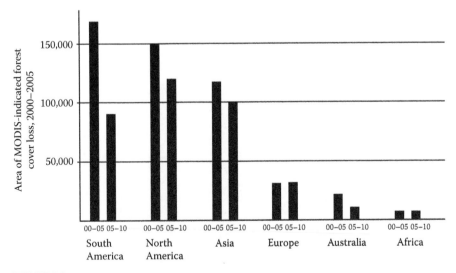

FIGURE 6.8
MODIS-indicated forest cover loss totals per continent for the 2000–2005 and 2005–2010 epochs.

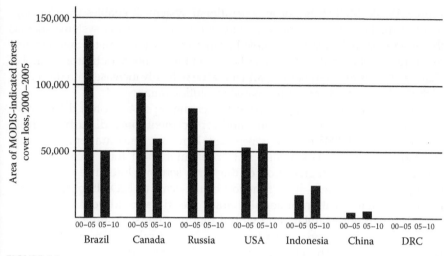

FIGURE 6.9
MODIS-indicated forest cover loss totals per country for the 2000–2005 and 2005–2010 epochs (only countries with greater than 1,000,000 km² of forest cover in 2000).

The results, as shown in Figure 6.4, have significant errors of omission, mainly related to the coarse scale of observation, as stated previously. Obvious commission errors are associated largely with two environmental dynamics. First, residual haze and cloud cover impact the metric space and lead to noise-related commission errors in a few humid tropical regions referred to earlier. Second, wetlands are very dynamic in their patterns of spectral change as floods arrive and recede along with attendant vegetation responses. Wetland formations are another source of forest change commission error. Finally, the uncertainty regarding the radiometric stability of the Terra instrument could significantly impact the repeated use of a single model over the two 5-year intervals. Further study is required to resolve the impact of Terra's radiometric degradation on the observed forest extent changes of this study, particularly between the two 5-year epochs.

6.5 Conclusion

The combined high-temporal observation frequency and moderate spatial resolution of MODIS data enable global forest change indicator mapping. The ability to synoptically characterize forest disturbance at the global scale allows for direct comparison of change rates through time and across space. The continuous acquisition of multispectral observations at the global scale for 10+ years illustrates the value of operational systems in quantifying environmental dynamics. As noted, such analyses are dependent on a stable radiometric data source. While MODIS is not an operational system, it enables the development of methods that can be implemented with operational systems such as the recently launched VIIRS (Visible Infrared Imager Radiometer Suite) instrument (Justice et al. 2010). This is a critical monitoring tool of indicators of global change, such as forest dynamics, and its value will only increase with the length of the high-temporal, moderate spatial resolution data record.

Our results document a pervasive and changing global forest disturbance dynamic. Overall, a reduction in stand-replacement forest disturbance from 2000 to 2005 and 2005 to 2010 was found. However, the data represent only indications of forest cover loss, not an estimation of total area, and may also be affected by degradation of the Terra sensor. Differences in epochal change illustrated here are a function of the scale of MODIS observations. Definitive quantification of aerial change over time could be different than that observed with MODIS and would require finer scale time-series imagery for either direct forest area loss estimation or calibration of the MODIS indicator product. The clearest reduction in forest cover loss occurred in Brazil and is related to policy and enforcement efforts to improve regulation of forest clearing in the Brazilian Amazon. Forest cover loss related to fire appeared to decline over the two epochs as well. The drivers of global forest change are many, and the spatial patterns seen in the MODIS change products capture four principle drivers: (1) agroindustrial scale clearing related to land use conversions and forestry practices, (2) fire, (3) disease, and (4) storm damage. Attributing each identified change pixel to a specific driver would greatly enhance the utility of the data for a host of land use and biogeochemical cycle modeling applications.

The ability to quantify both forest cover extent and change independent of land use designations is important in generating a consistent narrative of global forest change. Global observing systems such as MODIS enable such quantifications, but are limited in area estimation. As the discipline moves forward, high-temporal observations will be needed at finer resolutions in order to generate global forest cover extent and change maps that can be used directly in estimating area change. Landsat data, which have included a global acquisition strategy (Arvidson et al. 2001) and are now freely available (Woodcock et al. 2008), will be the data source to extend the methods developed using MODIS to finer spatial scales.

Acknowledgment

Support for this work was provided by the United States National Aeronautics and Space Administration's Land Cover and Land Use Change, MEASURES, and Science of TERRA and AQUA programs under Grants NNG06GD95G, NNX08AP33A, and NNX11AF38G.

About the Contributors

Matthew C. Hansen is a professor in the Department of Geographical Sciences at the University of Maryland, College Park, Maryland. He has a bachelor of electrical engineering from Auburn University, Auburn, Alabama. His graduate degrees include a master of engineering in civil engineering and a master of arts in geography from the University of North Carolina at Charlotte, North Carolina, and a doctoral degree in geography from the University of Maryland, College Park, Maryland. His research specialization is in large-area land cover monitoring using multispectral, multitemporal, and multiresolution remotely sensed data sets. He is an associate member of the MODIS Land Science Team and a member of the GOFC-GOLD Implementation Working Group.

Peter Potapov received a diploma in botany from the Moscow State University, Moscow, Russia, in 2000, and a PhD in ecology and natural resources from the Russian Academy of Science, Moscow, Russia, in 2005. From 1998 to 2006, he was a Geographical Information System (GIS) specialist at Greenpeace (Russian office) where he took part in developing an approach for mapping and monitoring forest degradation using the intact forest landscapes (IFL) method. He was a postdoctoral researcher from 2006 to 2011 at South Dakota State University, Brookings, South Dakota, and currently he is a research associate professor at the University of Maryland, College Park, Maryland. His current research is focused on establishing a global operational forest monitoring algorithm using integration of Landsat and MODIS data.

Svetlana Turubanova received a BSc in geography from the Komi Pedagogical Institute, Syktyvkar, Russia (1996), an MSc in ecology from the Pushchino State University, Pushchino, Russia (1998), and PhD in ecology from the Russian Academy of Science, Moscow, Russia (2002). She started her career as a GIS specialist at Greenpeace Russia in 1999. Later, she was a postdoctoral researcher at South Dakota State University (from 2008 to 2011), Brookings, South Dakota. Currently she is a research associate at the University of Maryland, College Park, Maryland. Her current research interests are focused on national-scale Landsat-based forest mapping, deforestation monitoring, and forest degradation assessment.

References

Arvidson, T., Gasch, J., and Goward, S.N., Landsat 7s long-term acquisition plan—An innovative approach to building a global imagery archive. *Remote Sensing of Environment*, 78, 13, 2001.

Achard, F., et al., Determination of deforestation rates of the world's humid tropical forests. *Science*, 297, 999, 2002.

Arino, O., et al., GlobCover—A global land cover service with MERIS. In *Proceedings of Envisat Symposium 2007*, Montreux, Switzerland, 2007.

Bartholomé, E. and Belward, A.S., GLC2000: A new approach to global land cover mapping from earth observation data. *International Journal of Remote Sensing*, 26, 1959, 2005.

Breiman, L., Bagging predictors. *Machine Learning*, 26, 123, 1996.

Breiman, L., Friedman, J., Olshen, R., and Stone, C. *Classification and Regression Trees*. Monterey, CA: Wadsworth, 1984.

Broich, M., et al., Time-series analysis of multi-resolution optical imagery for quantifying forest cover loss in Sumatra and Kalimantan, Indonesia. *International Journal of Applied Earth Observation and Geoinformation*, 13, 277, 2011.

Carroll, M., et al., Vegetative cover conversion and vegetation continuous fields. In B. Ramachandran, C. Justice, and M. Abrams (Eds.), *Land Remote Sensing and Global Environmental Change: NASA's EOS and the Science of ASTER and MODIS*. New York: Springer, 2010.

Clark, L.A. and Pergibon, D., Tree-based models. In T.J. Hastie (Ed.), *Statistical Models in S*. Pacific Grove, CA: Wadsworth and Brooks, 1992.

DeFries, R., Hansen, M., and Townshend, J., Global discrimination of land cover types from metrics derived from AVHRR Pathfinder data. *Remote Sensing of Environment*, 54, 209, 1995.

Freidl, M.A. and Brodley, C.E., Decision tree classification of land cover from remotely sensed data. *Remote Sensing of Environment*, 61, 399, 1997.

Friedl, M.A., et al., Global land cover mapping from MODIS: Algorithms and early results. *Remote Sensing of Environment*, 83, 287, 2002.

Hansen, M.C. and DeFries, R.S., Detecting long term global forest change using continuous fields of tree cover maps from 8 km AVHRR data for the years 1982–1999. *Ecosystems*, 7, 695, 2004.

Hansen, M.C., DeFries, R.S., Townshend, J.R.G., and Sohlberg, R., Global land cover classification at 1 km spatial resolution using a classification tree approach. *International Journal of Remote Sensing*, 21, 1331, 2000.

Hansen, M., Dubayah, R., and DeFries, R., Classification trees: An alternative to traditional land cover classifiers. *International Journal of Remote Sensing*, 17, 1075, 1996.

Hansen, M.C., Stehman, S.V., and Potapov, P.V., Quantification of global gross forest cover loss, *Proceedings of the National Academy of Sciences*, 107, 8650, 2010.

Hansen, M.C., et al., Global percent tree cover at a spatial resolution of 500 meters: First results of the MODIS vegetation continuous fields algorithm. *Earth Interactions*, 7(10), 15, 2003.

Hansen, M.C., et al., Estimation of tree cover using MODIS data at global, continental and regional/local scales. *International Journal of Remote Sensing*, 26, 4359, 2005.

Hansen, M.C., et al., Humid tropical forest clearing from 2000 to 2005 quantified using multi-temporal and multi-resolution remotely sensed data. *Proceedings of the National Academy of Sciences*, 105, 9439, 2008.

INPE (Instituto Nacional de Pesquisas Espaciais), *Monitoramento da cobertura florestal da Amazônia por satélites: Sistemas PRODES, DETER, DEGRAD E QUEIMADAS 2007–2008*. São José dos Campos, SP, Brazil: Instituto Nacional de Pesquisas Espaciais, 47 p, 2008.

Justice, C.O., et al., The moderate resolution imaging spectroradiometer (MODIS): Land remote sensing for global change research. *IEEE Transactions on Geoscience and Remote Sensing*, 4, 1228, 1998.

Justice C.O., et al., The evolution of U.S. moderate resolution optical land remote sensing from AVHRR to VIIRS. In B. Ramachandran, C.O. Justice, and M.J. Abrams (Eds.), *Land Remote Sensing and Global Environmental Change: NASA's EOS and the Science of ASTER and MODIS*. New York: Springer, 2010.

Loveland, T., et al., Development of a global land cover characteristics database and IGBP DISCover from 1 km AVHRR data. *International Journal of Remote Sensing*, 21, 1303, 2000.

Michaelson, J., et al., Regression tree analysis of satellite and terrain data to guide vegetation sampling and surveys. *Journal of Vegetation Science*, 5, 673, 1994.

Potapov, P., Turubanova S., and Hansen M.C., Regional-scale boreal forest cover and change mapping using Landsat data composites for European Russia. *Remote Sensing of Environment*, 115, 548, 2011.

Potapov, P., et al., Combining MODIS and Landsat imagery to estimate and map boreal forest cover loss. *Remote Sensing of Environment*, 112, 3708, 2008.

Potapov, P.V., et al., Quantifying forest cover loss in Democratic Republic of the Congo, 2000–2010, with Landsat ETM+ data. *Remote Sensing Environment*, in press.

Reed, B.C., et al., Measuring phenological variability from satellite imagery. *Journal of Vegetation Science*, 5, 703, 1994.

Shimabukuro, Y.E., et al., The Brazilian Amazon monitoring program: PRODES and DETER projects. In F. Achard and M.C. Hansen (Eds.), *Global Forest Monitoring from Earth Observations*. Taylor & Francis: Boca Raton, FL, 2012.

Vermote, E.F., El-Saleous, N., and Justice, C.O., Atmospheric correction of MODIS data in the visible to middle infrared: First results. *Remote Sensing of Environment*, 83, 97, 2002.

Wang, D., et al., Impact of sensor degradation on the MODIS NDVI time series. *Remote Sensing of Environment*, 119, 55, 2012.

Woodcock, C., et al., Free access to Landsat imagery. *Science*, 320, 1011, 2008.

7

Use of a Systematic Statistical Sample with Moderate-Resolution Imagery to Assess Forest Cover Changes at Tropical to Global Scale

Frédéric Achard, Hans-Jürgen Stibig, and René Beuchle
Joint Research Centre of the European Commission

Erik Lindquist and Rémi D'Annunzio
Food and Agriculture Organization of the United Nations

CONTENTS

7.1 Introduction

This chapter presents an operational remote sensing approach for monitoring forest cover at continental and global levels, based on a statistical sampling design and on satellite imagery from optical sensors of moderate spatial resolution (30 m × 30 m resolution).

There are two main approaches to forest characterization and monitoring with remotely sensed data (Achard et al. 2010): analyses that cover the full spatial extent of the forested areas, termed "wall-to-wall" coverage, or those that select a statistical sample of forested areas for careful analysis and extrapolate the findings to the entire area of interest. Wall-to-wall mapping has long been done with relatively coarse spatial resolution satellite data and, currently, moderate spatial resolution wall-to-wall analyses are possible (see following Chapters 9 to 13 for examples of wall-to-wall analyses). However, spatially exhaustive analyses are challenging to operationalize on frequent time intervals and over very large, heterogeneous areas. Statistical sampling approaches, therefore, serve an important role in providing cost-effective, timely, repeatable estimates of forest characteristics over large areas and at frequent time intervals (e.g., Brink and Eva 2009; Broich et al. 2009; Duveiller et al. 2008; Eva et al. 2010). A sampling procedure that adequately represents deforestation events (e.g., through a sufficiently dense systematic or stratified sample in space and time) can capture deforestation trends.

Whichever overall approach is chosen, sampling or wall-to-wall, the spatial unit of analyses or minimum mapping unit (MMU) must also be decided upon. There are two main choices for this. In pixel-based approaches, the smallest unit of analysis is the individual image pixel. Object-based approaches use pixel clustering algorithms to create spectrally homogenous pixel groupings, which are thereafter treated as individual units for analysis.

For the Global Forest Resources Assessment 2010 (FRA 2010), the FAO (Food and Agriculture Organization of the UN) has extended its global and continental monitoring of forest cover changes to include analysis of remotely sensed land cover and land use as a complement to standard national reporting. The survey applies object-based image analysis methods to a globally distributed, systematic sample of moderate-resolution satellite imagery to estimate forest land cover and land use change for the periods 1990–2000 and 2000–2005. The FAO has produced estimates of tropical forest cover changes as part of past assessments (FRA 1990, 2000), but the remote sensing survey (RSS) of FRA 2010 has been extended to all lands (FAO et al. 2009). This survey has been conducted by a partnership between FAO and its member countries, the European Commission Joint Research Centre (JRC) as the main scientific partner, South Dakota State

University, the United States Geological Survey (USGS), and the U.S. National Aeronautics and Space Administration (NASA). Over 200 national experts from 106 countries have participated in the survey.

This chapter presents the scientific and technical methods that have been developed for monitoring forest cover changes in the framework of this global survey.

7.2 Sampling Strategy

The grid system selected for the global systematic sample is a rectilinear grid, based on degrees of geographical latitude and longitude (Figure 7.1), that enables a straightforward implementation, and easy location and understanding (Mayaux et al. 2005). Although stratified sampling is generally preferable for improving the efficiency of land cover change

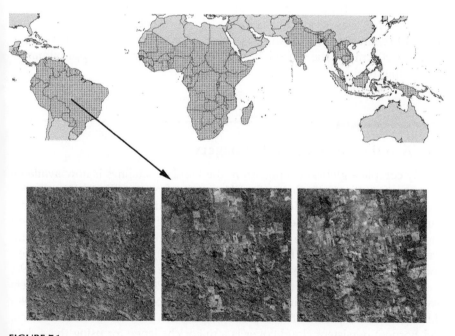

FIGURE 7.1
(See color insert.) Example of time series (for years 1990, 2000, and 2005) of Landsat satellite imagery over one sample site in the Amazon Basin (20 km × 20 km size). Forests appear in dark green, deforested areas (agriculture and pastures) appear in light green or pink.

estimation (Stehman et al. 2011), a systematic, nonstratified sampling has been implemented because:

1. This sampling scheme is intended to be also used for future time periods (for year 2010 and later), and it is impossible to reliably predict where deforestation "hot spots" will be located in future years.

2. The systematic sample scheme can be easily intensified for specific purposes, in particular for assessment at a national level or for a particular ecosystem. Indeed, a number of countries supported by FAO are already carrying out national forest assessments based on an intensification of the global sampling scheme (http://www.fao.org/forestry/nfma).

The global systematic sampling approach has already been tested against wall-to-wall reference data over the Brazilian Amazonia basin (Eva et al. 2010). It has also been intensified and tested for the Congo River basin region for the 1990–2000 period (Duveiller et al. 2008) and for the French Guiana territory (Eva et al. 2010), demonstrating its potential to estimate forest cover changes from continental to regional levels (Broich et al. 2009).

Globally, the survey involved 13,690 sample sites. Sampling has not been performed for latitudes higher than 75° north or south. At most sites, the area surveyed was 10 km × 10 km, which represents approximately 1% of the world's land surface. In the tropics, the area surveyed for each site was 20 km × 20 km for the period 1990–2000, which represents approximately 3.6% of the tropics.

7.3 Acquisition of Satellite Imagery

Nearly complete global coverage from the Landsat satellites is now available at no cost from the Earth Resources Observation Systems (EROS) Data Center (EDC) of the USGS (http://eros.usgs.gov/). A recent product, called the Global Land Survey (GLS), represents a global archive of good quality, orthorectified and geodetically accurate image acquisitions from Landsat Multispectral Scanner (MSS), Landsat Thematic Mapper (TM), and Landsat Enhanced Thematic Mapper (ETM+) sensors focused on the epochs ca. 1975, ca. 1990, ca. 2000, mid-2000s, and ca. 2010 (Gutman et al. 2008). These GLS data sets play a key role in establishing historical deforestation rates (Masek et al. 2008), although in some parts of the tropics (e.g., Western Colombia, Central Africa, and Borneo) persistent cloud cover is a major challenge for using these data (Ju et al. 2009; Linquist et al. 2008). For these regions, the GLS data sets can be complemented by remote sensing data from other satellite sensors with similar characteristics, in particular, optical sensors of moderate spatial resolution. The GLS data sets are described with full details in Chapter 4.

For each sample location of the systematic grid, the available Landsat data (from TM or ETM sensors) were sought from the GLS database (primary data source). These data were downloaded at full resolution (30 m × 30 m). Image subsets of 20 km × 20 km covering the sample sites were extracted in UTM projection (Potapov et al. 2011). The sample site target size is 10 km × 10 km, but a 5 km buffer has been used for data extraction and processing in order to keep contextual information. In the event of the data being unacceptable (due to cloud cover or artifacts from visual screening assessment), replacement data were sought from different sources with the help of the GEOSS (Global Earth Observing System of Systems) Land Surface Imaging Constellation. In particular, for the 4,016 sample sites covering the tropics, 2,868 suitable image pairs were found for the period 1990–2000 from the GLS data sets, representing 71.6% of the tropical sample (Beuchle et al. 2011). Better alternatives could be found for 26.6% of these 4,016 sites, substituting cloudy or missing GLS data sets at one or the other epoch or both (GLS-1990 or GLS-2000). Gaps were filled from the USGS Landsat archives (1,070 samples), data from other Landsat archives (e.g., GISTDA, ACRES, INPE; 53 samples), or with alternatives to Landsat, i.e., 15 samples from SPOT (Satellite Pour l'Observation de la Terre). This increased the effective number of sample pairs to 3,945, representing 98% of all target samples. No suitable image pairs were found for 71 confluence points, which were not randomly distributed, but mostly concentrated in the Congo basin, where around 15% of the region remains unsampled. There is a higher number of missing sites in the second period assessed (2000–2005) in particular for tropical regions, due to the malfunctioning of the line scanner on the Landsat 7 ETM sensor after June 1, 2003, which corrupts around 25% of each image acquisition (Maxwell 2004). The missing sites in the tropics for the 2000–2005 period are mainly located in Central America, Ecuador, the Colombian Choco, the Guianas, the southern ridge of West Africa, the western part of Congo basin (South Cameroon, Equatorial Guinea, Gabon, and Western Congo), Central Democratic Republic of Congo, Eastern Tanzania, and Indonesia (Kalimantan, Sulawesi, and Irian Jaya).

7.4 Preprocessing of Satellite Imagery

For each sample site, satellite image subsets (from 1990, 2000, and 2005) were preprocessed for geometric control, radiometric calibration and normalization, segmentation, and classification. Prior to the object segmentation and classification steps, radiometric correction to a common radiometric scale is required in order to apply standard supervised classification algorithms to the full imagery data set, making use of spectral training data of representative vegetation types. Acquisition errors and irrelevant data (e.g., clouds and cloud shadows) must also be removed in the preprocessing phase. A robust approach applicable to a large amount of multidate and multiscene Landsat imagery has been developed to convert all images into normalized radiometric values (Bodart et al. 2011). The different preprocessing steps were (1) conversion to top-of-atmosphere (ToA) reflectance, (2) cloud and cloud shadow removals, (3) haze correction, and (4) image radiometric normalization. The conversion to ToA reflectance was achieved by first converting raw digital numbers (DN) into at-sensor spectral radiance for each band and subsequently the at-sensor radiance was converted into ToA reflectance. The remaining clouds and cloud shadows in the selected images were masked in two steps. The first step was to detect all potential cloud and cloud shadow pixels using an automatic spectral rule-based mapping approach followed by a second step that consisted of a sequential application of a postprocessing algorithm based on morphological and topological methods designed to create a refined mask for images where clouds were visually identified. Image contamination by haze is relatively frequent in tropical regions (semitransparent clouds and aerosol layers that alter the spectral signatures of objects, especially in the visible bands). Partially contaminated images were corrected on the basis of the method using the fourth component of the tasseled cap transformation (TC4) computed from the six reflective bands of Landsat imagery. The applied image radiometric normalization is a relative normalization of multitemporal imagery covering different areas. Relative normalization adjusts the spectral values of all images to the values of one reference image. Dense evergreen forest pixels have been considered as pseudo-invariant features (PIF), i.e., stable targets between dates, assuming that reflectance differences in these stable targets are due to atmospheric perturbations. This normalization algorithm, referred to as "forest normalization," has been applied to each sample image with significant presence of dense evergreen forests (i.e., more than 2,000 pixels in the image). The median forest value parameter was extracted from a forest mask based on empirically determined thresholds of NDVI and bands 4 and 5 from Landsat imagery from years 1990 and 2000 and intersected with a 250 m forest map derived from the vegetation continuous field (VCF) product (Hansen et al. 2003). For those sites with a lower proportion of dense evergreen forests (i.e., less than 2,000 pixels in the image), a relative normalization has been performed whenever possible by visually selecting an area that did not change between the two dates, using the image of year 2000 as the reference image.

The haze correction algorithm improved the visual appearance of the image and significantly corrected the digital numbers for Landsat visible bands. The normalization procedures (forest normalization and relative normalization) improved the correlation between the spectral values of the same land cover in multidate images. The image subsets from the year 2000 were taken as the reference for geometric and radiometric controls. The preprocessed multitemporal data set constituted the basis for an automatic object-based supervised classification.

7.5   Segmentation of Satellite Imagery

After preprocessing, the image subsets were segmented so as to identify homogenous land units that can then be classified for each date (Raši et al. 2011). This approach comprises two automated steps of multidate image segmentation and object-based land cover classification (based on a supervised spectral library), followed by an intense phase of visual control and expert refinement. Image segmentation is done at two spatial scales, introducing the concept of an MMU via the automated selection of a site-specific scale parameter. The automated segmentation of land cover polygons and the pre-classification of land cover types mainly aim at avoiding manual delineation and at reducing the efforts of visual interpretation of land cover to a reasonable level, making the analysis of 13,000 sample sites feasible.

Several segmentation algorithms were tested. Based on technical performance and visual assessment of the object delineation, the eCognition software (Trimble) was chosen as most suited for our specific purpose. In particular, this software can process large amounts of data and classify objects in one common processing chain. For the purpose of forest cover monitoring, a multidate segmentation approach has been preferred to two separate, single-date image segmentations. Multidate segmentation integrates from the very beginning of the temporal aspect into the generation of spatially and spectrally consistent mapping units. For the tropical 4,000 sites, the segmentation process was initially implemented on two-date imagery (1990 and 2000) in a single operation. The Landsat TM or ETM+ spectral bands 3, 4, and 5 (ToA reflectance values) of both reference years (1990 and 2000) were therefore used as a common input to the segmentation procedure, assigning equal weights for all six bands. The weights of two other parameters in the eCognition software—referred to as "spectral" and "shape"—had to be determined for segmentation. Based on a series of tests with varying settings, the main weight of 0.9 has been empirically assigned to the "spectral" parameter, i.e., the spectral homogeneity accounts for 90% of the merging decision rules. The resulting weight for the "shape" parameter of 0.1 (as sum of the two weights = 1) proved to be sufficient for avoiding very irregular and fringed objects.

The main parameter controlling the size of objects is referred to as the scale parameter. The higher the scale parameter, the larger the average size of image objects, and in particular the maximum object size. We developed a process that automatically determines a specific scale parameter for each sample site in order to reach the desired MMU. This is achieved by increasing the scale parameter through iterative segmentations, until a size threshold for the smallest polygons is reached: the iterative process is stopped when the largest object among the 5% smallest objects reaches the desired MMU, i.e., when at least 95% of the remaining objects in the sample site are larger than the MMU. An initial MMU of 1 ha was set for the segments. This is a compromise between not having segments that are too small, and avoiding segments with mixed land covers. The segments of the individual image subsets are then classified using an automated supervised classification. In a second phase, these classified segments are aggregated into segments of 5 ha by increasing the scale parameter through iterative segmentations. In a final step, the number of the remaining small polygons below 5 ha size was reduced by merging each object smaller than 3 ha (corresponding to ca. 33 Landsat TM pixels) with the object it shared the longest common borderline with. The image objects resulting from the multidate segmentation conform to a standard MMU and exhibit similar spectral characteristics in time and in space. This 3 ha MMU size enables a feasible visual assessment of the classification by local experts.

7.6   Definition of Land Cover and Land Use Classes

Four main land cover categories were defined for labeling the 1 ha MMU segments: "tree cover" (TC), "other wooded land" (OWL), "other land" (OL), and "water" (WA). TC comprises all tree cover where canopy density can be expected to be ≥10% and tree heights to be ≥5 m. Included are natural forests and forest plantations, but also tree cover outside forests, such as in parks or on agricultural lands. OWL comprises all woody vegetation of lower height (<5 m), mainly shrub land, but also shrub-like agricultural crops, vegetation regrowth, or plantations with small trees. OL includes all nonwoody land cover (e.g., herbaceous cover, pastures, nonwoody crops, burnt areas, bare soils, settlements), except for water. The water class consists of rivers and in-land water bodies. The definition for tree cover has been chosen to be compatible with the FAO "forest" definition (FAO 2010). From the spectral and textural information of the moderate-resolution satellite imagery used in this study, one can only infer approximate tree density and broad height categories. The class thresholds served therefore rather as guidance for interpretation and for selection of training areas.

Land cover is the observed biophysical properties of the land surface, whereas land use is defined by the human activities and inputs on a given land area. Four main land use categories have been defined: "forest," "other wooded land," "other land use," and "water." Treating forest as a land use is consistent with the forest definition used in FAO's Global FRA country reports and national reports to the United Nations Framework Convention on Climate Change (UNFCCC). Forest land use may include periods during which the land is devoid of tree cover, for example, during cycles of forest harvesting and regeneration. In such cases, a land use is considered to be forest land use when management or natural processes will, within a reasonable time, restore tree cover to the point where it constitutes a forest.

7.7 Supervised Classification of Segmented Satellite Imagery for the Tropical Sample Sites

Spectral signatures were collected from the preprocessed Landsat ETM+ data of the year 2000 from one common set of training areas representing the main land cover classes within a region (Raši et al. 2011). For the first level classification at 1 ha, a large number of spectral classes were required to cover the variability of spectral reflectance within any particular land cover class, e.g., the TC class consists of 15 spectral classes including dense evergreen forests, degraded evergreen forests, dry deciduous forests in different phenological phases, mangrove, and swamp forest. Only homogeneous land cover units were selected as training areas, using additional references like fine-resolution satellite data (e.g., Google Earth). The number of pixels ultimately used for establishing the spectral signature of a subclass was generally higher than 1,000. Spectral signature statistics (means and standard deviations) were calculated at the level of subclasses.

A generic supervised classification of the 1 ha level segmentation objects was performed uniformly for all sample sites. The classification was based on membership functions established from the spectral signature of each subclass for the Landsat TM/ETM+ spectral bands 3, 4, and 5. The membership functions of each subclass were defined as an approximation of the class probability distribution, represented by isosceles triangles in the feature space of each spectral band. The top of the triangle corresponds to the class mean (m) and represents the spectral value of highest probability for class assignment. The two triangle legs descend from that position up to a spectral distance of $m \pm 3$ sd (sd = standard deviation), linearly decreasing the probability of class assignment to a value of "0" at the positions $m \pm 3$ sd.

The classification process compares the object spectral mean values to the membership functions defined for all subclasses. An object was assigned to the class displaying the highest membership probability for the object spectral mean values. We applied these membership functions to the imagery of all reference years, having performed previous spectral calibration to ToA reflectance values, haze correction, as well as normalization of the satellite imagery. The subclasses resulting from supervised classification served only for the mapping of the four main land cover classes.

The 1 ha level classified segments were automatically aggregated to 5 ha level into the five broad land cover classes based on the proportion of tree cover. The supervised classification result obtained for the 1 ha objects served as direct input to the thematic aggregation done at the second-level segmentation (5 ha MMU). The labeling of the second-level objects was performed by passing through a sequential list of classification criteria, with a main emphasis on tree cover proportions within second-level objects, e.g., TC class is defined as containing more that 70% tree cover within the 5 ha segment. As a consequence of merging objects from a finer scale (1 ha MMU), a "tree cover mosaic" class has been introduced for objects containing partial tree cover at the second level (objects containing an area portion of 40%–70% tree cover).

7.8 Visual Verification and Refinement of the Land Cover Classifications

The resulting land cover multitemporal classifications are then interdependently visually controlled by national experts. A dedicated graphical user interface has been developed for the visual verification and potential reassignment of land cover labels (Simonetti et al. 2011). For a selected sample site, the tool displays simultaneously the pair of image subsets (e.g., of 1990 and 2000) and the corresponding digitally classified land cover maps. The tool offers an optimized set of commands including image enhancement, simultaneous zoom of displayed data, single or multiobject selection and relabeling, specific class selection, and highlighting. The graphical user interface is available in English, Spanish, French, and Russian.

Visual control and refinement of the digital classification results at the 5 ha MMU level were implemented using, whenever available, very high-resolution satellite imagery (e.g., through Google Earth), but also existing vegetation maps and field knowledge as supplementary references: a revision of the mapping results was then carried out by forestry experts from the tropical countries who contributed local forest knowledge to improve the interpretation. During a final phase of regional harmonization, an experienced image interpreter performed a control of the interpretation consistency across the region, applying final corrections where necessary. Figure 7.2 shows a simplified example of the main steps used in visual verification and refinement of the land cover and land cover changes between 1990 and 2000.

The phase of visual control and refinement has been designed as a crucial component for correcting classification errors and for implementing the change assessment. The importance of visual control and correction can be perceived when comparing to the initial automatic classification result: e.g., in South East Asia about 20% of the polygon labels were changed through expert knowledge by visual interpretation (Raši et al. 2011). More than 120 experts from tropical countries have been involved in this verification and refinement phase of the survey.

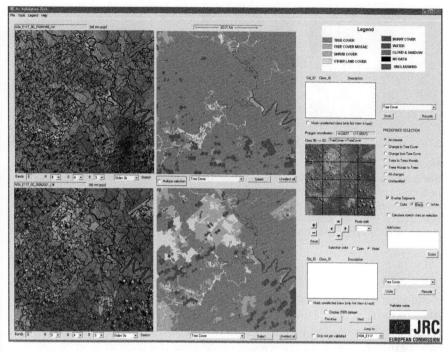

FIGURE 7.2
(See color insert.) Visualization tool used for the process of verification and correction of multitemporal classifications. *Left column*: Segmented Landsat imagery displayed (top: year 1990, bottom: year 2000). *Right column*: Land cover maps produced from satellite imagery.

7.9 Conversion of the Land Cover Maps into Land Use Maps

Land cover maps were first converted automatically into land use maps, and then the conversion results were reviewed through visual control by national experts. The automatic conversion of land cover maps into land use maps uses the following systematic rules:

- Classes TC and tree cover mosaic are converted to forest
- Class OWL remains as OWL
- Class OL is renamed other land use
- Class WA remains as WA

Because a direct translation possible from land cover to land use is not always possible, a visual interpretation and refinement of the land use classifications must be carried out by national experts. For example, when a forest has been

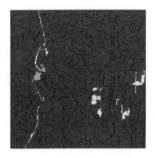

FIGURE 7.3
(See color insert.) The 20 km × 20 km multi-spectral Landsat image (left) for a sample site in the boreal forest showing, for the central 10 km × 10 km portion (red box), the classification of land cover (center) and land use (right). Land cover is classified as TC (green), tree cover mosaic (light green), OWL (orange), and other land cover (yellow). Land use is classified as forest (green), OWL (orange), and other land use (yellow).

clear-cut and is temporally unstocked, the land cover derived from any kind of automatic classification or visual interpretation will indicate something other than tree cover. However, the land use will remain as forest for a temporary clearing caused by timber harvest or fire, and this information can only be inferred by local knowledge of the land use context (Figure 7.3).

7.10 Production of Transition Matrices and Correction to Reference Dates and for Missing Data

For each sample site, land area transition matrices are produced for each period (1990–2000 and 2000–2005) and for both land cover and land use transitions (Table 7.1).

It was not possible to acquire all images at the exact reference date, with acquisitions ranging from 1984 to 1992 for the first reference year (1990), 1997 to 2003 for the second reference year (2000), and 2004 to 2009 for the third reference year (2005) (Beuchle et al. 2011). Each sample site's transition matrix was then adjusted to the baseline dates of June 30, 1990, 2000, and 2005; this was done by assuming that the land cover change rates are constant during the given period. We, therefore, linearly adjusted the land cover change matrices to the three reference dates.

Cloudy areas were considered as an unbiased loss of data and assumed to have the same proportions of land cover as noncloudy areas within the same site. This is achieved by converting the transition matrices 1990–2000 and 2000–2005 to area proportions relative to the total cloud-free land area of the sample site. For the missing sample sites in tropical regions, we

TABLE 7.1

Example of Land Cover Transition Matrix for Site [North 2°; West 074°] (areas in km²)

Year 2000/Year 1990	Tree Cover (TC)	Tree Cover Mosaic (TCM)	Other Wooded Land (OWL)	Other Land Cover (OLC)	WA	Total Year 1990
TC	44.9	4.4	2.8	9.8	0	61.9
TCM	0	3.4	1.7	5.4	0	10.5
OWL	0	0.6	4.1	3.4	0	8.1
OLC	0	0.3	1.6	17.9	0	19.8
WA	0	0	0	0	0	0
Total year 2000	44.9	8.7	10.2	26.5	0	100.2

used a local average from surrounding sample sites as surrogate results. The following weights ($\delta_{jj'}$) were applied for the local average of missing sites:

$$\delta_{jj'} = \frac{1}{d(j,j')} = \frac{1}{(d(\text{lat}))^4 + (d(\text{long}))^4} \tag{7.1}$$

where the differences in latitude and longitude between two sample sites (j and j') is used with a power of 4.

Small differences may appear between land cover proportions of year 2000 obtained from the successive transition matrices [1900–2000] and [2000–2005] due to the linear temporal extrapolation to the reference dates. To correct for potential inconsistencies for the common year 2000, the land cover proportions of year 2000 from the change matrices for period 2000–2005 are "calibrated" to the land cover proportions of year 2000 from the [1990–2000] transition matrix through a linear adjustment for each sample site.

7.11 Production of Statistical Estimates

For the statistical estimation phase, the sample sites are weighted in relation to their probability of selection (Eva et al. 2012). Indeed the sampling frame, although systematic, does not give equal probability because the distance between sites along a parallel is not the same as the distance along a meridian. All sample units were given a weight, equal to the cosine of the latitude, to account for this unequal probability. The impact of these weights is moderate in tropical areas. The sample sites that contain a proportion of sea compensate for unselected sample sites that contain a proportion of land (when the center of the site is located in the sea) because they were considered as full sites.

The area change proportions of all sample sites are then extrapolated to the study area using the Horvitz–Thompson direct expansion estimator. The estimator for each area class transition is the mean proportion of that change per sample site, given by Equation 7.2:

$$\bar{y}_c = \frac{1}{m}\sum_{i=1}^{n} w_i \cdot y_{ic} \tag{7.2}$$

where y_{ic} is the proportion of area change for a particular class transition in the ith sample site. The weight of the sample unit is w_i and m is the sum of the sample weights. The total area of change for this class transition Z_c is obtained from:

$$Z_c = D \cdot \bar{y}_c \tag{7.3}$$

where D is the total area of the study region.

The usual variance estimation of the mean is known to have a positive bias. Alternative estimators based on a local estimation of the variance have been shown to reduce the bias. We use an estimator of the standard error based on local variance estimation:

$$s^2 = (1-f)\frac{\displaystyle\sum_{j \neq j'} w_{jj'}\delta_{jj'}(y_j - y_{j'})^2}{2\displaystyle\sum_{j \neq j'} w_{jj'}\delta_{jj'}} \tag{7.4}$$

where

f is the sampling rate

weight $w_{jj'}$ is an average of the weights w_j and $w_{j'}$

$\delta_{jj'}$ is a decreasing function (7.1) of the distance between j and j'.

The standard error is then calculated from this local variance using the total number of available sample sites, i.e., not accounting for the missing sites even if they are replaced by a local average.

The observations (source data sets) that are used to produce these results are derived from satellite interpretations. These surrogates to ground observations may be subject to uncertainty (bias). The use of such surrogate data for assessing area change is inevitable in many areas of the tropics where no ground observations exist and where large areas of inaccessible forests can only be monitored at affordable costs by using satellite data.

7.12   Perspectives

An operational system for processing and analysis of a global sample of moderate-resolution satellite imagery has been developed to produce maps and estimates of forest area changes in the periods 1990–2000 and 2000–2005 at tropical to global scale (Figure 7.4).

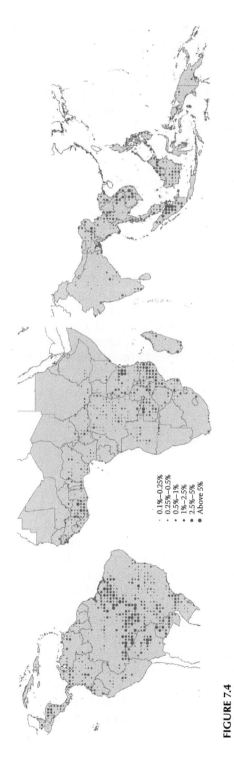

FIGURE 7.4

(See color insert.) Annual rate of gross forest cover loss during the period 2000–2005 for the tropical sample units of the global systematic sample.

The preliminary findings of an in-depth analysis of forest land-use change globally (FAO and JRC 2011) can be summarized as follows:

- The area in forest land use declined between 1990 and 2005, with global mean rates of loss between 1990 and 2000 of 2.7 (±0.9) million ha/year, rising to a mean annual loss of 6.3 (±1.4) million ha/year between 2000 and 2005.
- Just over half the world's forests are in tropical or subtropical climatic domains.
- There were important regional differences in forest loss and gain. In particular, forest loss was highest in the tropics going from –5.7 (±0.8) million ha/year in the 1990s to –9.1 (±1.2) million ha/year between 2000 and 2005.

The methods developed through the survey will be used to improve the measurement and reporting of forest area and change in forest area over time as part of the continual improvement of the FAO FRA process.

These results can be an important input to national and international reporting processes where forest area and change statistics are needed, such as the Convention for Biological Diversity and the emerging initiative for Reducing Emissions from Deforestation and Forest Degradation in Developing countries (REDD+) under the UNFCCC.

About the Contributors

Frédéric Achard is a senior scientist at the JRC, Ispra, Italy. His current research interests include the development of earth observation techniques for global and regional forest monitoring and the assessment of the implications of forest cover changes in the tropics and boreal Eurasia on the global carbon budget. Achard received his PhD in tropical ecology and remote sensing from Toulouse University, France, in 1989. He has coauthored over 50 scientific peer-reviewed papers in leading scientific journals.

Hans-Jürgen Stibig is a senior scientist at the JRC, Ispra, Italy. He received his PhD in forestry and remote sensing from the University of Freiburg, Germany, and has then been working in the field of forest inventory, mapping, and monitoring in several tropical countries, particularly in Southeast Asia. His current research interests is on global and regional forest monitoring by remote sensing, analyzing the impact and drivers of forest change in the tropics, with a geographical focus on Southeast Asia.

Erik Lindquist is a remote sensing specialist with the Food and Agriculture Organization. He is the coordinator of remote sensing activities within the Global FRA team based in Rome, Italy. He is completing his PhD in geospatial science and engineering from South Dakota State University, Brookings, South Dakota.

René Beuchle is a scientist at the JRC, Ispra, Italy. He has specialized in remote sensing, geoinformation systems, and cartography. He received a degree in cartography from the Karlsruhe University of Applied Sciences, Karlsruhe, Germany, in 1991. His current research activities are related to the assessment of tropical forest cover change by remote sensing with a geographical focus on South and Central America.

Rémi d'Annunzio is an forestry officer at the Food and Agriculture Organization (FAO) of the United Nations where he has been working since 2009, within the remote sensing team of the FRA team. He has an MSc in image analysis and segmentation and a PhD in forest science from AgroParistech, Paris, France.

References

Achard, F. et al., Estimating tropical deforestation. *Carbon Management*, 1, 271, 2010.

Beuchle, R. et al., A satellite data set for tropical forest area change assessment. *International Journal of Remote Sensing*, 32, 7009, 2011.

Bodart, C. et al., Pre-processing of a sample of multi-scene and multi-date Landsat imagery used to monitor forest cover changes over the tropics. *ISPRS Journal of Photogrammetry and Remote Sensing*, 66, 555, 2011.

Brink, A. and Eva, H.D., Monitoring 25 years of land cover change dynamics in Africa: A sample based remote sensing approach. *Applied Geography*, 29, 501, 2009.

Broich, M. et al., A comparison of sampling designs for estimating deforestation from Landsat imagery: A case study of the Brazilian legal Amazon. *Remote Sensing of Environment*, 113, 2448, 2009.

Duveiller, G. et al., Deforestation in Central Africa: Estimates at regional national and landscape levels by advanced processing of systematically distributed Landsat extracts. *Remote Sensing of Environment*, 112, 1969, 2008.

Eva, H.D. et al., Monitoring forest areas from continental to territorial levels using a sample of medium spatial resolution satellite imagery. *ISPRS Journal of Photogrammetry and Remote Sensing*, 65, 191, 2010.

Eva, H.D. et al., Forest cover changes in tropical South and Central America from 1990 to 2005 and related carbon emissions and removals. *Remote Sensing*, 4, 1369, 2012.

FAO, *Global Forest Resources Assessment 2010: Main Report*, FAO Forestry Paper 163, FAO, Rome, 2010.

FAO and JRC, *Global Forest Land-Use Change from 1990 to 2005*, FAO, Rome, 2011 (available at: http://www.fao.org/forestry/fra/remotesensingsurvey/en/).

FAO, JRC, SDSU, and UCL, *The 2010 Global Forest Resources Assessment Remote Sensing Survey: An Outline of the Objectives, Data, Methods and Approach*, Forest Resources Assessment Working Paper 155, FAO, Rome, 2009.

Gutman, B.G. et al., Towards monitoring land-cover and land-use changes at a global scale: The global land survey 2005. *Photogrammatic Engineering and Remote Sensing*, 74, 6, 2008.

Hansen, M.C. et al., Development of 500 meter vegetation continuous field maps using MODIS data. *International Geoscience and Remote Sensing Symposium*, 1, 264, 2003.

Ju, J., The availability of cloud-free Landsat ETM+ data over the conterminous United States and globally. *Remote Sensing and Environment*, 112, 1196, 2008.

Lindquist, E.J., et al., The suitability of decadal image data sets for mapping tropical forest cover change in the Democratic Republic of Congo: Implications for the global land survey. *International Journal of Remote Sensing*, 29, 7269, 2008.

Masek, J. et al., North American forest disturbance mapped from a decadal Landsat record. *Remote Sensing and Environment*, 112, 2914, 2008.

Maxwell, S., Filling Landsat ETM+ SLC-off gaps using a segmentation model approach. *Photogrammatic Engineering and Remote Sensing*, 70, 1109, 2004.

Mayaux, P. et al., Tropical forest cover change in the 1990s and options for future monitoring. *Philosophical Transactions of the Royal Society B: Biological Science*, 360, 373, 2005.

Potapov, P. et al., The global Landsat imagery database for the FAO FRA remote sensing survey. *International Journal of Digital Earth*, 4, 2, 2011.

Raši, R. et al., An automated approach for segmenting and classifying a large sample of multi-date Landsat-type imagery for pan-tropical forest monitoring. *Remote Sensing and Environment*, 115, 3659, 2011.

Simonetti, D., Beuchle, R., and Eva, H.D., *User Manual for the JRC Land Cover/Use Change Validation Tool*. EUR 24683 EN, Luxembourg: Publications Office of the European Union, 2011.

Stehman, S.V. et al., Adapting a global stratified random sample for regional estimation of forest cover change derived from satellite imagery. *Remote Sensing and Environment*, 115, 650, 2011.

8

Monitoring Forest Loss and Degradation at National to Global Scales Using Landsat Data

Peter Potapov, Svetlana Turubanova, and Matthew C. Hansen
University of Maryland

Ilona Zhuravleva and Alexey Yaroshenko
Greenpeace Russia

Lars Laestadius
World Resources Institute

CONTENTS

8.1 Introduction

Information on the extent and change of forest cover at the national to global scale is important for many reasons. At the national level, it provides a basis for terrestrial carbon accounting, land use management, monitoring of forest resources, and conservation planning. Many international processes use it too. It helps improve the forest cover change reporting of the United Nations Food and Agriculture Organization (FAO), which serves as the baseline reference for global-scale environmental accounting and modeling. It provides keystone variables for international initiatives to reduce deforestation, such as the process of reducing emissions from deforestation and degradation in developing countries (REDD+) of the United Nations Framework Convention on Climate Change (UNFCCC), which requires developing countries to have robust and transparent national forest monitoring systems. It is important to assess the status and threats for biological diversity as required by the Programme of Work on Forest Biological Diversity within the United Nations Convention on Biological Diversity. Environmental nongovernmental organizations such as World Wide Fund, Conservation International, and Greenpeace depend on forest degradation data to design forest conservation campaigns and combat illegal logging.

Ideally, such information should be comprehensive and consistent across the relevant space and time. Currently, the primary source of global forest cover extent and change is data from national forest inventories (NFIs), which are aggregated by FAO to form a series of Global Forest Resources Assessments (FRA). The usefulness of these assessments is reduced, however, by a number of factors that are inherent in the aggregation approach: (1) NFI data from different countries differ in terms of quality and age (update rates), and data from developing countries are often incomplete and inconsistent; (2) despite the efforts of FAO, countries de facto apply different definitions of forest cover and use, different forest accounting and change detection methods, thus making it difficult to synthesize results; (3) forest cover and change information are only provided in a tabular numerical format without any spatial disaggregation. The FRA process has started to incorporate remotely sensed data through the remote sensing survey, a sample-based assessment of global and biome-level forest extent dynamics (FAO 2009). However, for many applications, a spatially exhaustive map product is required.

Satellite remote sensing provides a viable data source to supplement NFIs and global forest monitoring initiatives. Forest cover extent and timely change estimates can be successfully retrieved from medium spatial resolution optical satellite data (Williams et al. 2006). These data are invaluable for the quantification of forest cover within the vast extent of remote and inaccessible forest landscapes, as well as for developing countries where lack of transportation infrastructure coupled with political instability often limit data collection and forest mapping on the ground.

During the last decade, a number of forest monitoring projects have been developed and implemented at the national level using satellite data. Major timber-producing countries, such as Finland (Tomppo 1993), Sweden (Willén et al. 2005), and Canada (Wulder et al. 2008), use optical satellite imagery as a standard source of information to supplement and extrapolate field plot measurements and to monitor forest management. Among developing countries, the Brazilian system on mapping annual deforestation (PRODES) is the largest and most robust operating forest monitoring system (INPE 2002). However, to expand these efforts to the biome and global scales, three major problems need to be solved: (1) methodological consistency must be improved (so that the results obtained at the national scale are directly comparable); (2) cost-effective monitoring methods must be developed (so that the cost of source data and data analysis will be low enough to allow national- to global-scale implementation); and (3) open data access must be ensured (so that various international and nongovernmental organizations and experts are able to analyze, review, and validate the monitoring results).

There are two main strategies for satellite-based forest monitoring at a large scale: sampling and wall-to-wall mapping. Several sample-based approaches have been successfully implemented during the last decade at biome (Achard et al. 2002) and global levels (FAO 2009; Hansen et al. 2010). Different sampling designs were used to select classified imagery subsets, including regular sampling (FAO 2009) and stratified sampling (Achard et al. 2002; Hansen et al. 2010). Both approaches, however, are challenged by low estimate precision due to the uneven distribution of change within forest landscapes (Tucker and Townshend 2000), and neither produces a spatially explicit result. This limits their usefulness for many applications.

Wall-to-wall coverage of satellite data with sufficient spatial resolution differs from sample-based approaches in that it allows for direct mapping of forest cover and change and for a spatially complete quantification of forest dynamics at the national scale. Low spatial resolution data of the kind produced by the MODIS or MERIS sensors are inadequate for direct estimation of forest change, as much of it occurs at subpixel scales (Jin and Sader 2005). Medium spatial resolution data, such as that produced by the Landsat sensor, do allow for accurate forest cover and change area measurement (Williams et al. 2006). The use of medium spatial resolution data for national-scale forest monitoring has been limited until recently by the high data costs, the difficulty of handling large data volumes, and data analysis problems in regions with persistent clouds, such as the humid tropics. Recently, however, changes in data distribution policies and data-processing algorithms have enabled fast and cost-effective national-scale forest cover and change assessment.

Undoubtedly, the most important enabling factor for large-scale satellite-based forest monitoring is free-of-charge data availability. While low spatial resolution data (AVHRR, MODIS) were freely available for decades, medium-resolution data have been costly until recently. In January 2008, the U.S. Geological Survey (USGS) implemented a new Landsat data distribution policy that provides Landsat data free of charge. The free-of-charge data allows financially constrained developing countries to use it for wall-to-wall forest mapping. For example, purchasing the 2000–2010 Landsat data for a country like the Democratic Republic of the Congo (DRC) would have cost more than 6 million U.S. dollars at the pre-2008 price. These resources can now be spent on data processing, analysis, and validation of results. Medium-resolution Landsat imagery provides the best balance between acquisition cost and spatial resolution, despite the fact that it is inadequate for the detection of small-scale forest change (e.g., low-intensity selective logging). Even when a complete national coverage of higher spatial resolution imagery is available, the high data cost will restrict its use by developing countries for national monitoring purposes.

Another important factor increasing the feasibility of using national wall-to-wall medium-resolution imagery for forest monitoring is the progress in computing capacity and data-processing algorithms. Modern computing hardware allows for rapid processing of Landsat data at the national scale (from several weeks to a month). Recent progress in automated Landsat data processing and mosaicing has made it possible to produce cloud-free annual or epochal composite images for persistently cloudy areas (Hansen et al. 2008; Potapov et al. 2011). Nonparametric classifiers (e.g., *k*-nearest neighbor, decision tree, support vector machines, and neural networks) allow for fast and precise mapping and change detection of heterogeneous land cover types such as forest cover (Hansen et al. 1996).

The rapid development in the quality and access to satellite imagery has widened the circle of actors that can monitor forests beyond national forest administrations, thereby enhancing transparency. Civil society, private industry, and researchers can now monitor forests in support of conservation, business, science, and other forest resource assessment and management applications. NFI and monitoring data provided by national governments can be validated by in-country and international nongovernmental organizations and expert groups, highlighting any data quality issues. This creates a competitive environment that stimulates the improvement of governmental policies and NFI methods. Forest monitoring transparency, however, requires that the source satellite data remain in the public domain and can be freely redistributed. Currently, only a few image data providers, including USGS and INPE, deliver satellite imagery under liberal licensing conditions that allow for sharing and redistribution of the data and derived monitoring products.

Our approach to national-scale forest cover loss monitoring is an evolution of an algorithm developed by Hansen et al. (2008). Data from the MODIS sensor were used to preprocess Landsat time-series images that in turn were used to characterize forest cover extent and loss. Our approach is based on a fully automated Landsat data processing, including scene selection, per-pixel quality assessment (QA), and normalization. The Landsat data archive was exhaustively mined, and all data that satisfied our selection criteria were used for the analysis. Individual Landsat images were normalized using MODIS-derived surface reflectance target and used to derive multitemporal metrics and time-sequential composites. These metrics, along with the MODIS data time series, were used as independent variables to build supervised decision tree models for mapping forest cover and change. Mapping and monitoring forest degradation, which include assessment of low-intensity disturbance and fragmentation, required an alternative method based on manual interpretation of time-sequential Landsat image composites following an approach developed by Potapov et al. (2008).

The objective of the forest assessment and monitoring method presented in this chapter is to provide regular national forest cover updates at 5- and 10-year intervals. The same algorithm can be used to produce results at finer temporal steps (e.g., annually), assuming that enough cloud-free observations are available; however, providing annual forest cover updates was beyond the objectives of this study. Further evaluation and evolution of the system will allow for more rapid updating of continental and global forest cover in the near future.

The forest cover loss and degradation assessment algorithms have been applied to different forest biomes, testing and illustrating their capability to be implemented at the global scale. Mapping and monitoring results have been published online along with Landsat image composites for use by national governmental and civil society organizations (European Russia data: http://globalmonitoring.sdstate.edu/projects/boreal/; the DRC data: http://congo.iluci.org/carpemapper/; Intact Forest Landscapes data: http://intacforests.org).

8.2 Landsat Data Processing

The Landsat remote sensing satellite program operated by the USGS provides free-of-charge data with a medium spatial resolution (30 m/pixel for reflective bands) suitable for the full spectra of forest monitoring studies from a local to the global scale (Williams et al. 2006). The Landsat program is unique due to its global image acquisition strategy, allowing land cover monitoring over the last three decades. Landsat ETM+ reflective spectral bands, which include visible (band 1, 450–515 nm; band 2, 525–605 nm; band 3, 630–690 nm), near infrared (band 4, 760–900 nm), and short infrared (band 5, 1,550–1,750 nm; band 7, 2,080–2,350 nm), provide a sufficient spectral profile for vegetation-type mapping and land cover change detection. The thermal infrared data (band 6, 10,400–12,500 nm) enable automatic cloud cover detection. One of the main advantages of the Landsat spectral bands is its radiometric consistency and continuity between Landsat sensors (TM, ETM+, and future LDCM) and with the MODIS sensor, allowing intercalibration of Landsat and MODIS datasets.

The complete global Landsat data archive is available through the USGS National Center for Earth Resources Observation and Science (EROS) from their Web portals: GLOVIS (http://glovis.usgs.gov) and Earth Explorer (http://earthexplorer.usgs.gov). The Earth Explorer data portal allows users to perform advanced archive inventory search as well as bulk Landsat data order and download. Image metadata browsing and selection is guided by the Worldwide Reference System-2 (WRS2) of path (ground track parallel) and row (latitude parallel) coordinates defining scene footprints.

In our study, to reduce computational time for Landsat data processing, only images having less than 50% cloud cover for any scene quarter, as estimated by the automatic cloud cover assessment (ACCA), were selected. However, the cloud cover threshold has been expanded to include images with 70%–80% cloud cover for scene footprints with low numbers of 50% cloud-free images. For boreal regions, only growing season images were selected. The annual growing season start/end dates were established for each Landsat WRS2 footprint using annual time series of MODIS-derived NDVI over a MODIS-derived forest cover mask. Image metadata analysis, scene selection, and bulk data ordering were performed using an automated metadata search tool.

The Landsat images are normally processed as Level 1 terrain (L1T) corrected data by the USGS EROS. The L1T corrected data product provides systematic geometric accuracy by incorporating ground control points and a digital elevation model (DEM) for topographic accuracy. However, if insufficient ground control points or elevation data necessary for terrain correction were available, images can be delivered as Level 1 systematic correction (L1G). Because L1G data often feature low geometric accuracy and require further geocorrection, only images processed as L1T, ensuring a high geolocation precision, were used for the subsequent data processing save for the few coastal scene footprints where L1T corrected data were not available at all.

Our fully automated Landsat data process included two main steps: (1) per-image processing including image resampling, applying at-sensor calibration, per-pixel observation QA, and radiometric normalization and (2) per-pixel observation coverage analysis, production of image composites, and derivation of multitemporal metrics for forest extent and change mapping (Figure 8.1).

To facilitate image processing and enable per-pixel compositing, all image data for the nation (region) were resampled to a predefined pixel grid. The pixel grid was specified separately for each continent in equal-area map projections chosen to reduce geometric distortion. The following examples of national-scale forest monitoring were prototyped using pixel grids with 60 m spatial resolution to reduce data volumes and computation time. The 30 m spatial resolution pixel grid will be used for future continental- to global-scale processing.

At-sensor calibration was applied to convert raw image digital numbers to top-of-atmosphere (TOA) reflectance (for reflective bands) and brightness temperature (for thermal infrared band) in order to minimize differences in sensor calibration, between sensors (TM, ETM+, and MODIS), in the sun–earth distance, and in the elevation of the sun. To calculate TOA reflectance and brightness temperature, we used the approach described in Chander et al. (2009), with coefficients taken from image metadata.

The purpose of per-pixel observation QA was to select cloud-free and cloud shadow-free land and water observations for subsequent image compositing. To automatically map clouds and cloud shadows, we used a set

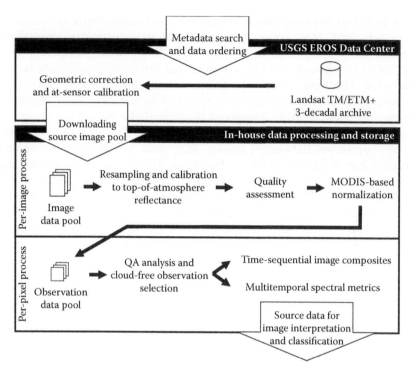

FIGURE 8.1
Landsat data-processing workflow.

of cloud, haze, shadow, and water detection models. The models correspond to a set of classification tree models (Breiman et al. 1984) derived from training data that were collected from a large sample of Landsat imagery. The Landsat training data and derived QA models are biome specific (separate models are used for tropical, temperate, and boreal forests). Training images were manually classified into land, water, cloud, haze, and shadow classes. From these images, 10% samples were randomly selected, aggregated for all images, and used to create generalized classification tree models. Each model was applied per Landsat image, yielding class probability values. Based on these values, a QA code was assigned to each pixel reflecting the probability of the pixel to be a land or water cloud-free observation, using the method described in Potapov et al. (2011).

Relative radiometric normalization of Landsat imagery was used to reduce reflectance variations between image dates due to atmospheric conditions and surface anisotropy. Only reflective bands used for image compositing (bands 3, 4, 5, and 7) were normalized. The shortwave visible bands (bands 1 and 2) were not used due to their sensitivity to atmospheric haze and water vapor, precluding correct normalization. The thermal infrared band 6 was used for the cloud screening model but was not included in the final image composite. The atmospheric correction of Landsat-derived TOA reflectance using time-synchronous atmospheric data and 6S radiative transfer code is a state-of-the art method (Masek et al. 2006) that should be implemented for obtaining consistent surface reflectance. However, simple techniques for relative image normalization using radiometrically consistent sets of moderate spatial resolution data could be successfully employed to facilitate image compositing over large regions (Olthof et al. 2005; Hansen et al. 2008). Our approach relied on the correlation between Landsat TOA and MODIS atmospherically corrected top-of-canopy (TOC) reflectance. MODIS normalization target reflectance data were collected from 2000 to 2009 (10-year) global Terra/MODIS 250 m data 16-day composites (MOD44C, collection 5), provided by the University of Maryland. The MODIS spectral bands 1, 2, 6, and 7 were chosen to correspond with Landsat bands 3, 4, 5, and 7. To reduce the presence of clouds and shadows, the mean surface reflectance corresponding to the three highest NDVI values from observations with the lowest cloud probability over the 2000–2009 interval were used as the normalization target. We calculated a mean bias between MODIS TOC and Landsat TOA reflectance for each spectral band over the land area and used it to adjust Landsat reflectance values. A simple empirically derived reflectance difference threshold was used to avoid areas of rapid land cover or phenological change. For tropical areas where the surface anisotropy effect significantly hindered image interpretation (Hansen et al. 2008), an additional correction for surface anisotropy was implemented. A simple linear regression between the MODIS/Landsat reflectance bias (dependent variable) and distance from orbit ground track (independent variable) was derived for each reflective band and then applied to correct band reflectance values within the entire Landsat image. Image normalization was performed independently for each spectral band and Landsat image. This fully automated image processing approach allowed us to use parallel computing methods, reducing the average image processing time to 12 s/image.

Our approach for image time-series analysis integrates the classic, multidate image compositing method (Holben 1986), with the novel approach of using multitemporal metrics to characterize reflectance variation within a given time interval (Hansen et al. 2003). Image time series were analyzed at per-pixel level using all processed Landsat observations for the entire time interval. For decadal forest monitoring, two sets of metrics were created for two 5-year time intervals: 2001–2005 and 2006–2010. To facilitate data management and to allow parallel computing, compositing was performed independently for a set of rectangular tiles dividing the entire area of analysis. To create an observation "data pool" from which time-sequential composites and spectral metrics could be derived, we preferentially selected observations with the least cloud/shadow contamination within the growing season. Growing season images are more appropriate for forest cover mapping than imagery captured during senescence or dormant periods. Preferential growing season boundaries can be defined either on a per-scene basis (Potapov et al. 2011) or on a per-pixel basis using MODIS-derived annual NDVI profiles. To create a "data pool," we analyzed QA flags for all available observations for the pixel. A set of criteria were designed to identify observations with the least cloud/shadow contamination to be included in the "data pool." Because the cloud shadow classification model was not tuned to water bodies, pixels with high water probability were selected separately. For land pixels, the number of growing season cloud/shadow-free observations for each 5-year interval (for decadal analysis) was calculated. If no cloud-free observations were found for any 5-year time interval, search boundaries were extended first to out-of-season observations, then to observations with moderate cloud/shadow probabilities. After the "data pool" pixels were selected, all other data (flagged as having higher cloud/shadow probability or out of season) were excluded from further processing.

The time-sequential image composites derived from the "data pool" observations represent start/end points for forest cover monitoring analysis and have been used for ca. year 2000 forest mapping, for change detection (for boreal regions), and for visual image interpretation and mapping of forest degradation. Several approaches for image compositing have been tested, including single-date compositing and multidate compositing using mean (or median) value or NDVI (or selected band reflectance) value ranking (Hansen et al. 2008; Potapov et al. 2011). We found that different approaches are appropriate for different applications. For change detection, the first/last single-date observation compositing was found to be the most suitable as it represents the land cover status for the first and last cloud-free image date in the "data pool." For visual interpretation, on the other hand, multidate composites were found to be more suitable due to low noise levels and consistent reflectance values within the area of analysis (Potapov et al. 2011). Our current automatic image compositing method produces a set of different time-sequential composites for use as classification metrics and for visual analysis.

While the time-sequential image composites are invaluable for visual image interpretation and for creating classification training sets, they are inadequate for forest cover change monitoring in tropical forests. This is because the rapid establishment of regrowth obscures the change signal over decadal and mid-decadal time intervals. To highlight reflectance variation within the analyzed time interval, a set of spectral metrics were created from the "data pool" observations. These metrics were designed to capture a generic feature space that facilitates regional-scale mapping and have been used extensively with MODIS and Landsat data (Hansen et al. 2003, 2008, 2010). Three groups of per-band metrics were created: (1) reflectance values representing 6-year maximum, minimum, and selected percentile values (10%, 25%, 50%, 75%, and 90% percentiles); (2) mean reflectance values for observations between selected percentiles (for the min-10%, 10%–25%, 25%–50%, 50%–75%, 75%–90%, 90%–max, min–max, 10%–90%, and 25%–75% intervals); and

(3) the value of the slope of a linear regression of band reflectance versus image date. Multitemporal metrics were used for forest cover and change classification, and selected metrics were employed for visual image analysis and creation of training data.

8.3 National-Scale Forest Cover Extent and Loss Mapping

Forest cover mapping and change detection was carried out on the basis of wall-to-wall image composites using a single national-scale supervised classification model. The classification model was built using an extensive set of training data collected within the entire area of analysis. This approach helped to avoid the problems that arise when a classification model based on local training data is extrapolated to neighboring images (Wulder et al. 2008). The classification and regression tree (CART) algorithm was used as the main tool for image classification and change detection. CART is a nonparametric supervised classification model constructed to predict the class membership by recursively splitting the feature space into a set of nonoverlapping subsets and then reporting the class probability within each subset. The CART algorithm has been shown to have a high precision for land cover mapping (Hansen et al. 1996). To improve the CART model stability and accuracy, a bootstrap aggregation (bagging) algorithm was used that corresponds to a set of trees created using random training data subsamples and taking the median class likelihood as the final result. Bagged classification tree models for forest cover and change mapping were generated using the training data as the dependent variable and multitemporal metrics plus time-sequential image composites as independent variables.

For the purpose of the regional-scale monitoring examples described below, forest was defined as having 30% or greater canopy cover for trees of 5 m or more in height. Forest cover and forest types were mapped for the year 2000, the first year of monitoring. All events resulting in stand replacement at the 60 m pixel scale within the analyzed time interval, including clearings (even if followed by forest regrowth within the same time interval), logging, fire, flooding, and storm damage, were mapped together as a gross forest cover loss class. Forest cover loss was mapped within the year 2000 forest mask. For the decadal monitoring, forest cover loss was mapped independently for each 5-year interval. To build the classification tree models for forest cover extent and forest cover loss mapping, a training set was manually created by visual interpretation of the region-wide time-sequential image composites. A number of additional datasets, including freely available QuickBird images from GoogleEarth™ and expert information, were used as reference materials to aid interpretation.

Two examples of region-wide Landsat forest cover mapping and change detection projects are briefly described below: one within the boreal and temperate forests of European Russia, another within the humid tropical forests and dry tropical woodlands of the DRC.

8.3.1   European Russia Forest Cover and Change Mapping

The forest cover change analysis from 2000 to 2005 was performed within the northern and central administrative regions of European Russia. The area of analysis spans from the northern forest–tundra ecotone to the forest–steppe boundary in the south and includes a variety of boreal and temperate forest types. A total of 7,227 Landsat ETM+ images from 1999 to 2007 were selected based on cloud cover and growing season date criteria. Landsat image normalization was performed using a MODIS-derived pan-boreal coniferous forest mask as the normalization target. Normalized Landsat images were used to create time-sequential image composites for 2000 and 2005 and a set of spectral metrics describing reflectance variability within ±1 year of the target composite date. For places with persistent cloud cover and/or a limited number of observations, images that were acquired more than 1 year before or after the target year were used for compositing and metrics. To create the image composite, all selected cloud-free observation dates for each pixel were ranked based on band 4 values. The image date corresponding to the band 4 median was chosen as the composite date, and all reflective bands from this date were used to create a final ca. 2000 or ca. 2005 image composite. In addition to the band 4 median value composites, a set of spectral metrics was created on the basis of a band 5 ranking meant to capture reflectance variation within the growing season. Owing to the time-preferential compositing rule, more than 95% of the composite areas for the ca. 2000 and 2005 could be created from images acquired within ±1 year of the target year. Less than 0.5% of the total composite area had to be excluded from analysis due to lack of cloud-free observations. Due to the relatively slow reforestation within the boreal and temperate forests, we concluded that using the composite difference would be sufficient for 5-year forest cover loss mapping (Figure 8.2).

Forest cover for the year 2000 was mapped using Landsat composites and metrics for ca. year 2000 supplemented with pixel latitude and MODIS annual metrics. The MODIS annual metrics included mean red reflectance and NDVI value for the growing season and annual highest red and NIR reflectance representing the extent of snow cover during the winter. Forest cover within European Russia is generally easily defined and mapped as most of the natural or managed forests have high canopy densities. Additional MODIS metrics helped improve the forest/nonforest classification within wetland forests, the forest–tundra, and the forest–steppe interface. Gross forest cover loss from 2000 to 2005 was mapped within the resulting year 2000 forest mask. All stand-replacing events, whether caused by logging, road/pipeline construction, wind throws, stand-replacement forest fires,

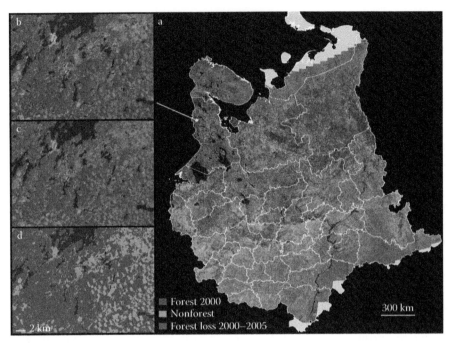

FIGURE 8.2
(See color insert.) Forest cover loss monitoring in European Russia. (a) The ca. 2000 region-wide Landsat ETM+ image composite. (b–d) Zoom-in example of forest cover and change mapping in the Republic of Karelia: b—the ca. year 2000 image composite; c—the ca. year 2005 image composite; d—classification result.

or severe insect outbreaks, were mapped together without any attempt to discriminate among them. Within low-intensity selective logging sites, only areas with significant forest impact (roads and clearings) were mapped.

The total forest area within analyzed regions of European Russia was estimated to be 150,228 thousand ha at the time around year 2000. The area of forest cover loss from 2000 to 2005 is 2,210 thousand ha, which represents a 1.5% of the year 2000 forest cover. Our forest extent estimate is within 1% difference with the latest available official forest cover area assessment for year 2003 (ROSLESINFORG 2003). At the regional level, our forest area estimates are well correlated (R^2 of 1.00) with official statistics. A per-pixel validation with independently derived forest cover mapping results for 23 blocks 20 km × 20 km in size within the boreal and temperate forests showed good agreement, with an overall forest cover accuracy of 89% (kappa of 0.78) and overall change detection accuracy of 98% (kappa of 0.71). A comparison at the individual sample block level, however, indicated relatively high forest cover classification uncertainty along the boreal forest's northern limit (overall accuracy of 87%) and low forest cover loss producer's accuracy (58%) within southern temperate forests featuring small-scale logging.

Forest cover loss was distributed unevenly within the administrative regions, reflecting several forest management issues. More than 60% of the total forest cover loss was found within the largest northern forest regions including Arkhangelsk, Kirov, Leningrad, and Vologda Oblast, Komi, and Karelia Republics. While regional forest cover loss is linearly related to forest area (R^2 of 0.84), the Leningrad region had the largest residual value, indicating a much higher rate of forest cover loss than the general trend within the area of study. One-third of the analyzed regions have a percent forest cover loss above the average and represent areas of intensive forest harvesting and frequent wildfires. These regions are located in the western and central parts of European Russia, close to large industrial cities and the Finnish border. Regions of eastern European Russia, the Urals, and northern forest–tundra transition have the lowest proportional gross forest loss. The three regions with the highest proportional forest cover loss are Vladimir, Leningrad, and Moscow Oblast (forest loss 3.7%, 3.5%, and 3.1% of year 2000 forest cover, respectively) (Figure 8.3).

The high forest cover loss within Leningrad region is thought to be a consequence of intensive forest harvesting. This is confirmed by official Russian forest use statistics for annual timber harvesting. The Leningrad region had

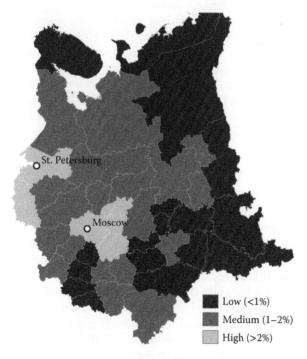

FIGURE 8.3
Forest cover loss intensity in European Russia (percent forest loss 2000–2005 of forest cover for year 2000 per administrative region).

the highest rate of timber removal of all analyzed administrative regions in the period from 2000 to 2005 (ROSSTAT 2008). The intensive felling in the Leningrad region and the neighboring Karelia Republic (gross forest cover loss 1.9% of year 2000 forest cover) is stimulated by the demand from the Nordic countries, particularly Finland, for timber from these border regions. The extensive gross forest cover loss due to industrial logging near the Russian–Finnish border could result in forest resource depletion and consequent environmental and social problems if not compensated by forest restoration.

While the gross forest cover loss in the Leningrad region was connected mainly with industrial timber harvesting, the forest loss in the Moscow and Vladimir regions is a consequence of several factors, including logging (partly illegal), insect outbreaks, human-caused fires, and expansion of settlements. The single largest forest cover loss event within these regions was due to the forest fires of year 2002. While in general wildfires play a comparatively small role in the forest dynamics within European Russia, severe drought conditions and human-induced fires led to extensive forest loss within the central regions of European Russia during the extreme fire season of 2002. According to official data, the area of burned forest in the Moscow region in 2002 was roughly 10 times higher than the mean annual burned area from 1992 to 2005 (ROSSTAT 2008). Another cause of forest cover loss around large cities is urban sprawl. For example, the expansion of settlements and industrial facilities around the city of Moscow led to the conversion of about 58 thousand ha of former forest and agriculture lands from 1998 to 2008 (Karpachevskiy et al. 2009). The forests that remain around large industrial cities provide ecological services (e.g., water and air purification, natural species refugee, recreation) that are important to urban populations. Our results raise concerns about the fate of the remaining forests in the most populated regions of European Russia.

8.3.2 Forest Cover Monitoring in the DRC

Information on forest cover extent and change is sparse or lacking for the DRC due to the vast extent of intact forest landscapes (IFLs), the lack of transportation infrastructure, and the continued political instability, all of which limit the possibilities to collect data on the ground. Satellite images are currently the only viable data source for national level mapping. We employed wall-to-wall Landsat imagery to map forest cover for the year 2000 and the gross forest cover loss between 2000 and 2010. The analysis was performed in partnership with Observatoire satellital des forêts d'Afrique central (OSFAC), a local nongovernmental organization supported by the Central Africa Regional Program for the Environment (CARPE) project of the United States Agency for International Development (USAID).

A total of 8,881 Landsat ETM+ images were selected, downloaded, and processed to create complete national-scale image composites and metrics.

About 99.6% of the country was covered by cloud-free Landsat observations. Gaps due to persistent cloud cover were located primarily in the coastal areas of the lower Congo River. The data gaps were mostly due to an insufficient number of cloud-free observations. Even though most of the available Landsat 7 observations (82%) were captured during the 11 years of observation in coastal areas, few of them were more than 50% cloud free for more than a quarter of a scene. This shows that data from a single sensor is often insufficient for monitoring forests in persistently cloudy tropical regions. A constellation of sensors with similar spectral and spatial resolution but varying overpass time and orbital cycle would be needed to provide sufficient observational coverage.

Forest cover and forest types were mapped for ca. year 2000. Forest cover classes included humid tropical forests (defined as having greater than 60% canopy cover) and woodlands (canopy cover between 30% and 60%). Humid tropical forests were additionally stratified into primary (mature) forests and secondary forests (regrowing after stand-replacement disturbance). A generic forest cover class category was mapped, and within this layer primary and secondary humid tropical forest classes were subsequently characterized. After mapping humid tropical forest classes, the remaining forest cover was assigned to the woodland class. Gross forest cover loss from 2000 to 2005 was mapped within the generic year 2000 forest mask, and forest cover loss 2005–2010 was mapped within the remaining forest area of 2005 (Figure 8.4).

The total forest cover extent in the DRC was estimated to be 159,529 thousand ha, which is within 1.5% of the FAO FRA estimate for year 2000. Primary and secondary humid tropical forests predominate (66% and 11% of total forest cover extent, respectively), with woodlands occupying the remaining 23%. The gross forest cover loss from 2000 to 2010 was 3,712 thousand ha or 2.3% of year 2000 forest area. About 57% of this loss occurred in secondary humid forests, 29% in primary humid forests, and 14% in woodlands. Secondary humid tropical forests experienced the most intensive loss (11.6% over 10 years), while the rate of loss in primary humid tropical forests and woodlands was considerably lower (1.0% and 1.4%, respectively). The gross forest cover increased by 14% between the 2000–2005 and the 2005–2010 periods. The increase was most prominent in primary humid tropical forests and woodlands (by 91% and 63%, respectively).

Visual examination of Landsat composite data suggests that almost all forest clearing was associated with the expansion of subsistence agriculture, local charcoal production, or mining. We found no evidence of major forest fires or windthrow events during the study period, with the exception of forest fires caused by the repeated eruptions of the Nyamuragira volcano. Clearings are common in secondary humid tropical forests due to the practice of rotational slash-and-burn agriculture. On the one hand, the fallow period between clearings (not quantified in this study) would be a useful indicator of land degradation. Clearing of primary forests, on the other hand, represents the expansion of agriculture into heretofore intact

FIGURE 8.4
(See color insert.) Forest cover loss monitoring in the DRC. (a) Nation-wide forest cover and change mapping result. (b–c) Zoom-in example of forest cover and change mapping around Buta: b—ca. year 2010 image composite; c—classification result.

forests, triggering changes in ecosystem dynamics and loss of floristic and faunal biodiversity. Clearing generally occurs in belts around secondary forests and roads due to the nearly continuous distribution of popula-tion along transportation infrastructure (Figure 8.4). Since forest clearing is mainly a consequence of small-scale subsistence farming, the change patches are small and have a mean area of 1.4 ha.

Most of the clearing occurred in areas with high population density and growth rates, such as Kinshasa, Kasai-Occidental, Sud-Kivu, and Kasai-Oriental provinces. Large industrial (Tshikapa, Mbuji-Mayi, Kolwezi, Lubumbashi) and artisanal mining areas (Kisangani, Beni, Buta) also exhib-ited intensive forest loss. The intensive forest loss along the boundaries of Virunga National Park (NP) in the North Kivu province is related to ongoing political unrest. The Virunga NP has the highest loss of primary forest of all national parks in the country (0.9%, compared to the mean of 0.4%), making it one of the most threatened natural protection areas. The loss of primary forest in protected areas increased by 64% from 2000–2005 to 2005–2010, highlighting the pressures and the need to improve the protection and management of nature reserves across the country.

8.4 Global- and National-Scale Forest Degradation Monitoring

It is well known that forest degradation, including fragmentation of natural landscapes, has a negative effect on global climate change and biodiversity (Harris 1984). However, forest degradation is a complex concept that is difficult to define and even more difficult to map. Unlike forest cover extent that can be quantified using straightforward biophysical parameters, assessing and monitoring forest degradation is a complicated task due to the great variability in the forms, factors, and degrees of human impact. In the late 1990s, a group of nongovernmental organizations including Greenpeace and the World Resources Institute developed a simple yet straightforward approach for assessment and monitoring of forest degradation called the IFL method (Potapov et al. 2008). An IFL is an unbroken expanse of natural ecosystems that shows no signs of significant human activity and is large enough to maintain all native biodiversity, including viable populations of wide-ranging species. The essence of the IFL method is to use medium spatial resolution satellite imagery to locate IFLs, establish their boundaries, and use them as a baseline for monitoring. The IFL method provides a simple and feasible way to cope with the complexity of the forest degradation concept by using changes in forest intactness as a proxy for forest degradation (Potapov et al. 2009). In this context, forest degradation is defined as a reduction in the ecological integrity of a forest landscape below a certain threshold due to human influence (e.g., conversion, alteration, and fragmentation), and forest landscapes are treated as being either intact (undegraded) or nonintact (altered or degraded).

An IFL boundary is defined using a sequence of two sets of criteria specifically developed for visual interpretation of medium spatial resolution satellite imagery. These criteria are globally applicable and easily replicable, allowing for repeated assessments over time as well as verification by independent assessments. The first set of criteria is used to eliminate lands with evidence of significant human-caused alteration from IFL status. Such alteration includes (1) settlements and industrial objects; (2) infrastructure used for transportation between settlements or for industrial development of natural resources; (3) agriculture and forest plantations; (4) industrial activities (including logging, mining, oil and gas exploration or extraction) during the last 30–70 years; and (5) stand-replacing wildfires during the last 30–70 years if located in the vicinity of infrastructure or developed areas. Some alterations, notably low-intensity human impacts that tend to occur in the vicinity of settlements and roads (e.g., selective logging and overhunting), are not visible in medium spatial resolution imagery. We, therefore, removed such areas by applying a buffer zone around settlements and transportation infrastructure, adapting the buffer width to the expected extent of human influence. For the global IFL method, a 1 km wide buffer was used. The second set of criteria is used to eliminate fragmented lands from IFL status by identifying patches of otherwise IFLs that are smaller or narrower than a selected threshold value. For the global analysis, a patch needed to meet the following criteria to qualify as an IFL: (1) minimal area of 500 km^2, (2) minimal width of at least 10 km (measured as the diameter of the largest circle that can be fitted inside the patch), and (3) at least 2 km wide in corridors or appendages to areas that meet the above criteria.

The IFL method was used to assess the ecological integrity of the world's forest landscapes. First, the current global extent of the forest zone was determined, defined as lands with at least 20% tree canopy cover (Hansen et al. 2003) and including treeless areas that occur naturally within forest ecosystems, such as wetlands. The area under consideration was then reduced by identifying and eliminating developed areas and infrastructure through visual interpretation of Landsat imagery. The global IFL mapping was done before the Landsat data archive was opened, and the GeoCover Landsat orthorectified image collection was therefore used. A global coverage of Landsat TM data (representing an average date of 1990) and ETM+ data (representing an average date of 2000) was used to systematically assess candidate IFL areas for human-caused alteration and fragmentation and to delineate IFLs. Fine-scale geospatial datasets on roads and settlements were used where available to facilitate interpretation. Infrastructure buffering was performed simultaneously with the visual image analysis. Altered and fragmented patches were eliminated from the area of study and remaining areas, if meeting the criteria, were classified as IFLs.

The current extent of the world's forest zone is 5,588 million ha. IFLs make up 23.5% of the forest zone (1,313 million ha). The remainder of the forest zone is affected by development or fragmentation and thus is either managed or degraded. The vast majority of the world's remaining IFLs are found within humid tropical and boreal forests (45.3% and 43.8% of the total IFL area, respectively). The distribution of IFL within these biomes is heterogeneous, reflecting differences in the history and intensity of economic development. Tropical IFLs are found mainly in the large forest massifs of the Amazon and Congo basins, and in insular Southeast Asia. More than half of the IFL area in the humid tropics is in the Amazon basin, while IFLs are largely absent in the lowlands of continental Asia. In the boreal region, the highest proportion of IFL is in the North. IFLs occupy more than half of the forest zone in Canada but have nearly disappeared in Europe due to the long history of intensive agriculture and forest management.

A particular strength of the IFL method is that it can easily be applied to different points in time, making it suitable for regular reassessments,

i.e., monitoring. The work is conducted through expert-based visual interpretation using the same criteria and the same type of data as in the baseline assessment (medium spatial resolution satellite imagery) but is much
less time consuming as only remaining IFLs need to be monitored. We
used the IFL method to assess change in IFLs from 2000 to 2010 (using
two 5-year steps) for the three largest tropical forest countries: Brazil, the
DRC, and Indonesia (Figure 8.5). For the DRC and Indonesia, national
reassessments were performed using Landsat time-sequential image composites (see Section 8.2), individual Landsat scenes, and ASTER imagery.
For Brazil, the forest cover loss monitoring results from PRODES (INPE
2002) were used to update the IFL map.

Our results show that a significant extent of intact areas has been lost within
all three countries after year 2000. The total proportion of IFLs lost was 5.2%,
1.9%, and 10.0% in Brazil, the DRC, and Indonesia, respectively. The IFL loss
in Brazil is mostly a consequence of agroindustrial development along the
forest/agriculture boundary of "arc of deforestation." In the DRC, the loss
of IFLs is unevenly distributed and located mostly within active timber

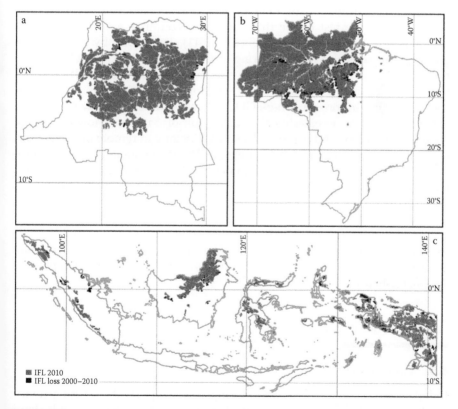

FIGURE 8.5
IFL decadal monitoring results for the DRC (a), Brazil (b), and Indonesia (c).

concessions (where selective logging is taking place) and in the vicinity of growing settlements (where subsistence agriculture, artisanal logging, and charcoal production are expanding). Conversion of IFLs to oil palm and timber plantations is common within the Indonesian lowlands of Sumatra and Kalimantan islands, while IFL loss in mountain areas is generally caused by selective logging.

While all analyzed countries experienced reductions of IFL area, the change trends are different, as approximated by IFL loss between 2000 and 2005 and 2006 and 2010. Brazil features a dramatic reduction in overall IFL loss from 4.1% during the first 5 years to 1.1% during the second half of the decade. In the DRC, the IFL loss rate was relatively stable (1.0% during 2000–2005 and 0.9% during 2006–2010). In contrast, the IFL loss rate in Indonesia increased from 4.2% to 5.8%. While no special analysis is available to explain these trends, we can speculate on their origins based on the global economy and the distribution of IFL loss. Undoubtedly, the global financial crisis that began in 2007 and followed by the recession during the end of the decade is a single most important factor behind the reduction of agroindustrial clearings and timber production worldwide. This crisis was more pronounced in Western countries but had consequences also for their main suppliers. Brazil was hit hardest of the three analyzed countries and experienced a negative GDP growth rate in 2009 (CIA 2011). The efforts by the Brazilian government to reduce forest clearing in the framework of the UN REDD+ program and the establishment of an effective deforestation monitoring system have likely also played a role. The situation was different in Asian countries, including Indonesia, where GDP either continued to grow or fell only slightly. Indonesia accelerated the conversion of remaining lowland forest areas to plantations and expanded selective logging in remote mountain forests, especially in the Papua island group. The IFL change dynamic is complicated in the DRC due to the combination of global economic drivers and local political instability. While economic stagnation and years of civil war have resulted in a low level of forest clearing in the country, an analysis of nature resources management (Endamana et al. 2010) highlighted that there was little change in conservation indicators in the Congo basin over the last decade. We may conclude that more favorable economic conditions may accelerate the loss of IFLs in the DRC, unless improved conservation policies are established.

8.5 Conclusion

Independent, satellite-based monitoring is an important tool for providing transparent information on forest change. Government officials, land managers, researchers, conservationists, and civil society groups can use such information to make better-informed decisions regarding the management of forest ecosystems. We have presented a novel, automated Landsat image processing approach that could be used for timely monitoring of forest cover change at national scales. This approach is a practical solution for examining trends in forest cover change at regional to national scales and could be implemented at a fraction of the cost of individual scene processing in terms of workload and processing time. Regional monitoring has the advantage of providing internally consistent, directly comparable results for assessing variation in the spatiotemporal trends of forest cover dynamics.

Landsat-based mapping of forest cover extent and change using supervised expert-driven classification is a well-established and accepted methodology, and reported accuracies for Landsat forest cover change detection range between 75% and 91% (Coppin and Bauer 1994). Our Landsat-based mapping algorithm has been tested for large forest regions, and our regional-scale Landsat forest cover change results are comparable with NFI data and individual scene supervised characterizations. The spatial accuracies of forest cover and change detection have not been rigorously validated, however, due to the lack of high spatial resolution imagery and field data. In the future, our approach can be validated using a series of high spatial resolution data sets. Our results can be used to target sampling with high spatial resolution imagery as part of a national-scale validation protocol.

The application of our forest monitoring approach in different biomes at the national/regional scales illustrate the possibility that it can be used also at the biome/global scales. Remaining challenges include possible gaps in future image availability, insufficient observation frequency for some areas, and the lack of a rigorous validation that uses high spatial resolution imagery along with field data. These concerns must be addressed before the proposed algorithm is implemented further. Yet having the technical ability to conduct satellite-based monitoring is not sufficient to detect and solve all environmental problems caused by inefficient and irresponsible forest management. First, only some components of ecosystem health can be monitored from space. Other components such as reductions in biodiversity due to overhunting and poaching, effects of chemical pollution, and global impact caused by human-induced climate change require a set of *in situ* measurements. Second, the forest management problems that are highlighted by monitoring data are sometimes a result of inadequate governmental control of natural resources exploitation and/or political and economic instability. Weak and/or corrupt governance precludes the maintenance of forest ecosystem services and protection of nature conservation areas. Integrating the drivers of forest cover change with satellite-based forest monitoring methods into national natural resource management systems and international conservational initiatives are important future steps for national-scale monitoring activities.

Acknowledgments

Support for this work was provided by the United States National Aeronautics and Space Administration's Land Cover and Land Use Change program and the USAID-supported CARPE project under Grants NNG06GD95G and NNX09AD26G.

About the Contributors

Peter Potapov received a diploma in botany from the Moscow State University, Moscow, Russia, in 2000, and a PhD in ecology and natural resources from the Russian Academy of Science, Moscow, Russia, in 2005. From 1998 to 2006, he was a GIS specialist at Greenpeace (Russian office) where he took part in developing an approach for mapping and monitoring forest degradation using the IFL method. He was a postdoctoral researcher from 2006 to 2011 at South Dakota State University, Brookings, South Dakota, and currently he is a research associate professor at the University of Maryland, College Park. His current research is focused on establishing a global operational forest monitoring algorithm using integration of Landsat and MODIS data.

Svetlana Turubanova received a BSc in geography from the Komi Pedagogical Institute, Syktyvkar, Russia (1996), an MSc in ecology from the Pushchino State University, Pushchino, Russia (1998), and PhD in ecology from the Russian Academy of Science, Moscow, Russia (2002). She started her career as a GIS specialist at Greenpeace Russia in 1999. Later, she was a postdoctoral researcher at South Dakota State University (from 2008 to 2011), Brookings, South Dakota. Currently she is a research associate at the University of Maryland, College Park, Maryland. Her current research interests are focused on national-scale Landsat-based forest mapping, deforestation monitoring, and forest degradation assessment.

Lars Laestadius holds a forest degree from the Swedish University of Agricultural Sciences, Uppsala, Sweden, and a PhD in forestry from the Virginia Polytechnic Institute and State University, Blacksburg, Virginia. He is a senior associate at the World Resources Institute, Washington, DC, and conducts forest policy research with a special interest in frontier forest landscapes assessment, forest landscape restoration, and the legality of forest management and trade. Prior to his current position, he served in the European Commission and was responsible for coordinating nationally funded European forestry and forest products research.

Alexey Yaroshenko received a diploma in botany in 1993 and a PhD in ecology and natural resources in 1997, both from Moscow State University, Moscow, Russia. From 1993 to 1997, he was a freelance researcher of forests of Central Russia and the Urals. Since 1997, he has been the head of the forest department in Greenpeace, Russia. His current work and main area of interest is related to forest policy and economics and boreal forest conservation. He is the founder (2004) and editor of the popular Russian portal for professional foresters and environmentalists—https://www.forestforum.ru.

Ilona Zhuravleva received her diploma in ecology from Tver State University, Tver, Russia in 2004. Since 2005, she has been working for Greenpeace Russia first as an assistant, and then as the head of the GIS unit. Her research interests are focused on IFL mapping, forest cover change analysis, forest fire monitoring, and conservation planning. Her current work is focused on developing Web applications to present deforestation mapping and IFL assessment results in tropics and boreal regions.

References

Achard, F. et al., Determination of deforestation rates of the world's humid tropical forests. *Science*, 297, 999–1002, 2002.

Breiman, L. et al., *Classification and Regression Trees*. Wadsworth and Brooks/Cole, Monterey CA, 1984.

Chander, G., Markham, B.L., and Helder, D.L., Summary of current radiometric calibration coefficients for Landsat MSS, TM, ETM+, and EO-1 ALI sensors. *Remote Sensing of Environment*, 113, 893–903, 2009.

CIA, *The World Factbook 2011*. CIA, Washington, DC, 2011 (https://www.cia.gov/library/publications/the-world-factbook/fields/2003.html). Accessed 12/14/2011.

Coppin, P.R. and Bauer, M.E., Processing of multitemporal Landsat TM imagery to optimize extraction of forest cover change features. *IEEE Transactions on Geoscience and Remote Sensing*, 32, 918–927, 1994.

Endamana, D. et al., A framework for assessing conservation and development in a Congo Basin Forest Landscape. *Tropical Conservation Science*, 3, 262–281, 2010.

FAO, *The 2010 Global Forest Resources Assessment Remote Sensing Survey: An Outline of the Objectives, Data, Methods and Approach*. FAO FRA Paper 155, FAO, Rome, 2009.

Hansen, M., Dubayah, R., and DeFries, R. Classification trees: An alternative to traditional land cover classifiers. *International Journal of Remote Sensing*, 17, 1075–1081, 1996.

Hansen, M., Stehman S., and Potapov P., Quantification of global gross forest cover loss. *Proceedings of the National Academy of Sciences of the USA*, 107, 8650–8655, 2010.

Hansen, M.C. et al., Global percent tree cover at a spatial resolution of 500 meters: First results of the MODIS vegetation continuous fields algorithm. *Earth Interactions*, 7, 1–15, 2003.

Hansen, M.C. et al., A method for integrating MODIS and Landsat data for systematic monitoring of forest cover and change and preliminary results for Central Africa. *Remote Sensing of Environment*, 112, 2495–2513, 2008.

Harris, L.D., *The Fragmented Forest: Island Biogeography Theory and the Preservation of Biotic Diversity*. The University of Chicago Press, Chicago, IL, 1984.

Holben, B., Characteristics of maximum-value composite images from temporal AVHRR data. *International Journal of Remote Sensing*, 7, 1417–1434, 1986.

INPE, *Deforestation Estimates in the Brazilian Amazon*. INPE, São José dos Campos, 2002.

Jin, S. and Sader, S.A., MODIS time-series imagery for forest disturbance detection and quantification of patch size effects. *Remote Sensing of Environment*, 99, 462–470, 2005.

Karpachevskiy, M.L. et al., *Natural Environment of Moscow Region: Losses 1992–2008 and Current Threats*. Biodiversity Conservation Center, Moscow, 2009.

Masek, J.G. et al., A Landsat surface reflectance dataset for North America, 1990–2000. *IEEE Geoscience and Remote Sensing Letters*, 3, 68–72, 2006.

Olthof, I. et al., Landsat ETM+ mosaic of northern Canada. *Canadian Journal of Remote Sensing*, 31, 412–419, 2005.

Potapov, P., Turubanova, S., and Hansen, M.C., Regional-scale boreal forest cover and change mapping using Landsat data composites for European Russia. *Remote Sensing of Environment*, 115, 548–561, 2011.

Potapov, P. et al., Mapping the World's intact forest landscapes by remote sensing. *Ecology and Society*, 13, 1–16, 2008.

Potapov, P. et al., *Global Mapping and Monitoring the Extent of Forest Alteration: The Intact Forest Landscapes Method*. FAO FRA Paper 166, FAO, Rome, 2009.

ROSLESINFORG, *Forest Fund of Russia (Data of State Forest Account, State by January 1, 2003)*. VNIILM, Moscow, 2003.

ROSSTAT, *Regions of Russia: Social and Economic Indicators*. ROSSTAT, Moscow, 2008.

Tomppo, E., Multi-source national forest inventory of Finland. *The Finnish Forest Research Institute. Research Papers*, 444, 52–60, 1993.

Tucker, C.J. and Townshend, J.R.G., Strategies for monitoring tropical deforestation using satellite data. *International Journal of Remote Sensing*, 21, 1461–1471, 2000.

Willén, E., Rosengren, M., and Persson, A., Multiresolution satellite data for boreal forest change detection mapping and monitoring. *Proceedings of ForestSat 2005*, 1–6, 2005.

Williams, D.L., Goward, S., and Arvidson, T., Landsat: Yesterday, today, and tomorrow. *Photogrammetric Engineering and Remote Sensing*, 72, 1171–1178, 2006.

Wulder, M.A. et al., Monitoring Canada's forests. Part 1: Completion of the EOSD land cover project. *Canadian Journal of Remote Sensing*, 34, 549–562, 2008.

9

The Brazilian Amazon Monitoring Program: PRODES and DETER Projects

Yosio Edemir Shimabukuro, João Roberto dos Santos,
Antonio Roberto Formaggio, Valdete Duarte, and
Bernardo Friedrich Theodor Rudorff
Brazilian Institute for Space Research

CONTENTS

9.1 Introduction

The Amazonia region comprises the greatest rain forest of our planet where the largest continuous remaining tropical forest can be found. In Brazil, an accelerated anthropization process began at the end of the 1960s in response to governmental policies to integrate the vast Amazonian region with the rest of the country. This was to be achieved mainly through road construction and incentivized transmigration policies that consequently expanded the Brazilian agriculture frontier. The anthropization process has been most intense in the so-called *arc of deforestation* where the Amazon ecosystem meets with the savanna (*cerrado*) ecosystem. Since 1973, Brazil has had access to remote sensing imagery from the series of Landsat satellites, enabling the quantification of natural resource extent and modification over the Amazon region. Based on the availability of these images, the Brazilian government began monitoring of the Amazon forest to quantify deforestation at multiyear intervals. Quantitative data on deforestation could then be used to assess the human impacts of the development policies, with the objective of minimizing the negative effects of the man–biome interaction on renewable and nonrenewable resources.

Since 1988, the Brazilian government has performed annual monitoring of the Amazon forest using Landsat-type imagery through the PRODES (monitoring of Amazon forest) project carried out by the Brazilian Institute for Space Research (INPE). PRODES has quantified approximately 750,000 km^2 of deforestation in the Brazilian Amazon through the year 2010, a total that accounts for approximately 17% of the original forest extent. PRODES data have revealed the annual deforestation rates to vary significantly in response to domestic political, economic, and financial policies as well as foreign market demands.

PRODES information is based primarily on Landsat imagery. Medium spatial resolution (30 m) data such as Landsat have a relatively low temporal resolution of 16-day repeat coverage, allowing for annual monitoring of deforestation. More rapid updating of forest disturbance is not possible with Landsat as the infrequent repeat coverage coupled with the persistent cloud cover of the humid tropical Amazon basin limits the number of viable land surface observations. This fact prevents the government and environment control agencies from making fast and adequate interventions to stop illegal deforestation activities.

Near-real-time deforestation monitoring is possible using the almost daily images of the MODIS (MODerate resolution Imaging Spectroradiometer) sensor on board the Terra and Aqua satellite platforms. Thus, a new methodology based on MODIS images was developed for rapid detection of deforestation in the Amazon region through the DETER (real-time detection of deforestation) project (Shimabukuro et al. 2006). While MODIS is a coarse spatial resolution sensor, and not viable for area estimation of deforestation, MODIS data can be valuable as a change indicator, or alarm product in the service of land management policies and enforcement.

This chapter presents an overview of the PRODES and DETER projects for annual and monthly monitoring of deforestation in the Brazilian Amazon, respectively. Initially, the Brazilian Amazon region is characterized in terms of its soil, biodiversity, climate, and vegetation followed by the deforestation history and the description of the methodology developed at INPE for the deforestation monitoring activities based on remote sensing image-processing and geographic information system (GIS) techniques. Results from more than three decades of monitoring are presented and discussed, illustrating the rapid deforestation that occurred during this period in the Amazon region. The results have quantified the magnitude and trends of deforestation in the Brazilian Amazon. Results provide an invaluable input to decision makers in establishing public policies and enforcing environmental governance in the critical ecosystems of the Brazilian Amazon.

9.2 Brazilian Amazon

The Amazon rainforest is located in South America and covers an area of 6.4 million km². Most of the Amazon rainforest (63%) is found in the Brazilian Legal Amazon (BLA) (Figure 9.1), with the remaining part being distributed among the countries of Peru, Colombia, Bolivia, Venezuela, Guiana, Suriname, Ecuador, and French Guiana. Much attention has been given to this region due to its relevance in terms of biodiversity as well its unique environmental services at the global scale.

The BLA is a geopolitical unit, established in 1966 by the Brazilian government. The BLA is located between 5° N, 20° S and 44° W, 75° W and covers an area of approximately 5 million km². It encompasses the whole states of Acre, Amapá, Amazonas, Mato Grosso, Pará, Rondônia, Roraima, Tocantins, and the western part (44° W) of the state of Maranhão (IBGE 2000). The BLA is included in the Amazon river basin except for the

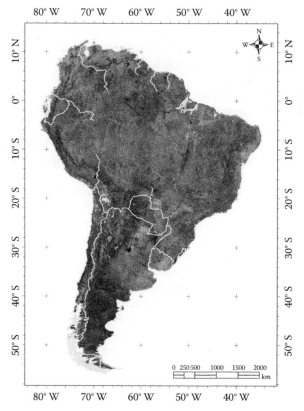

FIGURE 9.1
(See color insert.) The BLA (*red*) located in the South American continent.

southern part of Mato Grosso state (Paraguay river basin) and for part of Maranhão state (Parnaíba river basin).

Soils: The Amazon region includes varied soil classes formed under great geological diversity, exhibiting significant variation in relief and under the influence of high temperatures and precipitation typical for warm super humid or humid equatorial climates. The natural soil fertility is relatively low; however, the Amazon rainforest is a self-sustainable ecosystem due to its own nutrient cycles, making it vulnerable to anthropic interference (IBAMA 2009).

Biodiversity: The Amazon region comprises a large variety of ecosystems including upland forests (*terra firme*), swamp forests (seasonally flooded forest—*varzeas* and permanently flooded forest *igapós*), grasslands, and savannas (*cerrado*). An extremely rich biodiversity is found within the regions, including 1.5 million plant species; 3,000 fish species; 950 types of birds; and an enormous amount of insect, reptile, amphibian, and mammal species (IBAMA 2009).

Climate: The Amazon region is characterized by its enormous ability for water recycling. About 63%–73% of the water is lost through evapotranspiration, and approximately 50% of it is recycled within the region through precipitation (Salati 1985).

The average temperature varies from 25.8°C during the rainy season (May–September) to 27.9°C during the dry season (October–April). The duration of these seasons may vary due to the large extent of the Amazon region. The average annual precipitation is 2,250 mm, varying from 1,500 mm in the northern and southern regions to 3,000 mm in the northwestern region of the Amazon.

Vegetation: The Amazon region is covered by evergreen tropical rainforest comprised of three major classes of vegetation: (1) the evergreen tropical forest *stricto sensu*; (2) the semievergreen tropical forest; and (3) the semideciduous tropical forest (IBGE 1988). Evergreen tropical forests *stricto sensu* are mostly found in very moist regions where the annual precipitation is around 3,000 mm. They are composed of multilayered broadleaf evergreen trees that may reach 50 m in height, with a sparse substratum consisting mainly of herbaceous plants. Semievergreen tropical forests are spread along less humid areas, with annual precipitation varying from 2,000 to 3,000 mm. These forests are composed of three-layered formations of perennial and deciduous broadleaf trees, with the latter type being sparsely present and forming the top layer of the canopy. Semideciduous tropical forests differ from semievergreen ones by having a larger proportion of deciduous species.

The *cerrado* is a savanna-type ecosystem appearing mainly in the southern and eastern portions of the Amazon region. It is composed of broadleaf, semideciduous, or evergreen short trees typically growing in well-drained soils that are poor in nutrients, in a region where the average annual temperature ranges from 20°C to 26°C and annual precipitation ranges from 1,250 to 2,000 mm with marked influence of the austral winter dry season (May through September). In general terms, five structural types of *cerrado* are acknowledged to exist (Oliveira-Filho and Ratter 2002): *cerradão*—dominated by arboreous vegetation (8–12 m tall) whose canopy covers 50%–90% of the area; *cerrado* (*stricto sensu*)—dominated by trees and shrubs (3–8 m tall), with a more sparse canopy cover (above 30%); *campo cerrado*—formed by dispersed trees and shrubs, with a high density of herbaceous vegetation; *campo sujo*—dominated by herbaceous vegetation, with shrubs and small dispersed trees; and *campo limpo*—which is different from the *campo sujo* because it has no shrubs nor trees. *Cerrado* may also be associated with seasonally flooded areas. In total, the Amazon region has approximately 10%–15% of worldwide biomass (Houghton et al. 2001).

Deforestation in the BLA: Deforestation in the BLA has been a concern of several governmental and nongovernmental agencies, especially over the last three decades (Moran 1981; Skole and Tucker 1993). Although there is a longer history of human occupation in the BLA, nearly 90% of the deforestation for pasture and agriculture occurred between 1970 and 1988, as indicated by estimates based on satellite images (Skole et al. 1994).

Historically, the Brazilian territory was occupied along the coastline, with most of its population concentrated in this region. In an attempt to change this occupation pattern by increasing inland settlement, the federal capital was moved from the coast (Rio de Janeiro) to the Central region of Brazil (Brasília) in the mid-1950s (Mahar 1988). This occupation policy required major infrastructure investments to connect Brasília to the other regions of Brazil. The construction of the Belém-Brasília road (BR-010) in 1958 was the main factor that triggered major deforestation activities in the BLA (Moran et al. 1994; Nepstad et al. 1997). Subsequent events such as the construction of the BR-364 across the states of Mato Grosso, Rondônia, and Acre and the PA-150 in the state of Pará encouraged even more deforestation activities, converting forest into pasture and agriculture land (Moran 1993).

To introduce governance in the BLA, the SUDAM (Superintendência do Desenvolvimento da Amazônia) and the BASA (Banco da Amazônia) were established in 1966. Small producers were granted with incentives to invest in agriculture projects (Moran et al. 1994). Large producers were also granted tax incentives in exchange for converting forest to pasture land (Moran 1993). The incentives granted to large producers were the major drivers of deforestation; small producers had a lesser impact on deforestation due to the comparatively smaller scale practices of subsistence agriculture (Fearnside 1993).

Other activities with high economic value such as mining and selective logging also contributed to deforestation in the BLA (Cochrane et al. 1999).

Major deforestation in the BLA has been concentrated in the so-called *arc of deforestation,* located in the Southern and Eastern parts of the BLA from Acre to Maranhão states (Cochrane et al. 1999; Achard et al. 2002).

9.3   Deforestation Monitoring in the BLA

Since the late 1970s, INPE has performed deforestation assessments in the BLA using remotely sensed imagery. These assessments were carried out together with the former IBDF (Instituto Brasileiro de Desenvolvimento Florestal) that was later incorporated with IBAMA (Instituto Brasileiro do Meio Ambiente e dos Recursos Naturais Renováveis). The first assessments were carried out with the use of images acquired by the MSS sensor (four spectral bands with spatial resolution of 80 m) on board the Landsat-1, -2, and -3 satellites, during the periods of 1973–1975 and 1975–1978 using visual interpretation techniques (Tardin et al. 1980).

From 1988 onward, annual deforestation assessments were provided for the entire BLA using images from the TM sensor (six spectral bands with spatial resolution of 30 m) on board the Landsat-5 satellite, with improved mapping quality due to its improved spatial and spectral resolutions as compared to the MSS data. The methodology applied to map the deforested areas was based on visual interpretation of color composites (5R-4G-3B) of TM images in hard copy format at the scale of 1:250,000. The visually interpreted polygons of the deforested areas were summed up to compute the total deforested land for each state and presented in tabular format. This method, known as analog PRODES, was performed until 2001.

By the end of the 1990s, an automated methodology began to be developed and was named digital PRODES (Shimabukuro et al. 1998). However, the deforestation information provided by PRODES was not sufficient for the more frequent monitoring surveillance needs of various Brazilian government agencies. Therefore, the DETER project was developed based on the high temporal resolution images of the MODIS sensor to provide geospatial information on deforestation activities in near real time and has been in operation since 2004.

9.3.1   Digital PRODES Methodology

Digital PRODES is the world's largest remote sensing project for monitoring deforestation activities in tropical rain forests. It has the objective to survey all deforested areas within the 5 million km² of the BLA, an area covered by 229 Landsat scenes (Figure 9.2).

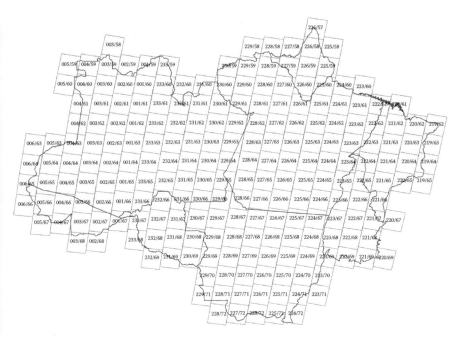

FIGURE 9.2
The BLA covered by 229 TM or ETM+/Landsat images for the annual survey of deforestation. (From INPE, *Monitoramento da cobertura florestal da Amazônia por satélites: Sistemas PRODES, DETER, DEGRAD E QUEIMADAS 2007–2008*, Instituto Nacional de Pesquisas Espaciais, São José dos Campos, SP, Brazil, 2008; Mahar, D., *Government Policies and Deforestation in Brazil's Amazon Region*, World Bank, Washington, DC, 1988.)

PRODES depicts deforestation within the BLA. A mask of nominally intact forest is annually updated by identifying new deforestation events to the exclusion of nonforest vegetation type and other change dynamics such as the clearing of secondary regrowth. Input Landsat TM images are selected from July, August, and September acquisitions. This period is within the *arc of deforestation's* local dry season and represents an atmospheric window where cloud-free images are typically available. These images are rectified using nearest neighbor sampling to a UTM projection, resulting in a cartographic product with 50 m internal error. For PRODES, TM 3 (red), TM 4 (NIR), and TM 5 (MIR) bands are used to generate the fraction images. The legend for the maps contains the following classes: forest, non-forest *cerrado arbustivo, campo limpo de cerrado, campinarana,* etc.), accumulated deforestation from previous years, deforestation from the current year, hydrography, and cloud.

Digital PRODES consists of the following methodological steps: (1) generation of per pixel vegetation-, soil-, and shade-fractional images; (2) segmentation based on growing regions' algorithm; (3) classification based on nonsupervised classifier; (4) mapping the classes based on the following legend: forest, nonforest (vegetation that is not characterized by a forest structure), deforestation (accumulated deforestation up to the previous year), hydrography, and clouds; and (5) editing of classified map based on visual interpretation to minimize omission and commission errors from the automatic classification to produce the final deforestation map in digital format. PRODES products are available at the official PRODES website (http://www.obt.inpe.br/prodes/index.html).

A linear spectral mixture model (LSMM) is used to produce fraction images of vegetation, soil, and shade applied to the TM spectral bands (Shimabukuro and Smith 1991). This method reduces data dimensionality and enhances the specific targets of interest. A vegetation-fraction image enhances the green vegetation, a soil-fraction image enhances bare soil, and a shade-fraction image enhances water bodies and burned land. The shade-fraction image was used to characterize the total previously deforested land in the BLA (Shimabukuro et al. 1998) up to 2001. The soil-fraction image is used to classify the annual deforested increment based on the contrast between forested and deforested land.

The LSMM can be written as:

$$r_i = a \times \text{vege}_i + b \times \text{soil}_i + c \times \text{shade}_i \times e_i,$$

where
   r_i is the response for the pixel in band i of TM image
   a, b, and c are the proportion of vegetation, soil, and shade in each pixel
   vege_i, soil_i, and shade_i correspond to the spectral responses of each
      component
   e_i is the error term for each band i

Landsat TM bands 3, 4, and 5 are used to form a linear equation system that can be solved by any developed algorithm (e.g., weighted least square). The resulting fraction images are resampled to a 60 m spatial resolution in order to minimize computer processing time and disk space, without losing information compatible with the 1:250,000 final product map scale.

Image segmentation is a technique to group the data into contiguous regions having similar spectral characteristics. Two thresholds are required to perform image segmentation: (a) *similarity*, that is the minimum value defined by the user to be considered as similar to form a region and (b) *area*, that is the minimum size given in number of pixels in order to be individualized. The unsupervised classification (ISOSEG) method is used to classify the segmented fraction images. It uses the statistical attributes (mean and covariance matrix) derived from the polygons of the image segmentation.

After the unsupervised classification, it is necessary to check the resulting maps. This task is performed by interpreters using interactive image editing tools. Color composites of Landsat bands 5, 4, and 3 are displayed in red–green–blue videos. Expert-identified omission and commission errors are manually corrected in order to improve the classification result. Then the individually classified images are mosaicked to generate the final maps per state and for the entire BLA. For the state mosaics, the spatial resolution is kept at 60 m and the scale for presentation is 1:500,000, while for the BLA the spatial resolution is degraded to 120 m and the scale for presentation is 1:2,500,000.

9.3.2 DETER Methodology

Starting in 2004, the DETER project was implemented to provide a near-real-time monitoring and detection of deforestation activities to support the Federal Government Action Plan for the Prevention and Control of Deforestation in the BLA. The procedure mimics the PRODES method but is meant to detect deforestation activities in near real time by exploiting the high temporal resolution of the MODIS sensor.

The first step in the method of the DETER project is to mask the intact forest based on the PRODES evaluation of the previous year. The map of intact forest is used as a reference for identifying new deforestation events in near real time throughout the current year. The monitoring activity with MODIS imagery begins in January, but becomes more active after March due to less cloud cover in the BLA. This does not significantly impact results as there is comparatively little deforestation occurring during the rainy season (November through March).

Daily MODIS images (surface reflectance—MOD09) used to identify deforestation spots are selected based on two criteria: (a) amount of cloud cover and (b) swath within sensor view zenith angle less than 35° (~1,400 km). The amount of cloud cover is evaluated based on quick-look images and, if deemed viable, a follow-on full spatial resolution assessment. The entire BLA is covered by 12 MODIS tiles from V09 to V11 and H10 to H13.

The images from the MOD09 product are delivered as HDF (hierarchical data format) files projected in a sinusoidal projection (WGS84 datum). All data are converted to a GeoTIFF format and reprojected to the geographic coordinate system for use in the SPRING software image-processing package.

From the set of seven reflective bands of the MOD09 product, bands 1 (red), 2 (NIR), and 6 (MIR) are used to generate the vegetation-, soil-, and shade-fraction images, respectively, using the linear spectral mixing model as previously described in the digital PRODES method. The soil-fraction images are then segmented, classified, mapped, and eventually edited by interpreters following the digital PRODES protocol.

The above procedure is carried out for every daily MODIS image acquired over the BLA. The results of the deforestation activities detected by DETER can be accumulated for different intervals such as weekly, biweekly, or monthly and are available in a digital format at the DETER website (http://www.obt.inpe.br/deter/index.html).

9.4   Results

9.4.1   Analog and Digital PRODES

Tardin et al. (1980) reported that deforestation in the BLA had reached a figure of 152,200 km^2 in 1978, which included the deforested land prior to 1960. Since that period, the average rate of deforestation has undergone significant changes. For example, from 1978 to 1988, the average deforestation rate was 21,130 km^2 year^{-1} while it gradually decreased to 11,130 km^2 in 1991. After 1991, it began to increase again, reaching a rate of 27,423 km^2 in 2004. However, an abnormally high rate of 29,059 km^2 was also observed in 1995. From 2004 on, a significant decrease in deforestation rates was observed, with a minimum rate of 7,000 km^2 in 2010 (Tables 9.1 and 9.2). This period is coincident with the implementation of the DETER project as part of the Federal Government Action Plan for the Prevention and Control of Deforestation in BLA.

Since the implementation of the digital PRODES method in 2002, the deforestation results are immediately provided to government agencies to implement policies that enforce the reduction of illegal deforestation. The PRODES results are available to the public at the Web site, and the main data on deforestation over the last 8 years are shown in Table 9.2.

Figure 9.3 illustrates the annual deforestation rates from 1988 to 2010 for the BLA.

Figure 9.4 presents the thematic map of the PRODES classes, showing the spatial distribution of the deforested areas up to 2010; note the concentration of forest loss in the *arc of deforestation*.

The remote sensing images acquired since the early 1970s proved to be an important tool for monitoring the deforestation in the entire BLA and largely coincide with enactment of policies by the Brazilian government to promote the occupation of the region. Spatiotemporal data on deforestation rates have significantly contributed not only to government policies in reducing illegal deforestation activities, but also to the scientific community and the study of human impacts on biodiversity, greenhouse gases emission, and regional and global climate change.

TABLE 9.1

Deforestation Estimates (km²) from the Analog PRODES Method from 1988 to 2001

States/Year	1988[a]	1989	1990	1991	1992	1993[b]	1994[b]	1995	1996	1997	1998	1999	2000	2001
Acre	620	540	550	380	400	482	482	1,208	433	358	536	441	547	419
Amazonas	1,510	1,180	520	980	799	370	370	2,114	1,023	589	670	720	612	634
Amapá	60	130	250	410	36	–	–	9	–	18	30	–	–	7
Maranhão	2,450	1,420	1,100	670	1,135	372	372	1,745	1,061	409	1,012	1,230	1,065	958
Mato Grosso	5,140	5,960	4,020	2,840	4,674	6,220	6,220	10,391	6,543	5,271	6,466	6,963	6,369	7,703
Pará	6,990	5,750	4,890	3,780	3,787	4,284	4,284	7,845	6,135	4,139	5,829	5,111	6,671	5,237
Rondônia	2,340	1,430	1,670	1,110	2,265	2,595	2,595	4,730	2,432	1,986	2,041	2,358	2,465	2,673
Roraima	290	630	150	420	281	240	240	220	214	184	223	220	253	345
Tocantins	1,650	730	580	440	409	333	333	797	320	273	576	216	244	189
Brazilian Amazon	21,050	17,770	13,730	11,030	13,786	14,896	14,896	29,059	18,161	13,227	17,383	17,259	18,226	18,165

[a] Average between 1978 and 1988.
[b] Average between 1993 and 1994.

TABLE 9.2

Deforestation Estimates (km²) from the Digital PRODES from 2002 to 2009

States/Year	2002	2003	2004	2005	2006	2007	2008	2009
Acre	883	1,078	728	592	398	184	254	211
Amazonas	885	1,558	1,232	775	788	610	604	406
Amapá	–	25	46	33	30	39	100	–
Maranhão	1,014	993	755	922	651	613	1,272	980
Mato Grosso	7,892	10,405	11,814	7,145	4,333	2,678	3,258	1,047
Pará	7,324	6,996	8,521	5,731	5,505	5,425	5,606	3,687
Rondônia	3,099	3,597	3,858	3,244	2,049	1,611	1,136	505
Roraima	84	439	311	133	231	309	574	116
Tocantins	212	156	158	271	124	63	107	56
Brazilian Amazon	21,394	25,247	27,423	18,846	14,109	11,532	12,911	7,008

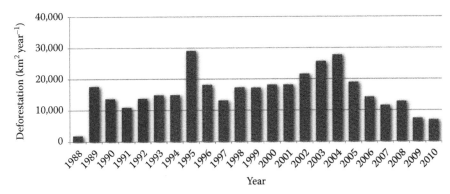

FIGURE 9.3

Variation of deforested areas during 1988–2010 time period for the Brazilian Amazonia region.

9.4.2 DETER Project

Figure 9.5 presents an example of the DETER monitoring results, showing the spatial distribution of the deforestation activities detected on a monthly basis for 2004.

The DETER system provides a near-real-time monitoring procedure to support the Federal Government Action Plan for the Prevention and Control of Deforestation in BLA since 2004, when a significant reduction in the deforestation rate started to be observed (Figure 9.3). DETER products are not used to estimate areas of deforestation but as an alarm to inform government agencies on potential illegal forest-clearing activities in the BLA. The availability of the high temporal resolution images from the MODIS sensor enables monthly reporting of forest loss alarms and has contributed to slowing illegal deforestation activities in the BLA.

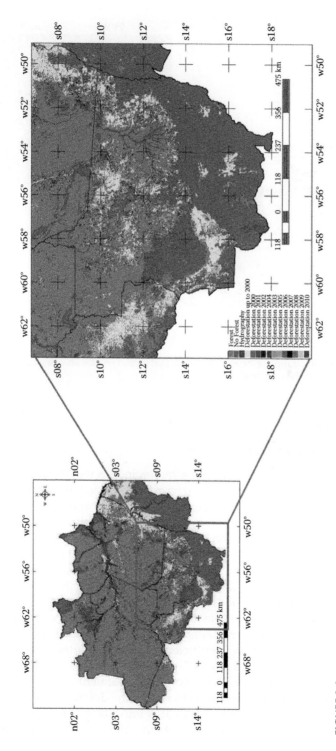

FIGURE 9.4
(See color insert.) Mosaic of digital PRODES mapping over the period 2000–2010.

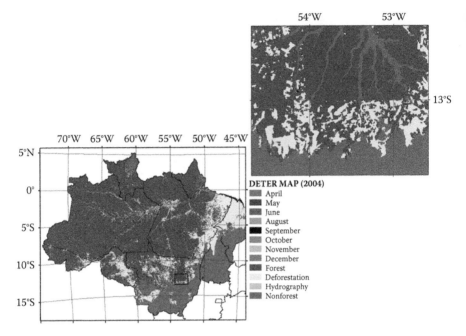

FIGURE 9.5
(See color insert.) Illustration of the example of DETER project results, showing the deforested areas detected during the year 2004.

9.5   Discussion and Conclusion

The initial monitoring of deforestation activities in BLA was performed by the analog PRODES product that was based on visual interpretation of hard copies of Landsat scenes at the scale of 1:250,000. This was an expensive and tedious procedure carried out by numerous interpreters on a yearly basis. However, it produced valuable information on deforestations rates until the 2001.

In 2002, the analog PRODES was replaced by the digital PRODES product that employs a semiautomatic method based on digital image-processing techniques and minor visual interpretation to correct for classification errors. The great advantage of digital PRODES is the provision of deforestation information in a compatible format for use in GIS for ecosystem and land use and cover change modeling. However, the annual frequency of deforestation estimates was insufficient to support other government needs, specifically that of reducing illegal deforestation activities.

As a consequence, the DETER project was implemented in 2004 to reinforce public policies that have helped to reduce the deforestation rates from 27,423 km² in 2004 to 7,000 km² in 2010. It is important to mention that the DETER does not replace but complements the digital PRODES monitoring procedure. The DETER detects deforestation activities in its initial stage without providing an area estimate, while the digital PRODES evaluates the total annual deforested area (INPE 2008).

The long-term history of the images acquired by the sensors on board the Landsat satellites proved to be an essential tool for monitoring the annual deforestation of the BLA. The Landsat record covers the majority of the period since the Brazilian government initially incentivized settlement of the BLA. The high temporal resolution of the MODIS sensor on board the Terra and Aqua platforms was also highly relevant to support government policies in stopping illegal deforestation. The result has been a consequent reduction of deforestation rates aided by the combined information from both the DETER and PRODES projects.

About the Contributors

Yosio Edemir Shimabukuro is a senior researcher at the Brazilian Institute of Space Research (INPE), São Paulo, Brazil. His fields of interest are use of satellite remote sensing data for studying vegetation cover and use of GIS techniques and models for environmental change detection over different biomes in Brazil.

João Roberto dos Santos is a senior researcher at the Brazilian Institute of Space Research (INPE), São Paulo, Brazil, and his fields of interest are optical and synthetic aperture radar (SAR) data for inventory and monitoring of forest typologies.

Antonio Roberto Formaggio is a researcher and postgraduation teacher/advisor in remote sensing for environment and agriculture at the Brazilian Institute of Space Research (INPE), São Paulo, Brazil. His research interests include impacts of agriculture in the Brazilian savannas and Amazonia biomes and environmental modeling and radiative transfer models for the retrieval of vegetation biophysical variables.

Valdete Duarte is a researcher in change detection and monitoring of forest by multisensor remote sensing satellite data. He contributed to establish the methodologies of PRODES (monitoring of Amazon forest) and DETER (real-time detection of deforestation) projects and is currently the principal investigator of the Panamazonia-II project to monitor the tropical forests in South America.

Bernardo Friedrich Theodor Rudorff is a researcher at the Remote Sensing Division of the Brazilian Institute of Space Research (INPE), São Paulo, Brazil, working with the assessment of agricultural production using remote sensing techniques and agrometeorological models. He is currently coordinating the soy moratorium project (monitoring of soy plantation in recent deforested land in the Amazon biome) and the Canasat project (monitoring of sugarcane for ethanol production).

References

Achard, F. et al., Determination of deforestation rates of the world's humid tropical forests. *Science*, 297, 999–1002, 2002.

Cochrane, M.A. et al., Positive feedbacks on the fire dynamic of closed canopy tropical forests. *Science*, 284, 1832–1835, 1999.

Fearnside, P.M., Deforestation in Brazilian Amazonia: The effect of population and land tenure. *Ambio*, 22(8), 537–545, 1993.

Houghton, R.A. et al., The spatial distribution of forest biomass in the Brazilian Amazon: A comparison of estimates. *Global Change Biology*, 7, 731–746, 2001.

IBAMA (Instituto Brasileiro do Meio Ambiente), Ecossistemas brasileiros: Amazônia. Disponível online: http://www.ibama.gov.br/ecossistemas/amazonia.htm. Consulta em 30 de novembro de 2009.

IBGE (Fundação Instituto Brasileiro de Geografia e Estatística), Brasil, Ministério da Agricultura, Instituto Brasileiro de Desenvolvimento Florestal, Secretária de Planejamento e Coordenação da Presidência da Republica, Mapa de Vegetação do Brasil, escala 1:5000000. Brasília, 1988.

INPE (Instituto Nacional de Pesquisas Espaciais), *Monitoramento da cobertura florestal da Amazônia por satélites: Sistemas PRODES, DETER, DEGRAD E QUEIMADAS 2007–2008*. São José dos Campos, SP, Brazil: Instituto Nacional de Pesquisas Espaciais, 47 p., 2008.

Mahar, D., *Government Policies and Deforestation in Brazil's Amazon Region*. Washington, DC: World Bank, 1988.

Moran, E.F., *Developing the Amazon*. Bloomington, IN: Indiana University Press, 1981.

Moran, E.F., Deforestation and land use in the Brazilian Amazon. *Human Ecology*, 21(1), 1–21, 1993.

Moran, E.F. et al., Integrating Amazonian vegetation, land-use, and satellite data. *Bioscience*, 44(5), 329–338, 1994.

Nepstad, D.C. et al., Land-use in Amazonia and the Cerrado of Brazil. *Ciência e Cultura Journal of the Brazilian Association for the Advancement of Science*, 49(1/2), 73–86, 1997.

Oliveira-Filho, A.T. and Ratter, J.A., Vegetation physiognomies and woody flora of the cerrado biome. In: P.S. Oliveira and R.J. Marquis (Eds.). *The Cerrados of Brazil*. New York: Columbia University Press, 91–120, 2002.

Salati, E., The climatology and hydrology of Amazonia. In: G.T. Prance and T.E. Lovejoy (Eds.). *Amazonia*. New York: Pergamon, 18–48, 1985.

Shimabukuro, Y.E. and Smith, J.A., The least-squared mixing models to generate fraction images derived from remote sensing multispectral data. *IEEE Transactions on Geoscience and Remote Sensing*, 29, 16–20, 1991.

Shimabukuro, Y.E. et al., Using shade fraction image segmentation to evaluate deforestation in Landsat Thematic Mapper images of the Amazon region. *International Journal of Remote Sensing*, 19(3), 535–541, 1998.

Shimabukuro, Y.E. et al., Near real time detection of deforestation in the Brazilian Amazon using MODIS imagery. *Ambiente e Água*, 1, 37–47, 2006.

Skole, D. and Tucker, C., Tropical deforestation and habitat fragmentation in the Amazon: Satellite data from 1978 to 1988. *Science*, 260, 1905–1910, 1993.

Skole, D.L. et al., Physical and human dimensions of deforestation in Amazonia. *Bioscience*, 44(5), 14–32, 1994.

Tardin, A.T. et al., Subprojeto Desmatamento. Convênio IBDF/CNPq/INPE, Rel. Técnico INPE-1649-RPE/103, 1980.

10

Monitoring of Forest Degradation:
A Review of Methods in the Amazon Basin

Carlos Souza, Jr.

Amazon Institute of People and the Environment

CONTENTS

10.1   Introduction

Forest degradation is an anthropogenic process that can lead to significant carbon loss from forests to the atmosphere. Measuring and mapping of forest degradation have become important tasks for advancing carbon payment negotiations through the reducing emissions from deforestation and degradation (REDD+) process (Herold et al. 2011). The forests of the Brazilian Amazon are significantly impacted by forest degradation due to three main processes: selective logging, forest fires, and forest fragmentation.

These degradation dynamics operate synergistically and recurrently, result-ing in the loss of original carbon stocks of intact forests. In extreme cases, forest degradation can lead to a complete conversion of forests to other land cover types (i.e., pasture or agriculture lands). However, it is more common for forests to remain nominally as forests, but with a reduced carbon stock and altered biodiversity and forest structure.

The annual area of selectively logged forest in the Brazilian Amazon is as large as that cleared by deforestation (Nepstad et al. 1999; Asner et al. 2005). Due to the significance of this disturbance dynamic to forest structure in the Amazon basin, several remote sensing techniques have been tested and developed to detect, measure, and map the areal extent of forest degradation (Souza and Barreto 2000; Asner et al. 2002; Souza et al. 2005a; Matricardi et al. 2007). Selective logging has also been studied in the Brazilian Amazon in terms of its ecological impacts, including changes in carbon stocks, biodi-versity loss, soil compaction, forest microclimate, and biogeochemical cycles (Verissimo et al. 1992, 1995; Johns et al. 1996; Pereira et al. 2002).

Forest fires (Cochrane et al. 1999; Alencar et al. 2004) and forest fragmenta-tion (Laurance et al. 2000, 2002) have also received great scientific attention, including studies of the synergism between these two processes (Cochrane 2001; Cochrane and Laurance 2002). The synergism between selective logging and forest fires is also well understood (Holdsworth and Uhl 1997; Nepstad et al. 1999). Remote sensing techniques to map forest fragments (FFs) have been developed since the early 1990s (Skole and Tucker 1993). However, map-ping burned area extent is more challenging as ground fires result only in degradation of forest understory. Moreover, fire is often related to forests that have been previously logged, further complicating their quantification and unique contribution to emissions.

A host of ecological and remote sensing studies of forest degradation have been conducted in the Brazilian Amazon, making the region a suitable area for a review and evaluation of optical remote-sensing techniques for REDD+ projects. Presenting a review of these remote-sensing techniques is the first objective of this chapter. By definition, REDD+ includes both forest conver-sion as well as forest degradation, and the Brazilian Amazon is the only tropical forest where both deforestation and forest degradation have been studied in great detail. The second objective of this chapter is to demonstrate how remote sensing techniques can be integrated with forest biomass field measurements to construct reliable baselines of carbon emissions associated with forest degradation. In order to achieve these objectives, the chapter is divided into three sections. The first section presents a summary of forest degradation processes and their impacts on forest carbon stocks and includes an evaluation of those attributes of forest degradation that can be quantified using remotely sensed data. In the second section, the optical remote sens-ing techniques available for detecting and mapping forest degradation are presented in detail, including a discussion of their strengths and limitations when applied to mapping changes in forest carbon stocks. The last session presents a framework for integrating deforestation and forest degradation monitoring activities in developing baselines for REDD+.

10.2 Field Characterization of Forest Degradation

10.2.1 Definition

Forest degradation is a temporary or permanent change in density, composition, or structure of natural forest attributes caused by anthropogenic factors. Forest degradation differs from forest changes caused by natural phenomena, such as natural tree falls, windthrows, and lightning strikes, as these changes in forest attributes are not human induced (Lambin 1999). Several ecological field studies conducted in the Brazilian Amazon have shown that selective logging, forest fires, and forest fragmentation are the main processes responsible for forest degradation (Verissimo et al. 1992; Barros and Uhl 1995; Holdsworth and Uhl 1997; Cochrane et al. 1999; Cochrane and Laurance 2002). Forest degradation processes operate at different intensities and time scales, creating a continuum from intact forests to degraded forests to complete stand replacement and conversion (Figure 10.1). Defining the types of forest attributes affected by degradation processes is important, as is assessing the capabilities of remote sensing in measuring changes to these attributes.

In the Brazilian Amazon, logging creates small clearings, known as log landings or logging decks, varying in size from 40 to 190 m². Log landings are connected by primary logging roads that can be 6–15 m wide and account for additional clearings of 60–567 m² per hectare. These roads give access to harvesting areas through secondary roads and/or skid trails. Tree fall gaps are commonly found in forest areas where commercial tree species are harvested, given that vine cutting is not a widespread practice in this region. High tree diameters (i.e., diameter at breast height [DBH] > 45 cm) are usually taken in the first harvesting cycle, but recurrent logging cycles can occur as smaller trees are successively harvested (i.e., 15 < DBH < 45 cm) (Figure 10.1). The harvesting intensity varies from 1 to 9 trees per hectare (Verissimo et al. 1992, 1995; Barros and Uhl 1995; Johns et al. 1996; Pereira et al. 2002).

It is well established that logging leads to favorable conditions for burning forests. Logging creates canopy gaps that allow penetration of more incoming solar radiation into the understory environment. As result, understory humidity is reduced, drying out remaining logging debris or slash. Agriculture fires can unintentionally escape to adjacent logged forests (Holdsworth and Uhl 1997). Similar to logging, forest fires can also reoccur in the same forest, creating a positive feedback in increasing forest degradation (Cochrane et al. 1999; Cochrane and Schulze 1999) (Figure 10.1).

Several logging cycles and fire events can drastically deplete forest carbon stocks to carbon density levels similar to those of a deforested area. However,

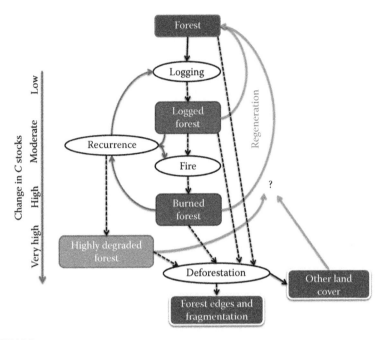

FIGURE 10.1
(See color insert.) Forest degradation processes and interactions commonly found in the Brazilian Amazon. Pristine forests can be subject to selective logging, creating favorable conditions for burning when fires from adjacent agriculture fields unintentionally escape. Logging and fires can be recurrent, creating highly degraded forests. Eventually, degraded forests can be converted by deforestation, increasing forest edges and landscape fragmentation. If degraded forests are not cleared, vegetation regeneration processes can prevail given the high resiliency of forests.

before this occurs, it is more common for degraded forests to be cleared. The fate of degraded forests in the Brazilian Amazon varies across the region. In areas close to deforestation frontiers, degraded forests are more likely to be cleared within 5–10 years, a process that increases forest edges and landscape fragmentation (Asner et al. 2005) (Figure 10.1). The degraded forests that are not converted by deforestation may regenerate, returning to their original carbon stocks after several decades. However, the original species composition may not be restored due to local extinctions (Figure 10.1).

10.2.2 Types of Degraded Forests

As discussed above, forest degradation creates a continuum from intact forest to clearings. But, for mapping purposes a typology of classes is required. Here, degraded forests are classified in terms of the processes and intensities associated with degradation (Souza et al. 2009). The first type of degraded forests in the Brazilian Amazon is logged forests. Three types of selectively logged forests have been identified in this region: nonmechanized logging (NML), managed logging (ML), and conventional logging (CL). Agricultural fires are more likely to burn forests that experienced CL. CL forests have favorable conditions for burning due to a greater amount of slash and collateral canopy damage. Fires in logged forests lead to a new class of forest degradation named burned forest (BF). Finally, forest patches of different sizes can be isolated due to landscape fragmentation. The resulting FF class has often been subject to logging and/or fire. Thus, a suitable classification scheme to characterize forest degradation in the Brazilian Amazon based on field ecological studies, associated with different processes and their interactions (Figure 10.1), and covering a spectrum of intensity, can be proposed as follows:

- Undisturbed forest (UF): Old-growth intact forest dominated by shade-tolerant tree species and original carbon stocks.
- NML: Logged forest without the use of heavy vehicles such as skidders and trucks, also known as traditional logging. Logging infrastructure (log landings, roads, and skid trails) are not built.
- ML: Planned selective logging where a tree inventory is conducted, followed by road and log landing planning to reduce harvesting impacts.
- CL: Conventional unplanned selective logging using skidders and trucks. Log landings, roads, and skid trails are built causing extensive canopy damage. Low-intensity understory burning may occur, but forest canopy is not burned.
- BF: Either NML or logged forests (ML and CL) where forest canopy has been intensively burned.
- FF: Isolated forest patches created by deforestation with abrupt changes on edges to pasture and agriculture lands, or with partial transitional edges to secondary forests. Fragments in the study area are usually subject to recurrent NML and fires.

10.2.3 Attributes of Degraded Forests Detectable Using Remote Sensing

At the field scale, logged forests are composed of three main environments: (1) forest islands that were not disturbed due to poor access imposed by difficult topography and rivers, or a lack of commercial timber species; (2) areas where the forest has been cleared to create roads for machine movements (skidders and trucks) and log landings to store the harvested timber; and (3) canopy-damaged forests (i.e., harvested areas and areas damaged by tree falls and machine movements) (Souza and Roberts 2005) (Figure 10.2). All of these environments can be found in the ML and CL classes, but the difference is that in ML, reduced impact logging practices are conducted to reduce direct and collateral damages (Johns et al. 1996; Pereira et al. 2002).

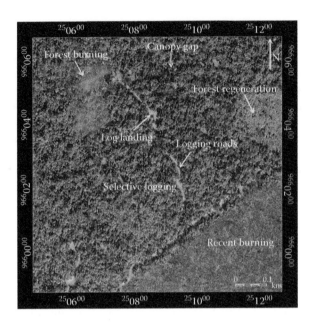

FIGURE 10.2
(See color insert.) Very high spatial resolution false-color infrared IKONOS image showing the different environments commonly found in logged and burned (LB) forests in the eastern Brazilian Amazon. At 1 m spatial resolution, log landings, logging roads, tree fall canopy gaps, and forest edges can be identified as well as "islands" of UFs and signs of regeneration. Signs of forest erosion along the edges between the LB forest and the recently slashed-and-burned forest can also be observed. (From Souza, C.M. and Roberts, D., *Int. J. Remote Sens.*, 26, 425, 2005.)

For these two classes, logging harvesting intensity varies from 30 to 40 m³ of logs per hectare (Verissimo et al. 1992; Johns et al. 1996). The NML class does not feature the various logging environments described above as no heavy machinery is used to harvest trees and a low harvest intensity is practiced (i.e., 5–10 m³ of logs per hectare). When fires penetrate logged forests, undetected damage under the canopy is expected. Prolonged and more intense fires can damage the tree canopy, exposing tree branches and trunks and making remote sensing detectability possible (Souza and Roberts 2005).

Tree inventories and forest impact measurements have been conducted to characterize forest degradation caused by selective logging (Verissimo et al. 1992; Johns et al. 1996; Pereira et al. 2002). Gerwing's (2002) was the first study in the Brazilian Amazon that proposed an all-encompassing approach to characterize the biophysical properties of a range of degraded forests. Slightly different forest degradation classes were proposed for this study. For example, repeated logging and burning were placed in separate classes. Our research group has adjusted Gerwing's method to characterize classes of forest degradation that can be easily integrated with remotely sensed measurements (Souza et al. 2005a, 2009).

The forest survey proposed by Gerwing (2002) consisted of measuring all trees with DBH >10 cm along transects of 10 m × 500 m (i.e., 0.5 ha).

Moreover, subparcels (10 m × 10 m; 0.1 ha) were established at every 50 m along transects, and all trees <10 cm DBH were surveyed. Logging and/or burning impacts were measured in the subparcels, including ground cover, and canopy gaps were estimated using a hemispherical lens and densitometer. Aboveground live and dead biomass pools were estimated for trees >10 cm DBH for each transect using tree inventory data and available allometric equations. Ancillary information about land use and disturbance history (i.e., time since last disturbance, number of times the area was disturbed) was collected during the field surveys. The forest transects were randomly defined in the field, and more than three must be conducted per class of degraded forest.

10.2.4 Ecological Impacts

Field ecological studies have provided the foundation for understanding the structural and compositional changes caused by forest degradation processes on pristine UFs. For remote sensing detection of forest degradation impacts, the following attributes are relevant: (1) ground cover comprised of intact vegetation, wood debris, and disturbed soils; (2) canopy cover; and (3) aboveground live biomass (AGLB). Our research group has conducted more than 100 transects in the Brazilian Amazon using an adaptation of Gerwing's methodology to link field measurements with remotely sensed data (Souza et al. 2005b, 2009). We have observed that for a single degradation event, intact vegetation and canopy cover decrease with an increase in forest degradation intensity by 20% and 60%, respectively. Conversely, soil disturbance and wood debris increase by 10% and 40%, respectively. However, when repeated degradation events are considered, these impacts tend to be more drastic. For example, repeated logging in the eastern Amazon region can disturb up to 70% of the original vegetation and deplete up to 40% of the original canopy cover (Gerwing 2002).

The forest structure changes caused by the forest degradation processes described above affect species composition and carbon stocks of UFs. The mean AGLB of UF obtained for our transect measurements was 377 Mg per hectare, with minimum biomass for the Ji-Paraná site (273 Mg per hectare) and maximum for Santarém (497 Mg per hectare). This result is compatible with field AGLB estimates using very large forest plots (Keller et al. 2001) and within the range of average values reported for the Brazilian Amazon region (Malhi et al. 2006; Saatchi et al. 2007). Using the mean AGLB obtained with our transects and assuming that carbon makes up 50% of the forest biomass, we can then demonstrate how carbon stocks vary with degradation intensity (Figure 10.3). A trend of reduced carbon stocks in pristine UF undergoing forest degradation processes has been observed. The more significant change is when UF is fragmented or burned, leading to respective 28% and 30% reductions in carbon stocks relative to original UF stocks. NML, ML, and CL degradation classes each experienced a <10% carbon loss. The carbon

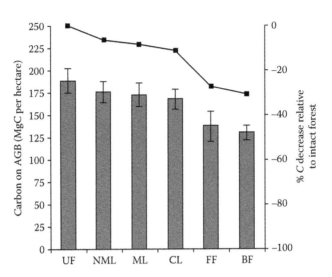

FIGURE 10.3
Change in aboveground live biomass as a function of degradation intensity. Bars represent standard error of the mean value and lines represent the percent change of C mean relative to intact forest. (From Souza, C. et al., Integrating forest transects and remote sensing data to quantify carbon loss due to forest degradation in the Brazilian Amazon. In *Case Studies on Measuring and Assessing Forest Degradation*. Forest Resources Assessment Working Paper 161, FAO, Rome, 20 p., 2009.)

stock changes presented in Figure 10.3 are for one event of forest degradation only. When considering recurrent forest degradation events, carbon stocks can be reduced by up to 50% (Gerwing 2002).

10.3 Remote Sensing of Forest Degradation

Detecting and mapping forest degradation with optical remotely sensed data is more complicated than mapping forest clearings by deforestation because degraded forest "pixels" are complex environments with mixtures of different land cover materials (i.e., vegetation, dead trees, bark, tree branches, soil, shade; Figure 10.1 [Souza and Roberts 2005]). Furthermore, signs of forest degradation disappear within 1–2 years due to rapid canopy closure and understory revegetation, making spectral characteristics of degraded forests similar to that of UFs (Stone and Lefebvre 1998; Asner et al. 2004a,b; Souza et al. 2005a, 2009).

The first attempts to map degraded forests in the Brazilian Amazon focused on detecting the processes responsible for degradation. Mapping selective logging received considerable attention, given its large extent and negative ecological impacts. The annual logged area in this region has been considered as large as the annually deforested area, with first estimates coming from socioeconomic field surveys (Nepstad et al. 1999) and the following ones based on satellite imagery (Asner et al. 2005; Matricardi et al. 2007). Techniques to map forest fire scars have also been developed, and forest fragmentation can be mapped with traditional techniques used to map deforestation. More recently, an all-encompassing approach for mapping forest canopy damage caused by these degradation processes has been proposed. Techniques for doing so are discussed in the following sections.

10.3.1 Remote Sensing Approaches to Mapping Selective Logging

Several remote sensing techniques were tested and applied to local and regional scale studies in the Amazon region to map selectively logged forests (Table 10.1). These techniques can be grouped in terms of mapping goals and methods utilized. In terms of mapping goals, some techniques were developed to map the total forest area affected by logging, which includes forest canopy damage and forest clearings created by log landings and roads, and to map intact forest islands surrounded by logging infrastructure and canopy-damaged areas. The second mapping goal focused on the mapping of areas with forest canopy damage only (i.e., intact forest islands were not included). In terms of methods for mapping logging, visual interpretation, semiautomated, and automated techniques have been tested (Table 10.1), and most of them can be applied to different spatial and spectral resolution sensors.

At high spatial resolutions (i.e., <5 m pixel size), images acquired by either space-borne or aerial platforms are viable for small-area analyses. Most of the features found in logging environments (i.e., roads, log landings, tree fall gaps, and UF islands) can be easily identified at this scale (Figure 10.1). Fusion techniques of panchromatic and multispectral images are commonly applied to enhance the imagery (Read et al. 2003; Souza and Roberts 2005), and visual interpretation is the most common mapping technique used. However, given the cost for image acquisition and interpretation, their use in mapping and monitoring logging is limited. For these reasons, the methods presented in the following sections focus only on medium spatial resolution imagery (i.e., 10–60 m pixel size). These data are freely available and are regularly acquired, unlike higher spatial resolution commercial data sets.

10.3.1.1 Visual Interpretation

Watrin and Rocha (1992) pioneered the use of satellite images to map selective logging in the Amazon region. Their work focused on Paragominas municipality, which was the most important logging center of the Brazilian Amazon from 1985 to 1995 (Verissimo et al. 1992). This study used printouts

TABLE 10.1

Remote Sensing Methods Most Often Used to Detect Forest Degradation Caused by Selective Logging in the Amazon Region

Mapping Approach	Studies	Sensor	Spatial Extent	Objective	Advantages	Disadvantages
Visual interpretation	Watrin and Rocha (1992)	Landsat TM	Local	Map total logging area	Does not require sophisticated image-processing techniques	Labor intensive for large areas and may be user biased to define the boundaries of logged forest
	Stone and Lefebvre (1998)	Landsat TM	Local			
	Matricardi et al. (2001)	Landsat TM	Brazilian Amazon			
	Santos et al. (2002)	Landsat TM	Brazilian Amazon			
Combining remote sensing and GIS (detection of logging landings + buffer)	Souza and Barreto (2000) Matricardi et al. (2001) Monteiro et al. (2003) Silva et al. (2003) Graça et al. (2005)	Landsat TM and ETM+	Local	Map total logging area (canopy damage, clearings, and undamaged forest)	Relatively simple to implement and satisfactory for estimating the total potential logging area	Logging buffers vary across the landscape and do not reproduce the actual shape of the logged area
Textural analysis	Matricardi et al. (2007)	Landsat TM	Brazilian Amazon	Map logging infrastructure	Implementation is straightforward and fully automated	Less sensitive to detect canopy damage created by tree falling
Decision tree	Souza et al. (2003)	SPOT 4	Local	Map forest canopy damage associated with logging and burning	Simple and intuitive classification rules	Has not been tested in very large areas, and classification rules may vary across the landscape

Method	Reference	Sensor	Scale	Objective	Advantages	Limitations
Change detection	Souza Jr. and Roberts (2002)	Landsat TM and ETM+	Local	Map forest canopy damage associated with logging and burning	Enhances forest canopy-damaged areas	Requires two pairs of images and does not separate natural and anthropogenic forest changes
Image segmentation	Graça et al. (2005)	Landsat TM	Local	Map total logging area (canopy damage, clearings, and undamaged forest)	Relatively simple to implement and satisfactorily estimate the total logging area. Free software available	Has not been tested in very large areas and segmentation rules may vary across the landscape
CLAS	Asner et al. (2005, 2006)	Landsat TM and ETM+	Five states in the Brazilian Amazon	Map total logging area (canopy damage, clearings, and undamaged forest)	Highly automated and standardized to very large areas	Requires high computation power and pairs of images for forest change detection
NDFI+CCA	Souza et al. (2005)	Landsat TM and ETM+	Local	Map forest canopy damage associated with logging and burning	Enhances forest canopy-damaged areas	Has not been tested in very large areas and does not separate logging from burning damages

of Landsat TM5 bands 4 and 5 acquired in 1988 to first visually identify and trace on overlay paper the boundaries of selectively logged areas. Next, the resulting polygons were hand digitized using a geographic information system (GIS) at 1:100,000 scale. The authors used the boundaries of forest scars created by roads, log landings, and canopy-damaged areas as the criteria for defining logged areas. Stone and Lefebvre (1998) also used visual interpretation of Landsat TM5 data to map logged forests in Paragominas for 1986, 1988, 1991, and 1995. In 2001, a large-scale study was conducted to map selective logging of the Brazilian Amazon using visual interpretation of Landsat TM5 digital imagery. In this study, Santos et al. (2001) mapped logged forests at a 1:250,000 scale and estimated an average of 1,580 km² per year for the period 1988–1998.

There are drawbacks to the use of visual interpretation for mapping selective logging. First, defining the boundary of logged and UFs is not always straightforward, even when using more detailed imagery such as IKONOS (Read et al. 2003; Souza and Roberts 2005). Second, there is some level of subjectivity in defining forest degradation created by logging and forest fires; none of the studies that used visual interpretation methods define rigorous criteria for separating these two causes of forest degradation. Third, visual interpretation is labor intensive and may be cost prohibitive for operational forest monitoring projects (Table 10.1).

10.3.1.2 Combining Remote Sensing and GIS

The need for a faster, cheaper, and replicable method to detect and map selective logging has driven the development of automated techniques. The first attempt combined automated detection of log landings from soil fraction derived from a spectral mixture analysis (SMA; covered in detail later) applied to Landsat images followed by the application of buffer regions (Souza and Barreto 2000). This technique requires field measurements to estimate harvesting radius from log landings in order to define the buffer radius. For tropical dense forest of the eastern Amazon and open forests of the central–southern region, buffer sizes were 180 m (Souza and Barreto 2000) and 350 m (Monteiro et al. 2003), respectively; both are considered local studies. Matricardi et al. (2001) used this buffer approach (with fixed radius of 180 m) to estimate selective logging impact over the Brazilian Amazon, differing with the use of texture measures applied to Landstat TM5 bands 3–5 to detect log landings. This large-scale study estimated an annual average area affected by logging of 4,690 km² per year for the period 1992–1999. This result is almost three times the one obtained by visual interpretation (Santos et al. 2001), though the product is at a more detailed scale (1:50,000) (Table 10.1).

The buffer technique for estimating logging areas also has limitations. Logging buffers are not fixed, and neither circular (Souza and Barreto 2000) nor squared buffers (Monteiro et al. 2003) adequately capture logged areas. The area affected by logging in most cases did not follow the contours of the buffer regions, resulting in commission and omission classification errors. To overcome this problem, a technique that uses region growth algorithms from log landings was proposed (Graça et al. 2005) to map canopy-damaged areas (Table 10.1).

10.3.1.3 SMA

Studies in the Brazilian Amazon have shown that Landsat reflectance data have limited the capacity for detecting logged forests, with bands 3 and 5 providing the best spectral contrast between logged and intact forests (Stone and Lefebvre 1998; Asner et al. 2002; Souza et al. 2005a). Vegetation indices and texture filters also showed some potential for detection of canopy damage created by logging (Asner et al. 2002; Souza et al. 2005a), but are more useful for enhancing logging infrastructure using Landsat band 5 (i.e., roads and log landings; Matricardi et al. 2007) (Table 10.1).

Alternatively, SMA has been proposed to overcome the challenge of using whole-pixel information to detect and classify logged forests. Landsat pixels typically contain a mixture of land cover components (Adams et al. 1995). In logged forests (and also in BF and forest edges), mixed pixels predominate and are expected to have a combination of green vegetation (GV), soil, nonphotosynthetic vegetation (NPV), and shade-covered materials. Therefore, fractional images derived from SMA analyses have the potential to enhance the detectability of logging infrastructure and canopy damage within degraded forests. For example, soil fractions enhance log landings and logging roads (Souza and Barreto 2000), while NPV fractions enhance forest-damaged areas (Cochrane and Souza 1998; Souza et al. 2003), and GV highlights forest canopy gaps (Asner et al. 2004a).

In SMA, the Landsat TM/ETM+ reflectance data of each pixel can be broken down into GV, NPV, soil, and shade fractions, which are the expected materials found in pixels within areas of forest degradation. The SMA model assumes that the image spectra are formed by a linear combination of n pure spectra, referred to as endmembers (Adams et al. 1995), such that:

$$R_b = \sum_{i=1}^{n} F_i R_{i,b} + \varepsilon_b \tag{10.1}$$

for

$$\sum_{i=1}^{n} F_i = 1 \tag{10.2}$$

where
R_b is the reflectance in band b
$R_{i,b}$ is the reflectance for endmember i, in band b
F_i the fraction of endmember i
ε_b is the residual error for each band

The SMA model error is estimated for each image pixel by computing the root mean square (RMS) error, given by:

$$\text{RMS} = \left[n^{-1} \sum_{b=1}^{n} \varepsilon_b \right]^{1/2} \qquad (10.3)$$

The identification of the nature and number of pure spectra (i.e., endmembers) in the image scene is an important step in obtaining correct SMA models. Two approaches have been proposed to define endmembers. First, reflectance spectra can be acquired at the field level with a handheld spectrometer (Roberts et al. 2002). The pure spectra measured on the ground are named reference endmembers and need to be well calibrated to the image data. The second approach uses image endmembers obtained directly from the images (Small 2004). This approach does not require spatial and radiometric calibration between endmembers and image data since their acquisition is from the same sensor and scale. SMA automation is also required to make this technique useful for monitoring large areas. A Monte Carlo unmixing technique using reference endmember bundles was proposed for that purpose (Bateson et al. 2000) and applied to map selective logging with Landsat images over the Brazilian Amazon (Asner et al. 2004a, 2005). An alternative approach using generic image endmembers (Small 2004) was implemented for the same application (Souza et al. 2005b), avoiding the need for collecting reference field spectra.

A novel spectral index applicable combines SMA fractions to derive the normalized difference fraction index (NDFI) (Souza et al. 2005b). The NDFI was developed to more accurately map selective logging. The NDFI is computed as:

$$\text{NDFI} = \frac{GV_{\text{Shade}} - (NPV + Soil)}{GV_{\text{Shade}} + NPV + Soil} \qquad (10.4)$$

where GV_{shade} is the shade-normalized GV fraction given by

$$GV_{\text{Shade}} = \frac{GV}{100 - Shade} \qquad (10.5)$$

NDFI values range from −1 to +1. For intact forests, NDFI values are expected to be high (i.e., about 1) due to the combination of high GV_{shade} (i.e., high GV and canopy shade) and low NPV and soil values. As forest becomes degraded, the NPV and soil fractions are expected to increase, lowering NDFI values relative to intact forest. Cleared forests are expected to exhibit low GV and shade, and high NPV and soil, making it possible to distinguish them from degraded forests as well (Figure 10.4).

Fraction images obtained with the subpixel estimation of forest endmembers through SMA enhanced the detection of forest degradation caused by

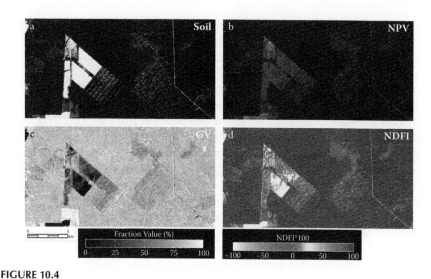

FIGURE 10.4
(See color insert.) Subset of a Landsat TM image showing fractions obtained from SMA and NDFI. (a) High soil fraction shows logging infrastructure (log landings and roads); (b) NPV shows higher fraction values for canopy-damaged areas along infrastructure relative to the surrounding intact forest; (c) canopy damage is also identified with lower GV fraction values (dark colors); and (d) all the fraction information are combined to enhance the detection of logged forest.

logging. As a result, spatial and contextual classifiers were developed and applied to fraction images improving detection and mapping of selectively logged forests. The techniques varied from simple GV change detection (Souza et al. 2002) and contextual–spectral classifiers (Souza et al. 2005b) to more sophisticated and computer-intensive spectral and spatial pattern recognition techniques (Asner et al. 2005) (Table 10.1). As a result, selective logging, initially considered cryptic to Landsat-like images (Nepstad et al. 1999), became visible and measurable over large forest areas of the Brazilian Amazon. Subsequent analyses proved that this type of degradation was affecting areas as large as those cleared by deforestation, as indicated by field survey estimates (Nepstad et al. 1999).

10.3.2 Classification of Forest Degradation

The remote sensing techniques described in Section 10.3.1 represent a considerable contribution toward mapping selective logging, which is one of the processes responsible for forest degradation. However, the application of these techniques has also revealed challenges in separating logging damage from that created by forest fires. For example, SMA fractions have been used to map fire scars of previously logged forests of the eastern Amazon (Cochrane and Souza 1998; Cochrane et al. 1999); the large-area mapping studies of selective logging did not take into account the associated fire impacts on forests (Asner et al. 2005; Matricardi et al. 2007), assuming that the forest damage was created only by logging. Therefore, new classification algorithms were needed to account for the different change dynamics created by logging and fires.

Morton et al. (2011a) proposed a technique, also applied to SMA fractions, to detect the spatial and temporal pattern of forest burn damage and recovery (BDR) in order to distinguish forest degradation from logging and forest fires. The BDR technique was applied to Landsat and MODIS data, with the latter more suitable for mapping large burn scars (i.e., >50 ha). This technique requires robust time series including a postdisturbance recovery signal, meaning that the result is always 1 year out-of-date. An alternative to this method is to use spatial–contextual classifiers to separate logged forest from BFs based on the size and shape of the forest damage (Souza et al. 2005b) or the burn scar index (BSI) (Alencar et al. 2011), which is an SMA fraction-based approach to map BFs. However, these methods do not eliminate all spatial and temporal overlaps between the different degradation processes. Therefore, it is more appropriate to map canopy damage without regard to the cause of forest degradation (either logging or forest fire), and then use contextual information to distinguish the process responsible for the impact.

For example, Figure 10.5 shows the result of a time-series (1984–2010) analysis of deforestation and forest degradation for a Landsat TM scene (226/68) covering Sinop municipality, in Mato Grosso state, southern Amazon region. A decision tree classifier was built and applied to fractions (GV, NPV, soil, and shade) and NDFI derived from SMA to map forest canopy damage caused by selective logging and forest fires every year. Then, forest degradation age and frequency were obtained from these annual maps. Moreover, a carbon emission simulator (CES) (Morton et al. 2011a) model was used to estimate carbon emissions associated with deforestation and forest degradation and associated uncertainty. Forest degradation frequency enables the CES model to keep track of carbon stock reduction; forest degradation age is important to track carbon sequestration due to forest regeneration.

Because CES is based on a Monte Carlo simulation approach, emission factors from deforestation and forest degradation and model parameters are defined as ranges of possible values. For example, forest carbon stock changes due to forest degradation in this region range from 10% to 30% (Figure 10.3). CES runs several times (i.e., at least 100 times), and in each simulation carbon stock changes associated with forest degradation can have any possible value between this range. Here, we assumed a uniform distribution since we do not have sufficient data to define the actual statistical distribution of carbon stock changes in degraded forests. Then, uncertainty of carbon emissions associated with deforestation and forest degradation can be estimated with CES.

The CES results showed that the carbon emissions for the 226/62 Landsat scene covering the Sinop region in Mato Grosso totaled 46.7–82 MgC (i.e., tons of C) from 1984 to 2010 (Figure 10.5). The average total carbon

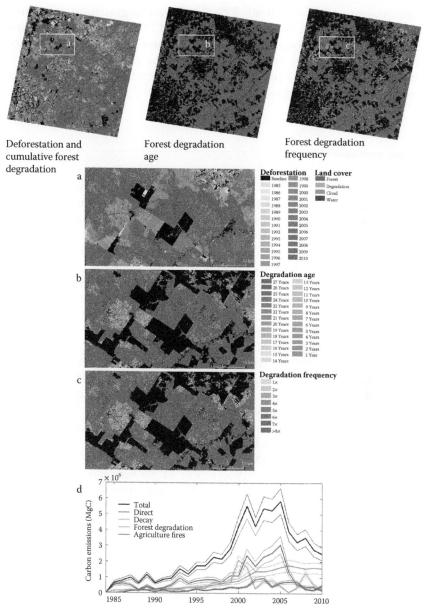

FIGURE 10.5
(See color insert.) In this example, a long time series (i.e., >25 years) of Landsat TM/ETM+ data from Sinop, Mato Grosso state, was used to track deforestation and forest degradation. Forest degradation age and frequency maps are obtained from the annual maps and used together with the forest degradation and deforestation maps in a CES model to estimate carbon emissions associated with these processes. More reliable and consistent baseline scenarios for REDD+ can be obtained with this type of model because information about forest degradation is included and associated uncertainty estimated.

emissions were 66.5 MgC (with 95% CI). Forest degradation contributed 19% (i.e., 8.7–16.3 MgC; average of 8.7 MgC) of the carbon emissions over this 26-year period. However, in 2000, 2007, and 2008, carbon emissions from forest degradation were higher than emissions from direct forest conversion. These results reinforce the need to measure carbon emissions associated with forest degradation (Figure 10.5).

10.4 Forest Monitoring for REDD+

In a recent study conducted in the forests of Mato Grosso state, the sources of uncertainties for carbon emission estimates from deforestation, forest degradation, and forest carbon stocks were identified for the period 1990–2008 (Morton et al. 2011b). The sources of deforestation data showed good agreement for multiyear periods (i.e., 5-year interval), but annual deforestation rates differed by >20%. Data sources of forest carbon stocks ranged more significantly, between 99 and 192 MgC per hectare. Even though there were several ecological studies of the impacts of forest degradation in this region and remote sensing techniques for mapping forest degradation were available, existing maps of forest degradation were scarce. Additionally, the available forest biomass maps did not account for changes in forest carbon stocks due to forest degradation. As a result, full carbon accounting for REDD+ is compromised. The remote sensing techniques described in this chapter can be used to reduce this uncertainty by quantifying annual transitions involving degraded forest and their relation to deforestation and reduction of forest carbon stocks (Figures 10.1 and 10.6).

Selective logging, forest fires, and forest fragmentation are the major sources of depletion of forest carbon stocks in the Amazon region through forest degradation, even though less carbon-impacting forest degradation processes have been recognized (Peres et al. 2006). Therefore, the lessons from the Amazon region regarding characterization of forest degradation through ecological and remote sensing measurements can be useful for establishing a framework for the spatially explicit estimation of carbon emissions and their sources of uncertainty for REDD+ (Figure 10.6). The proposed framework is that of the United Nations Framework Convention on Climate Change (UNFCCC) Approach 3 and Tier 3 forest area change and carbon stocks estimates (Herold et al. 2011).

First, the baseline period for the project must be defined. In our study in Mato Grosso, we concluded that a long (>15 years) historic assessment could help reduce uncertainty in remote sensing data sources. In the example provided in Figure 10.5, 1984 was defined as the baseline year for mapping forest changes. For mapping deforestation, there are several well-established remote sensing techniques and operational monitoring systems in place in

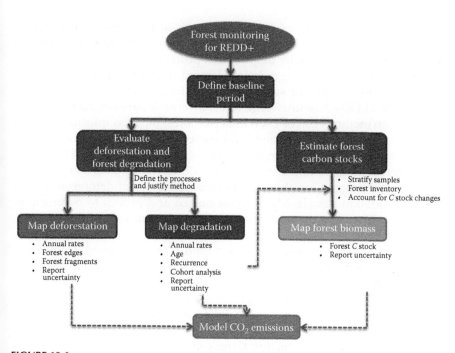

FIGURE 10.6
(See color insert.) Integrating deforestation and forest degradation information to estimate forest carbon stock changes for REDD+ projects.

the Amazon region. For forest degradation, Table 10.1 offers several options to map forest canopy-damaged areas. The reported map accuracy for the methods used to map logging and forest fires ranged from 89% to 93%. However, it is important to previously characterize the processes responsible for degradation in order to support the selection of the remote sensing method.

Deforestation maps over the REDD+ baseline period allow estimation of annual deforestation rates. Additionally, deforestation maps can also inform the length of forest edges and the extent of forest fragmentation. For example, in 1999 and 2002, more than 32,000 km and 38,000 km of new forest edges were created, respectively, as a result of deforestation and selective logging (Broadbent et al. 2008). Information on forest fragmentation and edge effects has not been taken into account in REDD+ projects, but can be a major source of carbon emissions (Numata et al. 2010, 2011). Forest degradation maps are important for providing information on annual rates of degradation and on forest degradation age and recurrence (i.e., frequency). Age and recurrence histories of forest degradation are necessary for updating forest carbon stock maps. Moreover, this information can aid in designing forest inventory sampling stratification schemes to estimate carbon stocks of degraded forests at field level. For example, forest inventories can be conducted in areas that have undergone several cycles of carbon depletion by degradation processes.

Annual maps of forest degradation derived from remote sensing offer a reliable spatiotemporal data set to account for forest carbon stock changes in preparing a REDD+ baseline. Once forest inventories are conducted, spatial interpolation methods can be used to derive forest biomass information over large areas. Kriging interpolation is an approach that has been successfully tested in the Brazilian Amazon to estimate spatially explicit unbiased averages of forest biomass and their associated uncertainty (Sales et al. 2007). Integration of krigged forest biomass maps with maps of deforestation and forest degradation has already been conducted and proven to be useful in reporting carbon emissions associated with these processes (Morton et al. 2011a; Numata et al. 2011).

These results are promising and support the proposed framework (Figure 10.6) for monitoring REDD+ projects. The challenges to applying this framework to other tropical forest regions include the lack of technical capacity for both remote sensing and forest inventory activities. However, options for monitoring forest degradation and deforestation going from a less to more rigorous approach/tier are available (Herold et al. 2011). Nonetheless, there is no technical reason to exclude carbon emissions estimates by forest degradation from REDD+ MRV activities.

10.5 Conclusions

Selective logging, forest fires, and forest fragmentation are the main processes responsible for forest degradation in the Brazilian Amazon. These processes can lead to significant reduction of forest carbon stocks, especially when recurrent forest degradation occurs. Additionally, significant change in forest structure also happens, allowing detection and mapping of forest degradation scars with optical remotely sensed data. A range of 1–30 m of spatial resolution imagery has been tested in the Amazon region for mapping forest degradation, using different techniques. But high spatial resolution imagery such as Landsat has been the most important source of data to map forest degradation in this region. Landsat imagery is important because it covers very large areas and allows to construct very long (i.e., >15 years) historical deforestation and forest degradation credible baseline for REDD+. In terms of techniques, subpixel information derived from SMA offers a better way to enhance forest degradation scars relative to whole-pixel classifiers or textural metrics (which is based on pixel neighborhood information). Moreover, forest change detection algorithms must be designed to track history and recurrent events of forest degradation to better estimate carbon emissions associated with these processes. Therefore, because of the large area affected and high impact on forest carbon stocks, baseline for REDD+ projects in the Amazon region must include annual forest area change and associated carbon emissions due to forest degradation, as demonstrated in this chapter.

Acknowledgments

My research activities on mapping forest degradation in the Brazilian Amazon have been funded by several organizations in the past 10 years. I am grateful to the LBA Research Program funded by NASA and the Brazilian government. I am also thankful to the Gordon and Betty Moore Foundation, the Packard Foundation, and the Avina Foundation for having supported my research and operational projects in Brazil. I would also like to thank Imazon for providing the work conditions to conduct this review. Antônio Victor Fonseca, Marcio Sales, and João Victor Siqueira helped to prepare some figures for this chapter and I would like to thank them.

About the Contributor

Carlos Souza, Jr. is a senior researcher at Imazon, Belém, Pará, Brazil. He received his PhD in geography from the University of California, Santa Barbara, in 2005, and got the Skoll Award for Social Entrepreneurship in 2010 for his contributions on monitoring deforestation and forest degradation in the Brazilian Amazon. His research focuses on development of methods for mapping forest changes with satellite images and modeling carbon emissions associated with forest and land use changes.

References

Adams, J.B. et al., Classification of multispectral images based on fractions of end-members: Application to land-cover change in the Brazilian Amazon. *Remote Sensing of Environment*, 52, 137, 1995.

Alencar, A. et al., Modeling forest understory fires in an eastern Amazonian landscape. *Ecological Applications*, 14, S139, 2004.

Alencar, A. et al., Temporal variability of forest fires in eastern Amazonia. *Ecological Applications*, 21, 2397, 2011.

Asner, G.P. et al., Remote sensing of selective logging in Amazonia: Assessing limitations based on detailed field observations, Landsat ETM+, and textural analysis. *Remote Sensing of Environment*, 80, 483, 2002.

Asner, G.P. et al., Canopy damage and recovery after selective logging in Amazonia: Field and satellite studies. *Ecological Applications*, 14, S280, 2004a.

Asner, G.P. et al., Spatial and temporal dynamics of forest canopy gaps following selective logging in the eastern Amazon. *Global Change Biology*, 10, 765, 2004b.

Asner, G.P. et al., Selective logging in the Brazilian Amazon. *Science*, 310, 480, 2005.

Asner, G.P. et al., Condition and fate of logged forests in the Brazilian Amazon. *Proceedings of the National Academy of Sciences (PNAS)*, 103 (34), 12947–12950, doi: 10.1073/pnas.0604093103, 2006.

Barros, A.C. and Uhl, C., Logging along the Amazon River and estuary: Patterns, problems and potential. *Forest Ecology and Management*, 77, 87, 1995.

Bateson, C.A. et al., End-member bundles: A new approach to incorporating end-member variability into spectral mixture analysis. *IEEE Transactions on Geoscience and Remote Sensing*, 38, 1083, 2000.

Broadbent, E.N. et al., Forest fragmentation and edge effects from deforestation and selective logging in the Brazilian Amazon. *Biological Conservation*, 141, 1745, 2008.

Cochrane, M.A., Synergistic interactions between habitat fragmentation and fire in evergreen tropical forests. *Conservation Biology*, 15, 1515, 2001.

Cochrane, M.A. and Laurance, W.F., Fire as a large-scale edge effect in Amazonian forests. *Journal of Tropical Ecology*, 18, 311, 2002.

Cochrane, M.A. and Schulze, M.D., Fire as a recurrent event in tropical forests of the eastern Amazon: Effects on forest structure, biomass, and species composition. *Biotropica*, 31, 2, 1999.

Cochrane, M.A. and Souza, C.M., Linear mixture model classification of burned forests in the Eastern Amazon. *International Journal of Remote Sensing*, 19, 3433, 1998.

Cochrane, M.A. et al., Positive feedbacks in the fire dynamic of closed canopy tropical forests. *Science*, 284, 1832, 1999.

Gerwing, J.J., Degradation of forests through logging and fire in the eastern Brazilian Amazon. *Forest Ecology and Management*, 157, 131, 2002.

Graça, P.M.L.A., Santos, J.R., Soares, J.V. and Souza, P.E.U., Desenvolvimento metodológico para detecção e mapeamento de áreas florestais sob exploração madeireira: estudo de caso, região norte do Mato Grosso. In: SIMPÓSIO BRASILEIRO DE SENSORIAMENTO REMOTO, 12. (SBSR), Goiânia. Anais, São José dos Campos: INPE, 1555–1562. CD-ROM, On-line. ISBN 85-17-00018-8. (INPE-12649-PRE/7941). Disponível em: <http://urlib.net/ltid.inpe.br/sbsr/2004/11.16.13.56>. Acesso em: 03 Jul. 2012, 2005.

Herold, M. et al., Options for monitoring and estimating historical carbon emissions from forest degradation in the context of REDD+. *Carbon Balance and Management*, 6, 13, 2011.

Holdsworth, A.R. and Uhl, C., Fire in Amazonian selectively logged rain forest and the potential for fire reduction. *Ecological Applications*, 7, 713, 1997.

Johns, J.S. et al., Logging damage during planned and unplanned logging operations in the eastern Amazon. *Forest Ecology and Management*, 89, 59, 1996.

Keller, M. et al., Biomass estimation in the Tapajos National Forest, Brazil: Examination of sampling and allometric uncertainties. *Forest Ecology and Management*, 154, 371, 2001.

Lambin, E.F., Monitoring forest degradation in tropical regions by remote sensing: Some methodological issues. *Global Ecology and Biogeography,* 8, 191, 1999.

Laurance, W.F., et al., Forest loss and fragmentation in the Amazon: Implications for wildlife conservation. *Oryx,* 34, 39, 2000.

Laurance, W.F. et al., Ecosystem decay of Amazonian forest fragments: A 22-year investigation. *Conservation Biology,* 16, 605, 2002.

Malhi, Y., et al., The regional variation of aboveground live biomass in old-growth Amazonian forests. *Global Change Biology,* 12, 1107, 2006.

Matricardi, E.A.T., Skole, D.L., Chomentowski, W., and Cochrane, M.A., Multi-temporal detection and measurement of selective logging in the Brazilian Amazon using Landsat data. BSRSI/RA03-01/w. Michigan State University, East Lansing, Michigan, 2001.

Matricardi, E. et al., Multi-temporal assessment of selective logging in the Brazilian Amazon using Landsat data. *International Journal of Remote Sensing,* 28, 63, 2007.

Monteiro, A.L. et al., Detection of logging in Amazonian transition forests using spectral mixture models. *International Journal of Remote Sensing,* 24, 151, 2003.

Morton, D.C. et al., Historic emissions from deforestation and forest degradation in Mato Grosso, Brazil: 1) source data uncertainties. *Carbon Balance and Management,* 6, 18, 2011a.

Morton, D.C. et al., Mapping canopy damage from understory fires in Amazon forests using annual time series of Landsat and MODIS data. *Remote Sensing of Environment,* 115, 1706, 2011b.

Nepstad, D.C. et al., Large-scale impoverishment of Amazonian forests by logging and fire. *Nature,* 398, 505, 1999.

Numata, I. et al., Biomass collapse and carbon emissions from forest fragmentation in the Brazilian Amazon. *Journal of Geophysical Research,* 115, G03027, 2010.

Numata, I. et al., Carbon emissions from deforestation and forest fragmentation in the Brazilian Amazon. *Environmental Research Letters,* 6, 044003, 2011.

Pereira, R. et al., Forest canopy damage and recovery in reduced-impact and conventional selective logging in eastern Para, Brazil. *Forest Ecology and Management,* 168, 77, 2002.

Peres, C.A. et al., Detecting anthropogenic disturbance in tropical forests. *Trends in Ecology & Evolution,* 21, 227, 2006.

Read, J.M. et al., Application of merged 1-m and 4-m resolution satellite data to research and management in tropical forests. *Journal of Applied Ecology,* 40, 592, 2003.

Roberts, D.A. et al., Large area mapping of land-cover change in Rondonia using multitemporal spectral mixture analysis and decision tree classifiers. *Journal of Geophysical Research-Atmospheres,* 107, D20, 2002.

Saatchi, S.S. et al., Distribution of aboveground live biomass in the Amazon basin. *Global Change Biology,* 13, 816, 2007.

Sales, M.H. et al., Improving spatial distribution estimation of forest biomass with geostatistics: A case study for Rondonia, Brazil. *Ecological Modelling,* 205, 221, 2007.

Santos, J.R. et al., *Dados multitemporarais TM/Landsat aplicados ao estudo da dinâmica de exploração madeireira na Amazônia.* Simpósio Brasileiro de Sensoriamento Remote, 2001.

Santos, J.R. et al., Change vector analysis technique to monitor selective logging activities in Amazon. Geoscience and Remote Sensing Symposium, IGARSS '03. *Proceedings, 2003 IEEE International,* 4, 2580–2582, doi: 10.1109/IGARSS.2003.1294515, 2003.

Skole, D. and Tucker, C., Tropical deforestation and habitat fragmentation in the Amazon: Satellite data from 1978 to 1988. *Science,* 260, 1905, 1993.

Small, C., The Landsat ETM+ spectral mixing space. *Remote Sensing of Environment,* 93, 1, 2004.

Souza, C. and Barreto, P., An alternative approach for detecting and monitoring selectively logged forests in the Amazon. *International Journal of Remote Sensing,* 21, 173, 2000.

Souza Jr., C. and Roberts, D., Multi-temporal analysis of canopy change due to logging in Amazonian transitional forests with green vegetation fraction images. *Second LBA Scientific Conference,* Manaus, Brazil, July 7–10, 2002 (http://lba.cptec.inpe. br/lba-conf-manaus02-en/accept_abstracts.htm), 2002.

Souza, C.M. and Roberts, D., Mapping forest degradation in the Amazon region with Ikonos images. *International Journal of Remote Sensing,* 26, 425, 2005.

Souza, C. et al., Mapping forest degradation in the Eastern Amazon from SPOT 4 through spectral mixture models. *Remote Sensing of Environment,* 87, 494, 2003.

Souza, C.M. et al., Multitemporal analysis of degraded forests in the southern Brazilian Amazon. *Earth Interactions,* 1–25, 2005a.

Souza, J. et al., Combining spectral and spatial information to map canopy damage from selective logging and forest fires. *Remote Sensing of Environment,* 98, 329, 2005b.

Souza, C. et al., Integrating forest transects and remote sensing data to quantify carbon loss due to forest degradation in the Brazilian Amazon. In *Case Studies on Measuring and Assessing Forest Degradation.* Forest Resources Assessment WorkingPaper161, FAO, Rome, 20 p., 2009.

Stone, T.A. and Lefebvre, P., Using multi-temporal satellite data to evaluate selective logging in Para, Brazil. *International Journal of Remote Sensing,* 19, 2517, 1998.

Verissimo, A. et al., Logging impacts and prospects for sustainable forest management in an old Amazonian frontier: The case of Paragominas. *Forest Ecology and Management,* 55, 169, 1992.

Verissimo, A. et al., Extraction of a high-value natural resource in Amazonia: The case of mahogany. *Forest Ecology and Management,* 72, 39, 1995.

Watrin, O.S. and Rocha, A.M.A., Levantamento da vegetação natural e do uso da terra no Município de Paragominas (PA) utilizando imagens TM/Landsat. *Belém, Boletim de Pesquisa,* n. 124, 40 p. (EMBRAPA/CPATU), 1992.

11

Use of Wall-to-Wall Moderate- and High-Resolution Satellite Imagery to Monitor Forest Cover across Europe

Jesús San-Miguel-Ayanz, Daniel McInerney, Fernando Sedano, and Peter Strobl
Joint Research Centre of the European Commission

Pieter Kempeneers
Flemish Institute for Technological Research

Anssi Pekkarinen
Food and Agriculture Organization of the United Nations

Lucia Seebach
University of Copenhagen

CONTENTS

11.1 Introduction

Forest resources are very relevant in the political agenda of the European Union, as forestry influences many sectorial policies dealing with environmental protection, renewable energy, and biodiversity, to name some. The design, implementation, monitoring and evaluation, and impact assessment of environmental policies at the European level require reliable, consistent, and updated information of forest resources.

Although several countries in Europe collect a considerable amount of forest-related information, this is often not spatially continuous and frequently not accessible, nonharmonized, scattered in remote databases, and encapsulated in diverse data formats. One critical aspect regarding forest information in Europe is the different forest definitions used by countries, which hampers the comparability of nationally collected forest information.

Remote sensing–based products are thus the most suitable source of consistent and up-to-date forest information over large areas. Remote sensing techniques have been widely used for mapping forest resources at local and national levels. Working over large areas poses additional logistic, technical, and managerial challenges that have limited the number of existing pan-European products. Large-area projects usually require a considerable data management capacity. They also require carefully planned processing chains, including consistent preprocessing of satellite and ancillary information and mapping methodologies to produce large-area products. In addition, these methodologies must be robust, reliable, and flexible to handle suboptimal data sets of images from several sensors.

Several remote sensing–based products exist that include forest information and have pan-European coverage. However, these products were derived from coarse-resolution satellite images (Bartholomé and Belward 2005; DeFries et al. 2000; Friedl et al. 2002; Häme 2001; Hansen et al. 2000; Schuck 2003) or are labor intensive (Corine Land Cover [CLC]). Furthermore, the lack of comprehensive validation schemes of these products limits their utility in a number of applications.

The recent availability of a wider selection of remote sensing data allows an improvement in spatial resolution over the existing products. It also allows exploiting the temporal domain of remote sensing data. This scenario enables the development of products with higher spatial detail and increased thematic information content.

In this context, the Joint Research Centre (JRC) of the European Commission has been working on the production of enhanced remote sensing–based forest products. Two pan-European forest maps with a ground sampling distance (GSD) of 25 m have been produced based on Landsat ETM+ imagery (Pekkarinen et al. 2009) and IRS LISS-III, SPOT 4-5, and Moderate Resolution Imaging Spectroradiometer (MODIS) remote sensing data

(Kempeneers et al. 2011). Besides these high-resolution products, the JRC is carrying out research to improve forest monitoring capabilities at 250 and 500 m GSD based on time-series analysis of remote sensing data. This chapter presents the methodologies used in the production of these maps and their accuracies and discusses future potential developments in forest monitoring at the pan-European level.

11.2   Materials and Methods

This section describes the materials used in the production of the forest maps for the years 2000 and 2006 (Figures 11.1 and 11.2).

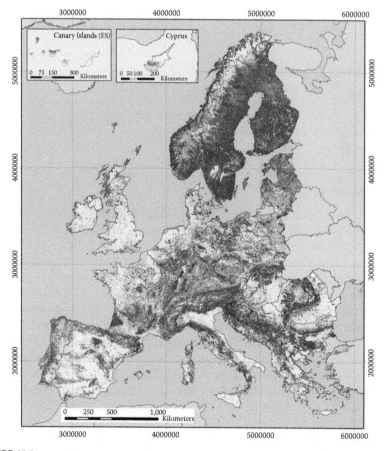

FIGURE 11.1
(See color insert.) JRC forest map 2000.

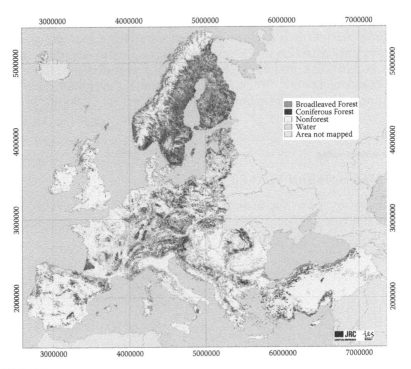

FIGURE 11.2
(See color insert.) JRC forest map 2006.

11.2.1 Materials

The forest/nonforest map for the year 2000 (FMAP2000) was derived from Landsat-7 ETM+ imagery. Scenes belonged to two different image datasets: the NASA Orthorectified Landsat Dataset (Tucker et al. 2004) available from the Global Land Cover Facility (GLCF) and the IMAGE2000 data set (JRC 2005). The two data sources were mixed in order to optimize cloud freeness and acquisition date. The target year for the scenes was 2000, but the acquisition window covered years from 1999 to 2002. The full data set included 415 scenes available as top of atmosphere (TOA) radiance: 285 of them from the GLCF and 130 from the IMAGE2000 data set. All images in the full data set were reprojected to the European Terrestrial Reference system 1989 and the Lambert Azimuthal Equal Area (ETRS89-LAEA) projection and resampled to 25 m rasters. In order to ensure consistent geometrical quality between scenes coming from the two different data sets, IMAGE2000 scenes were orthorectified taking GLCF scenes as a reference.

The forest/nonforest map for the year 2006 (FMAP2006) and the forest type map (FTYP2006) were derived from the IMAGE2006 data set. This data set includes TOA radiance IRS-LISS-3 scenes and additional SPOT 4 and

5 scenes for those regions in which cloud-free IRS-LISS-3 were not available. The scenes were orthorectified and geometrically corrected. As in the year 2000, all images were reprojected to ETRS89-LAEA projection and resampled to 25 m rasters.

In addition to the IMAGE2006 data set, the production of FTYP2006 required 12 (one per month) MODIS 16-day composites at 250 m spatial resolution. These composites were reprojected and resampled to 25 m to match the IMAGE 2006 data set.

11.2.2   Ancillary Data

11.2.2.1   Training Data

The CLC data set was used as training data. CLC includes 44 land cover (LC) and land use classes from which three correspond to forest classes (broadleaved, coniferous, and mixed forests). The CLC covers all EIONET countries, which includes the EU-27 Member States and neighboring countries. The CLC is available for the reference years 1990, 2000, and 2006. The corresponding data set was used for the production of each pan-European forest map.

11.2.2.2   Reference Data

The validation of the FMAP2000 was performed using two data sets. The first included field plot data from the land use/cover area frame statistical survey that was carried out in 2001 (LUCAS2001). LUCAS2001 is based on 94,984 sampling units, which consist of a circle with a 20 m radius. It is based on a seven LC classification nomenclature, with the forest class subdivided into broadleaved, coniferous, and mixed, but it also includes a land use component. The second data set was derived from the visual interpretation of sample points overlaid on very high-resolution satellite imagery from Google Earth. In total, 5,193 forest and nonforest points were collected from the interpretation of this data set and classified into forest and nonforest classes.

The FMAP2006 was validated using ground reference data that were derived from European National Forest Inventories (NFIs). NFI data are frequently collected by national authorities for the production and planning of forest resources at national and regional levels, but they are also needed to meet international reporting requirements to the FAO's Forest Resource Assessment (FAO 2010) and other requirements.

The NFI data used in this validation were managed in the so-called eForest platform. The eForest platform, established for the provision of data and services to the European Forest Data Center (EFDAC) of the European Commission, is the first step to produce a harmonized database of all European NFIs. It emerged from the work carried out by the COST Action E-43 that sought to develop methods, concepts, and definitions that would harmonize NFIs between countries (Tomppo et al. 2010). Of particular importance within

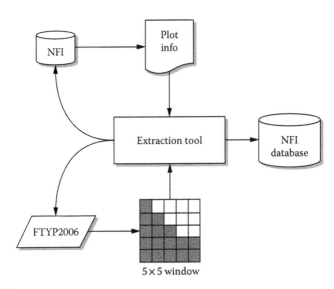

FIGURE 11.3
Pixel extraction tool.

this process was the harmonization of the definition of forest, which varies between NFIs.

The platform consisted of 1,080,829 NFI plots, distributed across 21 countries. However, the exact plot locations were not disclosed by the NFIs. For the validation, it was necessary to build a pixel extraction tool that was used by the data owners to extract the forest map data within a 5 × 5 window around the NFI plot coordinates (Figure 11.3). These data were used to compute the overall, producer, and user accuracies for the FMAP2006 at country and regional scales. Plots that were labeled as young stands or unstocked were removed from the eForest validation data set so that the accuracy assessment of the FMAP2006 focused on forest cover and nonforest use. It should be noted that unstocked forest areas are considered forests from a land use perspective, although they are not forests from an LC (remote sensing) perspective.

Additionally, the LUCAS2001 data were used to validate the FMAP2006 data set. The results of this validation process are described hereafter.

11.3   Methods

11.3.1   Data Preprocessing

The high spatial resolution scenes from IRS LISS-3 and SPOT4/5 were preprocessed by the German Aerospace Center (DLR). The scenes were orthorectified using rational polynomial functions (Lehner et al. 2005) and geometrically corrected using ground control points (GCPs) and a digital elevation model (DEM). The orthoimages were resampled to 25 m in the standard projection for Europe, using the ETRS89/LAEA projection (Annoni et al. 2003). The reported root mean square errors in both horizontal directions were less than a pixel. The images were only available as TOA radiances (not atmospherically corrected).

In the case of the MODIS, daily images were also preprocessed by DLR in the standard European projection. However, a geometric and atmospheric correction was performed to obtain ground reflectances for bands 1–7 at 250 and 500 m GSD.

A 16-day MODIS composite was created from the daily images. By not using the MOD13Q1 product (Huete 2002), a reprojection from sinusoidal to the standard European projection was not needed, avoiding an extra interpolation step. Unlike the MOD13Q1 product, our 16-day composite was not corrected for BRDF effects. Nevertheless, by selecting the median pixel value in the NIR band of all cloud-free observations within the 16-day window, some of the effects due to undetected clouds and extreme observation angles were alleviated.

11.3.2   Forest Mapping Approaches

A nonparametric supervised classification algorithm was used to obtain the forest maps FMAP2000 and FTYP2006. Supervised classification methods are preferable in cases where *a priori* information is available for the desired output classes and their spatial distribution (Cihlar 2000). With the CLC map, training data for forests (types) and nonforests were available in a consistent way for the entire area of interest (Europe).

Given the large geographic extent of the pan-European map, the interclass variance was expected to be high. For example, broadleaved forests in northern Europe have different spectral characteristics than those in southern Europe. Moreover, the digital numbers stored in the multispectral image bands represented TOA radiance and thus were not corrected for atmospheric effects. Consequently, image data were processed on a scene-by-scene basis, allowing the classifier to be trained for the specific conditions within each scene. The final output, the pan-European forest map, was then obtained by mosaicing the different scenes, using a composite rule where pixels did overlap. In the case of the FMAP2000, the composite rule was based on uncertainty information derived during the classification process. The number of overlapping scenes in the case of the FMAP2006 was larger (every pixel was observed at least twice but often three to four times). This allowed for a (weighted) maximum voting of the classified scenes. Weights were introduced based on seasonality. Summer scenes were weighted in favor of early spring or late autumn scenes.

The main requirements for the classification method were:

1. Consistency: The wall-to-wall pan-European forest map had to be produced in a homogeneous way.
2. Performance: Algorithms had to be fully automatic.
3. Robustness for deficiencies in the training and input data: The methods for the FMAP2000 and FTYP2006 showed some important differences in how this was achieved. This is explained in the following overview.

The FMAP2000 was mapped using a *k*-nearest-neighbor (*k*-NN) classifier (Tomppo et al. 2008). Instead of extracting spectral information for each Corine Land Use Land Cover patch, two key improvements in the classification approach were implemented to improve the performance (Pekkarinen et al. 2009): first, a segmentation prior to the classification step and second, an adaptive spectral representivity analysis (ASRA) (Pekkarinen et al. 2009). ASRA was developed to improve the training process and to minimize errors resulting from the relatively large minimum mapping unit of Corine. The segmentation was merely used to speed up the *k*-NN classification, which is known to be inefficient for processing large data sets. The ASRA was introduced after clustering the segments into spectral classes. It seeks to identify representative combinations of spectral and informational classes using a contingency table, derived from the cluster labels and CLC classes. For more details of the algorithm, the reader is referred to Pekkarinen et al. (2009).

The classification method for the FTYP2006 was based on an artificial neural network (ANN) (Rumelhart and McClelland 1986) that has been shown to combine two excellent classification properties: high accuracy (Chini et al. 2008; Licciardi et al. 2009) and robustness to training site heterogeneity (Paola and Schowengerdt 1995). Also important for the selection of the classifier was that the ANN, once trained, is very fast. Unlike for the production of the FMAP2000 method, a segmentation step was therefore not needed.

Another difference with the FMAP2000 is that forest types were introduced in the FTYP2006. To increase the potential of the classifier, multitemporal information was added to the multispectral information (data fusion). The multitemporal data were obtained from the MODIS sensor, using a 16-day composite for each month in 2006 at 250 m spatial resolution. The temporal aspect of the spectral reflectance can describe phenology, which is a potential indicator for LC types (DeFries et al. 1994; Hansen et al. 2005). The data fusion with this additional information source also increased the robustness of the classification process (Kempeneers et al. 2011).

However, fusing data from sensors at different spatial resolutions posed a challenge to retain the fine spatial resolution in the final LC map. A new data fusion method was therefore proposed, based on a two-step approach (Kempeneers et al. 2011). In step one, the classifier created a forest map, classifying forests and nonforests only. In step two, a new classifier refined forest into forest types, excluding the nonforested pixels from the classification process. The multitemporal data at medium spatial resolution were introduced only in step two.

The idea is that, as the classes are refined, the complexity of the classification increases. At this point, the classifier can benefit most from the added information obtained from data fusion. The forest/nonforest map was mapped using only the spectral information at fine spatial resolution and therefore retained the finest spatial resolution possible.

11.4   Results

The accuracy assessment of the forest cover maps was performed using three reference data sets that were previously described in Section 11.2.2. The overall accuracy (OA) of the FMAP2000 was 88.6% and 90.8% respectively for the VISVAL and LUCAS data sets, while the OA for the FMAP2006 was 88.0% and 84.0% based on the eForest and the LUCAS2001 data sets. The results for eForest and LUCAS2001 cannot really be compared due to a different coverage in both space and time (where LUCAS2001 can be regarded as outdated).

The calculation of the producer and user accuracies provided information on the performance of both maps for the forest and nonforest classes (Table 11.1). The producer's accuracy of the forest class was lowest for the FMAP2006 (75%) with respect to the eForest database, while it was slightly higher than FMAP2000 at 85.5% and 83.9% when compared to the VISVAL and LUCAS data sets. When compared to official statistics, the results demonstrated an overall underestimation of forest area in both forest maps, which was particularly emphasized in Ireland, Spain, Portugal, and Greece. This underestimation can be explained by the high rate of recent afforestation in Ireland, while in the Mediterranean countries, the forests typically have a very low percentage forest cover (e.g., 5% in Spain).

TABLE 11.1

FMAP2000 and FMAP2006 Accuracies with Respect to Validation Data Sets

	FMAP2000		FMAP2006	
Accuracies	VISVAL	LUCAS	eForest	LUCAS2001
OA%	88.6	90.8	88.0	84
Forest PA%	85.5	83.9	75	66
Forest UA%	77.66	85.8	87	85
Nonforest PA%	89.58	NA	94	94
Nonforest UA%	93.59	NA	88	84

The individual accuracies for the forest and nonforest classes were computed for the 3 × 3 window, and it was found that the producer accuracies improved by 1% for the eForest database (from 75% to 76%) and by 3% for the LUCAS data set (from 66% to 69%).

11.5   Applications

Harmonized spatial information on forest area is an important basis for environmental modeling and policy making at both national and international levels. Even if a majority of these data have been supplied by NFI statistics, detailed spatial distribution needed for modeling or further applications can mainly be provided by remote sensing–based products. Yet, reliability, consistency, and a high level of harmonization are important aspects to ensure comparability and enable the development of forest scenarios at an international level. The pan-European forest cover maps (FMAP2000 and FMAP2006) have the advantage to be produced under these prerequisites due to their harmonized approaches and, therefore, guarantee spatial consistency for further applications. Besides that, the medium resolution of the maps offers higher spatial details as previous pan-European LC products such as the CLC maps.

Most of the applications of the forest cover maps (FMAP2000 and FMAP2006) are related to the need for accurate and up-to-date estimates on the spatial distribution of forests as inputs into various models. Baritz et al. (2010) investigated the carbon concentrations and stocks in forest soils of Europe and located forested areas with the help of FMAP2000. Similarly, information on forest distribution was needed for a vulnerability study in the Alps and the Carpathian mountains (Casalegno et al. 2011). As the forest definition of the forest cover maps includes also urban parks in contrast to CLC, FMAP2000 could have been applied in a pan-European urban greening study, where growth of urban forest was investigated. In some of aforementioned studies, the initial medium resolution (25 m) was degraded down to 1 km resolution to speed up the process of the models, yet even with the degraded resolution of 1 km, FMAP2000 was found to be preserving the detailed forest spatial pattern of the original map (Seebach et al. 2011a). Besides applications at the pan-European level, the forest cover maps have been used in local or regional studies as the high resolution allows for detailed studies at that level. The large extent of Europe further enables potential reproducibility of regional studies using these maps as proposed by Lasserre et al. (2011) or Casalegno (2011). Another example of the same kind is the study of Chirici et al. (2011) that used FMAP2000 for a regional study in central Italy (Molise) as an initial forest mask for subsequent delineation of clearcuts based on very high-resolution imagery.

Another application of these maps apart from their indirect use as forest masks is, for example, the estimation of forest area at different units. Seebach et al. (2011b) investigated the applicability of FMAP2000 for reporting harmonized forest estimates for European countries. The comparison with official statistics derived from NFIs indicated an overall good agreement if uncertainties of both sources were taken into account; yet, discrepancies were found in areas with very low and fragmented forests or in mountainous regions. Another major driver of the remaining disagreements between official statistics and map-derived estimates originates from the common issue of land use versus LC. While official statistics reports are based on forest use definitions, estimates based on remote sensing products like FMAP2000/2006 will report land coverage with forest-like vegetation. The latter might become forest use maps only if extensive auxiliary data are available for their manipulation. A further direct application of the forest cover maps are their use for assessing change using postclassification comparison as both maps have been produced by a comparable and consistent approach. This was done for the European part of the FAO FRA 2010 Remote Sensing Survey (RSS), where both forest maps were used to detect reliable forest cover changes based on an enhanced postclassification approach. This approach accounts for potential misregistration errors and reduces the uncertainty of erroneous change detection due to classification errors (Seebach et al. 2010).

All in all, FMAP2000 has proved its ability to serve as a multipurpose product from direct use to downstream services. FMAP2006 and the associated FTYPE2006 have been recently released and are foreseen to be used in upcoming studies, where the differentiation of forest types is of high importance, for example, pan-European forest biomass estimation. Yet, care must be exercised for any application of these maps as every map inherits uncertainties, which need to be addressed depending on the intended use.

11.6 Conclusions and Future Aspects

The pan-European forest maps have been produced for the reference years 2000 and 2006 using optical satellite imagery and standardized methodologies with respect to preprocessing and classification. These maps have provided a baseline assessment of the spatial distribution and composition of forest resources in Europe and demonstrated improvements in terms of quality and production with respect to the CLC Project. In the frame of the Global Monitoring for Environment and Security (GMES) Initial Operations, the production of a new set of so-called high-resolution layers (HRLs) is foreseen, which will be coordinated by the European Environment Agency. Among these, HRLs will be a forest layer designed to closely resemble the JRC FMAP2000 and FMAP2006, but with a target reference year of 2012.

The mapping methods presented within this chapter were based on at-sensor radiances of the remote sensing sensors. Despite the fact that the applied methods are scientifically sound and practical, future mapping applications should be based on well-calibrated image data, from which the effects of the atmosphere have been removed. That would allow for the development of well-defined algorithms that could be applied to a range of different optical sensors, since these algorithms would be based on registered spectral responses of real-world objects. Recent advances in preprocessing algorithms and new European optical imaging sensors, such as RapidEye and ESA's Sentinel II, will hopefully facilitate future development of such mapping approaches.

It is evident that the demand for European level information on forest resources will increase in the future. We need to better understand the integrated role of forests in the protection of the environment, biodiversity, well being and recreation, timber and bioenergy production, as well as mitigation of climate change and monitoring compliance to international climate change agreements. In the future, other sources of Earth observation data should be further studied and used in large-scale mapping projects. For instance, interferometric SAR and space-borne LiDAR could be used to map land use and LC as well as being used to estimate other forest parameters, particularly by their combined use with field measurements and/or high-density airborne LiDAR data.

About the Contributors

Jesús San-Miguel-Ayanz is a senior researcher at the Institute for Environment and Sustainability of the European Commission's Joint Research Centre and leads the forest research activities of the JRC in Europe. He has a PhD and MSc in remote sensing and GIS from the University of California, Berkeley (1993 and 1989, respectively) and a degree in forest engineering (1987) from the Polytechnic University of Madrid.

Pieter Kempeneers received his MS in electronic engineering from Ghent University, Belgium, in 1994, and a PhD in physics from Antwerp University, Belgium, in 2007. He was a researcher with the Department of Telecommunications and Information Processing, Ghent University, and with Siemens (mobile communication systems). In 1999, he was with the Centre for Remote Sensing and Earth Observation Processes (TAP), Flemish Institute for Technological Research (VITO), as a scientist. From 2008 to 2011, he was a scientist with the Joint Research Centre, European Commission, Ispra, Italy. His research focus is on image processing, pattern recognition, and multi- and hyperspectral image analyses.

Daniel McInerney graduated from University College Dublin in 2002 with an honors degree in forestry; he also holds an MSc in geoinformation science and remote sensing from the University of Edinburgh (2004) and a PhD in remote sensing applied to forest inventories from University College Dublin. He is currently a postdoctoral researcher at the Institute for Environment and Sustainability at the European Commission's Joint Research Centre. His research interests include NFIs, forest mapping, and Web-based geoinformation systems.

Fernando Sedano holds a PhD from the University of California, Berkeley. He was a NASA Earth Science Fellow from 2005 to 2008 and a postdoctoral researcher with the Institute for Environment and Sustainability of the JRC. Previously, he also worked as a forest consultant in several tropical countries. Fernando Sedano's research focuses on the development of remote sensing methods and applications to monitor and understand forest dynamics.

Anssi Pekkarinen studied forest inventory, remote sensing, and GIS at the University of Helsinki, Finland, and holds a Dr Sc in agriculture and forestry. He worked at the Institute for Environment and Sustainability of the JRC-EC during the period 2005–2009 and is currently with the Food and Agriculture Organization of the United Nations. He has more than 15 years of experience in operational mapping and monitoring of forest resources with space and airborne remote sensing. His fields of specialty are forest inventory, remote sensing aided land use and LC mapping applications, and image processing.

Lucia Seebach received her diploma (masters equiv.) in geoecology from the University of Bayreuth, Germany, in September 2003. After her graduation, she worked until 2010 as a scientific officer at the Institute for Environment and Sustainability of Joint Research Centre of the European Commission in Ispra, Italy. Currently, she is a PhD fellow at the Department of Forest and Landscape, University of Copenhagen, Denmark. Her main areas of scientific interest are monitoring and modeling of forest resources, uncertainty analysis, and assessment of applicability of remote sensing–derived maps.

Peter Strobl graduated in geophysics from the University of Munich in 1991 and obtained a PhD in geosciences from the University of Potsdam, Germany, in 2000. His career history includes the University of Munich, the German Aerospace Centre (DLR), and the Joint Research Centre of the European Commission. He has worked on a wide range of remote sensing–related issues and currently focuses on preprocessing, analysis, and quality aspects of large multisensor, multitemporal data sets.

References

Annoni, A. et al., *Map Projections for Europe*. EUR 20120 EN, European Commission Joint Research Center, Ispra, Italy, 2003.

Baritz, R. et al., Carbon concentrations and stocks in forest soils of Europe. *Forest Ecology and Management*, 260, 262–277, 2010.

Bartholomé, E. and Belward, A., GLC2000: A new approach to global land cover mapping from Earth observation data. *International Journal of Remote Sensing*, 26, 1959–1977, 2005.

Casalegno, S., Urban and peri-urban tree cover in European cities: Current distribution and future vulnerability under climate change scenarios. In: Casalegno, S. (Ed.), *Global Warming Impacts: Case Studies on the Economy, Human Health, and on Urban and Natural Environments*. InTech, Croatia, 2011.

Casalegno, S. et al., Vulnerability of *Pinus cembra* L. in the Alps and the Carpathian mountains under present and future climates. *Forest Ecology and Management*, 259, 750–761, 2011.

Chini, M. et al., Comparing statistical and neural network methods applied to very high resolution satellite images showing changes in man made structures at rocky flats. *IEEE Transactions on Geoscience and Remote Sensing*, 46(6), 1812–1821, 2008.

Chirici, G. et al., Large-scale monitoring of coppice forest clearcuts by multitemporal very high resolution satellite imagery: A case study from central Italy. *Remote Sensing of Environment*, 115, 1025–1033, 2011.

Cihlar, J., Land cover mapping of large areas from satellites: Status and research priorities. *International Journal of Remote Sensing*, 21(6–7), 1093–1114, 2000.

DeFries, R. and Townsend, J., NDVI-derived land cover classifications at a global scale. *International Journal of Remote Sensing*, 15(17), 3567–3586, 1994.

DeFries, R. et al., A new global 1 km dataset of percentage tree cover derived from remote sensing. *Global Change Biology*, 6(2), 247–254, 2000.

FAO, *Global Forest Resources Assessment 2010*. Main report. FAO Forestry Paper 163. Rome, Italy, 2010.

Friedl, M. et al., Global land cover mapping from MODIS: Algorithms and early results. *Remote Sensing of Environment*, 83(1–2), 287–302, 2002.

Häme, T., AVHRR-based forest proportion map of the pan-European area. *Remote Sensing of Environment*, 77(1), 76–91, 2001.

Hansen, M. et al., Global land cover classification at 1km spatial resolution using a classification tree approach. *International Journal of Remote Sensing*, 21(6–7), 1331–1364, 2000.

Hansen, M. et al., Estimation of tree cover using MODIS data at global, continental and regional/local scales. *International Journal of Remote Sensing*, 26(19), 4359–4380, 2005.

Huete, A. et al., Overview of the radiometric and biophysical performance of the MODIS vegetation indices. *Remote Sensing of Environment*, 83(1), 195–213, 2002.

JRC, *IMAGE2000 and CLC2000 Products and methods*, Joint Research Centre of the European Commission, Ispra, Italy, 2005.

Kempeneers, P. et al., Data fusion of different spatial resolution remote sensing images applied to forest type mapping. *IEEE Transactions on Geoscience and Remote Sensing*, 49(12), 4799–4986, 2011.

Lasserre, B. et al., Assessment of potential bioenergy from coppice forests through the integration of remote sensing and field surveys. *Biomass and Bioenergy*, 35, 716–724, 2011.

Lehner, M. et al., DSM and orthoimages from QuickBird and IKONOS data using rational polynomial functions. In: *Proceedings of High Resolution Earth Imaging for Geospatial Information*. Hannover, Germany, 17–20 May 2005.

Licciardi, G. et al., Decision fusion for the classification of hyperspectral data: Outcome of the 2008 GRS-S data fusion contest. *IEEE Transactions on Geoscience and Remote Sensing*, 47(11), 3857–3865, 2009.

Paola, J. and Schowengerdt, R., A review and analysis of backpropagation neural networks for classification of remotely-sensed multi-spectral imagery. *International Journal of Remote Sensing*, 16(16), 3033–3058, 1995.

Pekkarinen, A. et al., Pan-European forest/non-forest mapping with Landsat ETM+ and CORINE Land Cover 2000 data. *ISPRS Journal of Photogrammetry and Remote Sensing*, 64(2), 171–183, 2009.

Rumelhart, D. and McClelland, J., *Parallel Distributed Processing: Explorations in the Microstructure of Cognition*, Vol. 1: *Foundations*. MIT Press, Cambridge, MA, 1986.

Seebach, L.M. et al., *Pilot Study in Europe for the Global Forest Remote Sensing Survey*. Joint Research Centre, Publication Office of the European Union, Luxembourg, EUR 24488 EN, 2010.

Seebach, L.M. et al., Identifying strengths and limitations of pan-European forest cover maps through spatial assessment. *International Journal of Geographic Information Science*, 25, 1865–1884, 2011a.

Seebach, L.M. et al., Comparative analysis of harmonised forest area estimates for European countries. *Forestry*, 84, 285–299, 2011b.

Schuck, A., Compilation of a European forest map from Portugal to the Ural Mountains based on earth observation data and forest statistics. *Forest Policy and Economics*, 5(2), 187–202, 2003.

Tucker, C. et al., NASA's global orthorectified Landsat dataset. *Photogrammetric Engineering Remote Sensing*, 70(3), 313–322, 2004.

Tomppo, E. et al., Combining national forest inventory field plots and remote sensing data for forest databases. *Remote Sensing of Environment*, 112(5), 1982–1999, 2008.

Tomppo, E. et al., *National Forest Inventories: Pathways for Common Reporting*. Springer, Dordrecht, 2010.

FIGURE 3.1
Global forest cover map derived from the GlobCover Land Cover map at 300 m resolution. Forested areas appear in green. (From Arino, O. et al., *ESA Bull.*, 136, 24, 2008; The GlobCover Land Cover map is available from the European Space Agency website at http://ionia1.estrin.esa.int/.)

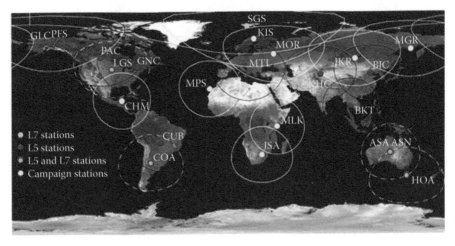

FIGURE 4.1
Active Landsat ground stations. (More details are available at http://landsat.usgs.gov/about_ground_stations.php.)

FIGURE 6.1
MODIS annual growing season image composite of shortwave, near-infrared, and red band, enhanced to appear as true color.

FIGURE 6.2
400 km × 400 km subset centered on 12° 4′ S, 55° 59′ W in Mato Grosso, Brazil. False-color composite of MODIS band 7 growing season metrics—*blue*: 2000 mean band 7 shortwave infrared reflectance from the three greenest 16-day composite periods, *green*: difference in the 2000 and 2005 mean band 7 shortwave infrared reflectance from the three greenest 16-day composite periods, and *red*: difference in the 2005 and 2010 mean band 7 shortwave infrared reflectance from the three greenest 16-day composite periods.

FIGURE 6.3
400 km × 400 km subset centered on 51° 45′ N, 72° 8′ W in Quebec, Canada. False-color composite of MODIS band 7 growing season metrics—*blue*: 2000 mean band 7 shortwave infrared reflectance from the three greenest 16-day composite periods, *green*: difference in the 2000 and 2005 mean band 7 shortwave infrared reflectance from the three greenest 16-day composite periods, and *red*: difference in the 2005 and 2010 mean band 7 shortwave infrared reflectance from the three greenest 16-day composite periods.

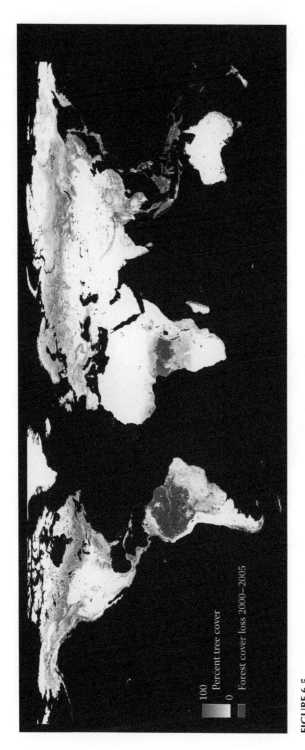

FIGURE 6.5
MODIS percent tree cover 2000 and indicated forest cover loss from 2000 to 2005.

Percent tree cover

100

0

Forest cover loss 2000−2005

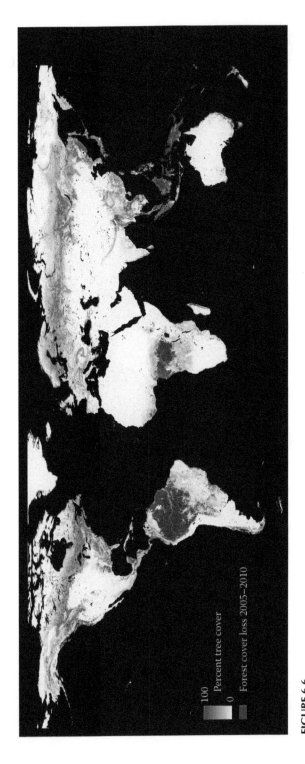

FIGURE 6.6
MODIS percent tree cover 2000 and indicated forest cover loss from 2005 to 2010.

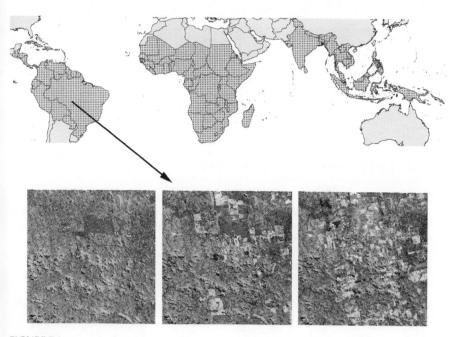

FIGURE 7.1
Example of time series (for years 1990, 2000, and 2005) of Landsat satellite imagery over one sample site in the Amazon Basin (20 km × 20 km size). Forests appear in dark green, deforested areas (agriculture and pastures) appear in light green or pink.

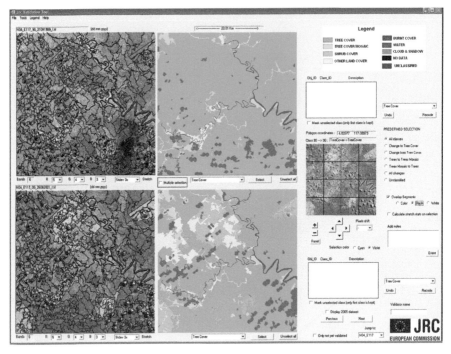

FIGURE 7.2
Visualization tool used for the process of verification and correction of multitemporal classifications. *Left column*: Segmented Landsat imagery displayed (top: year 1990, bottom: year 2000). *Right column*: Land cover maps produced from satellite imagery.

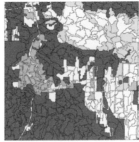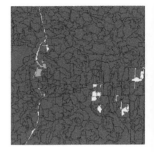

FIGURE 7.3
The 20 km × 20 km multi-spectral Landsat image (left) for a sample site in the boreal forest showing, for the central 10 km × 10 km portion (red box), the classification of land cover (center) and land use (right). Land cover is classified as TC (green), tree cover mosaic (light green), OWL (orange), and other land cover (yellow). Land use is classified as forest (green), OWL (orange), and other land use (yellow).

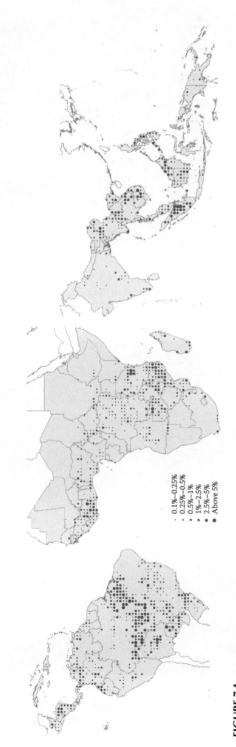

FIGURE 7.4
Annual rate of gross forest cover loss during the period 2000–2005 for the tropical sample units of the global systematic sample.

· 0.1%–0.25%
· 0.25%–0.5%
· 0.5%–1%
● 1%–2.5%
● 2.5%–5%
● Above 5%

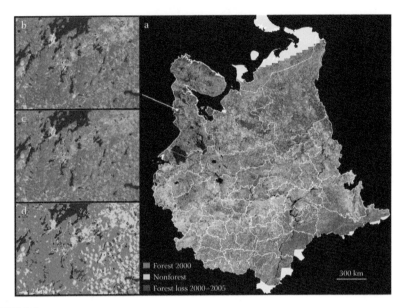

FIGURE 8.2
Forest cover loss monitoring in European Russia. (a) The ca. 2000 region-wide Landsat ETM+ image composite. (b–d) Zoom-in example of forest cover and change mapping in the Republic of Karelia: b—the ca. year 2000 image composite; c—the ca. year 2005 image composite; d—classification result.

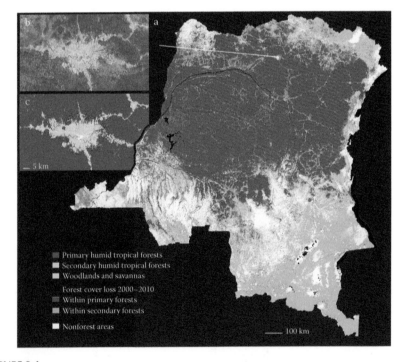

FIGURE 8.4
Forest cover loss monitoring in the DRC. (a) Nation-wide forest cover and change mapping result. (b–c) Zoom-in example of forest cover and change mapping around Buta: b—ca. year 2010 image composite; c—classification result.

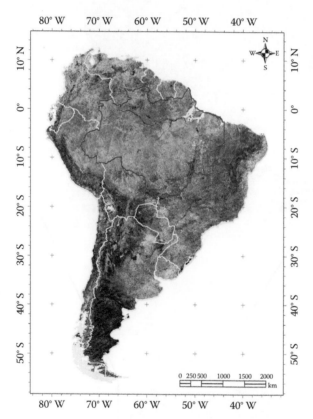

FIGURE 9.1
The BLA (*red*) located in the South American continent.

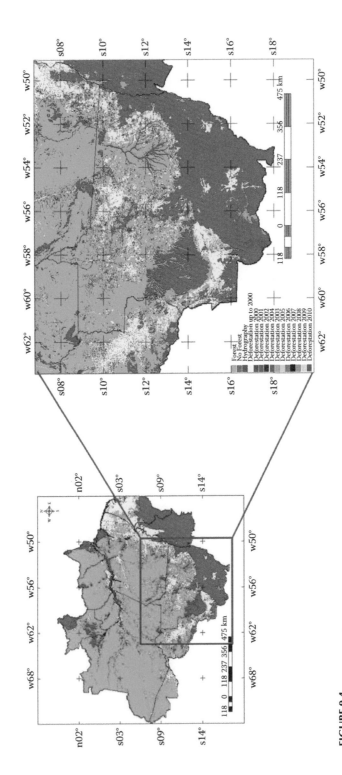

FIGURE 9.4
Mosaic of digital PRODES mapping over the period 2000–2010.

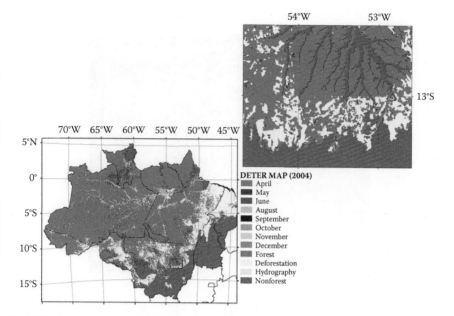

FIGURE 9.5
Illustration of the example of DETER project results, showing the deforested areas detected during the year 2004.

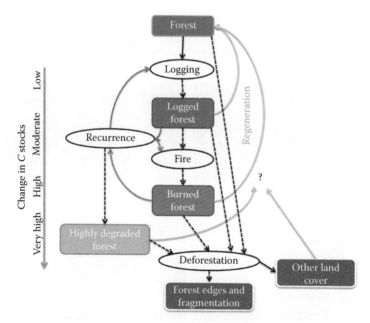

FIGURE 10.1
Forest degradation processes and interactions commonly found in the Brazilian Amazon. Pristine forests can be subject to selective logging, creating favorable conditions for burning when fires from adjacent agriculture fields unintentionally escape. Logging and fires can be recurrent, creating highly degraded forests. Eventually, degraded forests can be converted by deforestation, increasing forest edges and landscape fragmentation. If degraded forests are not cleared, vegetation regeneration processes can prevail given the high resiliency of forests.

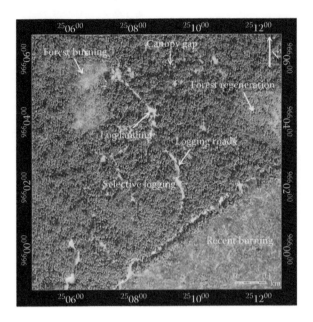

FIGURE 10.2
Very high spatial resolution false-color infrared IKONOS image showing the different environments commonly found in logged and burned (LB) forests in the eastern Brazilian Amazon. At 1 m spatial resolution, log landings, logging roads, tree fall canopy gaps, and forest edges can be identified as well as "islands" of UFs and signs of regeneration. Signs of forest erosion along the edges between the LB forest and the recently slashed-and-burned forest can also be observed. (From Souza, C.M. and Roberts, D., *Int. J. Remote Sens.*, 26, 425, 2005.)

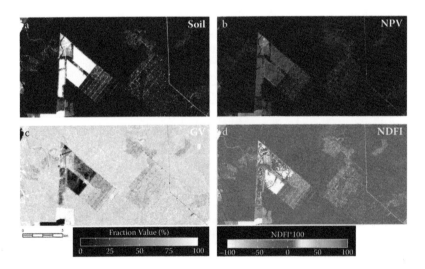

FIGURE 10.4
Subset of a Landsat TM image showing fractions obtained from SMA and NDFI. (a) High soil fraction shows logging infrastructure (log landings and roads); (b) NPV shows higher fraction values for canopy-damaged areas along infrastructure relative to the surrounding intact forest; (c) canopy damage is also identified with lower GV fraction values (dark colors); and (d) all the fraction information are combined to enhance the detection of logged forest.

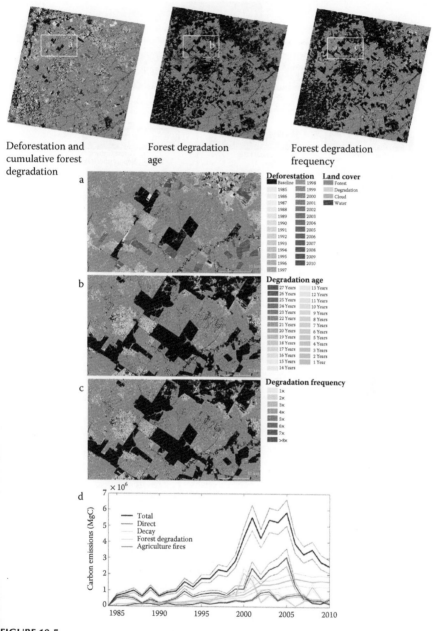

FIGURE 10.5

In this example, a long time series (i.e., >25 years) of Landsat TM/ETM+ data from Sinop, Mato Grosso state, was used to track deforestation and forest degradation. Forest degradation age and frequency maps are obtained from the annual maps and used together with the forest degradation and deforestation maps in a CES model to estimate carbon emissions associated with these processes. More reliable and consistent baseline scenarios for REDD+ can be obtained with this type of model because information about forest degradation is included and associated uncertainty estimated.

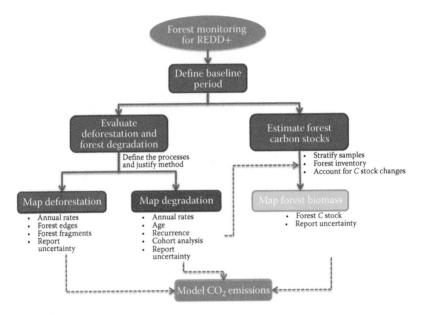

FIGURE 10.6
Integrating deforestation and forest degradation information to estimate forest carbon stock changes for REDD+ projects.

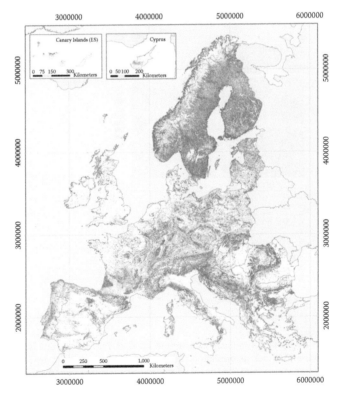

FIGURE 11.1
JRC forest map 2000.

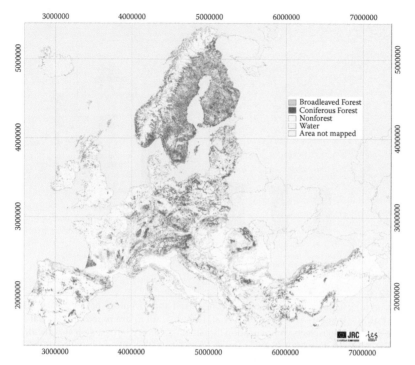

FIGURE 11.2
JRC forest map 2006.

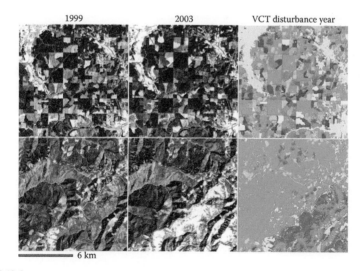

FIGURE 12.2
Examples of NAFD disturbance mapping from southern Oregon (Landsat path 46, row 30).
Top row: RGB imagery (bands 7, 5, 3) and VCT disturbance maps for an area of active harvest;
bottom row: RGB imagery and disturbance map for the northern edge of the 2002 Biscuit Fire.
The VCT maps shows permanent forest (green), permanent nonforest (gray), and the year of
mapped disturbance from 1985 to 2009 (other colors).

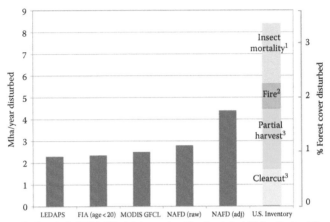

1. USDA Forest Service (2010) for USA—includes areas of mortality only, excluding defoliation without mortality.
2. U.S. EPA (2010)—includes Alaska fires.
3. Smith et al. (2009).

FIGURE 12.3
Comparison of disturbance rates among satellite-based and inventory-based studies. LEDAPS (Masek et al. 2008) and NAFD (Kennedy et al. in preparation) are based on Landsat change detection. NAFD (adj) reflects compensation for net omission errors based on visual validation. MODIS GFCL is based on MODIS gross forest cover loss (GFCL) (Hansen et al. 2010). The FIA (age < 20) is based on equating the area of young forestland from the FIA with an annualized turnover rate. The percent forest cover values are based on the area of forest land in the "lower 48" conterminous United States (~250 Mha).

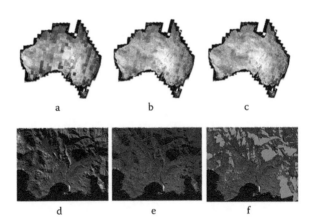

a   b   c

d   e   f

FIGURE 13.1
Image calibration (top) and normalization (bottom). Calibration: Landsat mosaic of Australia showing (a) uncalibrated, (b) TOA correction, and (c) TOA + BRDF correction. Normalization (From Wu et. al., 2004.): (d) uncorrected, (e) terrain illumination correction, and (f) estimated occlusion mask overlaid and shown in gray. (From Wu, X., et al., An approach for terrain illumination correction. Australasian Remote Sensing and Photogrammetry Conference, Fremantle, Western Australia, 2004.)

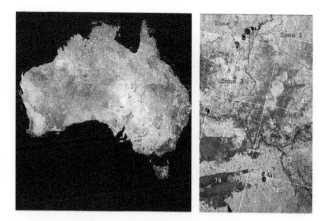

FIGURE 13.3
(Left) Graphical depiction of the location of high-resolution IKONOS data used in the derivation of classifier training information. (Right) Typically, samples are required by intersection of zone and image, though well-calibrated data can reduce this requirement by allowing extrapolation across scene boundaries in many cases.

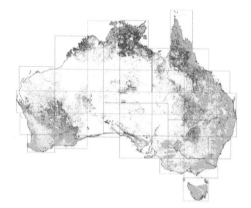

FIGURE 13.5
Map of Australia showing NCAS forest extent (green) and sparse extent (red).

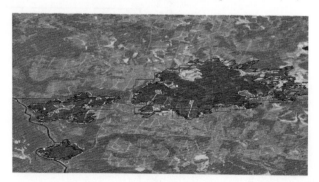

FIGURE 14.8
Example of burned area polygons derived from the three methods: red polygon, AFBA product; black polygon, SRBA product; yellow polygon, HRBA product. The results are displayed in the Web-service user interface with the Landsat-TM scene used for the HRBA product as a background image.

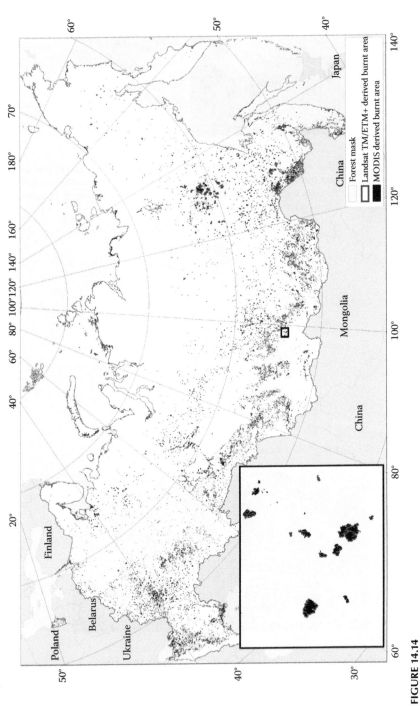

FIGURE 14.14
Burned areas for the year 2011 over Russian Federation as depicted by the SRBA product (black areas) and the HRBA product (red polygons).

PALSAR 10 m global mosaic 2009

©JAXA, METI Analyzed by JAXA

R:HH G:HV B:HH/HV

FIGURE 15.1
Global ALOS PALSAR color composite mosaic at 10 m pixel spacing (R: HH, G: HV, B: HH/HV). 95% of the data—a total of approximately 70,000 scenes—were acquired within the time period June–October 2009. (Courtesy of JAXA EORC, Tsukuba, Japan.)

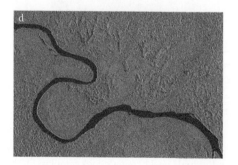

FIGURE 15.2
(d) A composite of HH data from two dates (September 12 and 15, 2011) and coherence (in RGB respectively; blue areas indicate deforested areas).

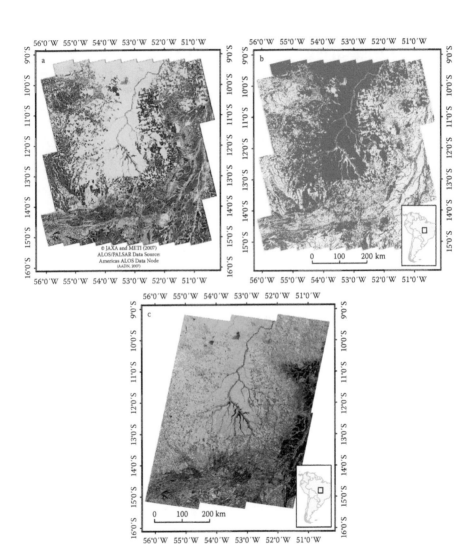

FIGURE 15.4
Satellite image mosaics produced for the Xingu River headwaters region. (a) ALOS PALSAR mosaic consisting of 116 individual Level 1.1 (single-look complex) fine beam, dual-polarimetric scenes (R/G/B = polarizations HH/HV/HH-HV difference). (b) Map of forest (green) and nonforest (beige) generated with an overall classification accuracy of 92.4% ± 1.8%. (c) Landsat 5 mosaic consisting of 12 individual Level 1G (Geocover) scenes (R/G/B = bands 5/4/3).

FIGURE 15.5
Multitemporal ALOS PALSAR L-band HV image generated from data acquired in 2007 (red), 2008 (green), and 2009 (blue) for a part of the Xingu watershed. Closed forest (white) is interspersed with fire scars (red tones) along the main stem of the Xingu River and tributaries (black).

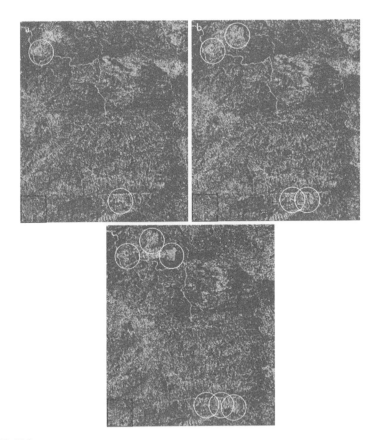

FIGURE 15.6
Forest degradation in Sarawak through selective logging observed through comparison of forest maps generated using ALOS PALSAR data for the years (a–c) 2007 through to 2009.

12

Monitoring U.S. Forest Dynamics with Landsat

Jeffrey G. Masek
NASA Goddard Space Flight Center

Sean P. Healey
U.S. Forest Service

CONTENTS

12.1 Introduction: U.S. Forest Dynamics in the Global Context

Forest dynamics in the United States differ substantially from those in the developing world and thus present unique monitoring requirements. While deforestation and conversion to semipermanent agriculture dominate tropical forest dynamics, the area of forest land in the United States has remained fairly constant for the last 50–60 years (Birdsey and Lewis 2003). Although the United States experienced rapid deforestation during the eighteenth and nineteenth centuries, much of the eastern clearing regrew during the twentieth century as marginal agricultural land was abandoned.

Recent inventory reports indicate very small rates of net forest cover change in recent decades, with the area of U.S. forests increasing by slightly more than one-tenth of 1% per year since 1987 (Smith et al. 2009).

Rather than land use conversion, forest dynamics in the United States are dominated by harvest, fire, and other temporary disturbance processes. These processes do not change the net area of forest land use, but dramatically affect the forest age structure, landscape ecology, carbon balance, and habitat suitability. It is thought that about 1.4% of forest land area is affected by harvest each year in the United States, and another 0.4% is affected by fire (Smith et al. 2009; U.S. EPA 2011). However, these disturbance rates are not static. Changes in forest management as well as recent climate change may be affecting contemporary disturbance rates relative to historic norms (e.g., van Mantgem et al. 2009).

The United States relies on its national forest inventory for domestic and international reporting of forest change. The U.S. Forest Inventory and Analysis (FIA) program collects data on a set of over 300,000 plots across the United States, with one plot per every ~2,430 ha. A range of attributes are collected in addition to stand volume, including stand age, species composition, and management practice. The key aspect of this design-based inventory is that the sampling error associated with any variable is well constrained, and thus robust estimates across broad areas can be made with known sampling uncertainty. Plots are remeasured on a 5- to 10-year cycle, depending on the state. Like other nations, the United States reports national forest carbon dynamics as part of the United Nations Framework Convention on Climate Change (UNFCCC). In this case, inventory data from the FIA and other agencies are collated and reported by the U.S. Environmental Protection Agency (EPA).

While the FIA is well suited for estimating national forest statistics, it is not designed to accurately capture local dynamics due to disturbance and other rare events. For example, while a difference between a 1% per year and 2% per year disturbance rate is truly significant from an ecological point of view, a very large number of random samples is needed to distinguish those two rates with any level of precision. Given the FIA plot spacing, this implies that disturbance rates cannot be accurately characterized below the scale of 100s of kilometers.

The desire for consistent, geospatial information on forest disturbance and conversion has invigorated the application of Landsat-type remote sensing technology for forest monitoring in the United States. This work builds on a significant legacy that dates back to the launch of Landsat-1 in 1972 (Cohen and Goward 2004). Early efforts at basic land cover mapping identified forests as a unique spectral region (the so-called badge of trees in red-near-IR space) that enabled reliable single-image mapping of forest cover. Studies during the 1980s and 1990s established the opportunity to use multidate Landsat imagery to characterize forest conversion, harvest, burned area, and insect damage. Recent increases in computing power, coupled with the gradual opening of the Landsat archive for free distribution, have resulted in researchers undertaking increasingly ambitious programs in large-area forest dynamics monitoring. Here we describe several of these efforts, focusing on national-scale work in the United States.

12.2 Overall Forest Disturbance: LEDAPS and NAFD Projects

The North American Carbon Program (NACP) is an ongoing interagency effort within the United States to constrain the North American carbon budget, improve process understanding, and forecast future scenarios. The NACP Science Strategy recognized at the outset that ecosystem disturbance was a critical but poorly known parameter required for more accurate assessments of ecosystem carbon flux. Accordingly, two Landsat-based projects were organized during 2004–2005 in order to meet NACP modeling needs (Goward et al. 2008).

The LEDAPS (Landsat Ecosystem Disturbance Adaptive Processing System) project was based on traditional two-date change detection, but across very broad spatial scales (Masek et al. 2008). The main objective was to map stand-clearing disturbance (primarily fire and clearcut harvest) across all forested land in the conterminous United States and Canada. At the start of the project, the Landsat archive was not yet free. Instead, the project chose to use the Global Land Survey (GLS) preprocessed Landsat data sets (Tucker et al. 2004). The GLS data sets consist of cloud-free imagery for epochs centered on 1975, 1990, 2000, 2005, and 2010. For the LEDAPS project, the focus was on estimating forest disturbance between 1990 and 2000.

The LEDAPS processing approach focused on establishing accurate surface reflectance values from each image, and then using those data to perform two-date change detection using a tasseled cap disturbance index. Considerable work went into establishing a sensor calibration and atmospheric correction approach suitable for use with the GLS data sets, including revising the Landsat-5 calibration look-up table based on invariant desert targets and adjusting the calibration of the older GLS data sets to reflect the new table. The development of a stand-alone atmospheric correction code for Landsat was a significant side benefit of the project.

Beyond sensor calibration, a number of other challenges were encountered during the disturbance mapping. First, the 10-year (1990–2000) change-detection span caused stands disturbed during the early part of the epoch to exhibit significant regrowth, resulting in high omission errors of 40%–50%. This issue has previously been documented (Jin and Sader 2005) and suggests that change detection on closer to annual timesteps is more appropriate for most forest monitoring applications. Statistical summaries reported in Masek et al. (2008) compensated for this issue by adjusting rates by the difference between omission and commission errors. Second, many of the GLS images from the 1990 and 2000 data sets were acquired during senescent parts of the growing season, confusing the change-detection approach. These images were replaced with new imagery purchased from U.S. Geological Survey (USGS).

The results of the continental mapping indicated that 2.3 Mha/year of U.S. forest land was affected by stand-clearing disturbance during the 1990s, representing a fractional disturbance rate of 0.9% per year, or an equivalent "turnover" period of 110 years. The highest disturbance rates were found in areas with significant harvest activity, including the southeastern United States, Maine/Quebec, and the Pacific Northwest. Rates in the mid-Atlantic and New England were lower, reflecting both less overall harvest activity and greater prevalence of partial harvest, which could not be reliably detected using the LEDAPS measurement period.

While LEDAPS focused on wall-to-wall assessment of disturbance at a coarse temporal timestep, the North American Forest Dynamics (NAFD) project took an alternate path: characterizing disturbance using a sparse geographic sample of Landsat imagery at annual temporal resolution (Goward et al. 2008). The NAFD originally began with a sample of 23 Landsat frames across the United States and later expanded to a set of 50 frames. For each frame, a set of biennial (later annual) Landsat imagery was assembled, and time-series analysis was used to map forest disturbance.

The NAFD geographic sample was designed to support robust characterization of national disturbance rates (eastern and western United States as separate estimates) based on an unequal probability sampling design. This sampling design was based on selecting across strata for U.S. forest types (Ruefenacht et al. 2008) while also accommodating the inclusion of fixed sites from earlier phases of the work. The decision to increase the number of samples from 23 to 50 reflected the desire to reduce the national sampling error to less than 10% (Figure 12.1).

Aligned with several other recent studies (Kennedy et al. 2007), the NAFD disturbance-mapping effort relied on detecting anomalies in per-pixel spectral time series. The specific algorithm, the vegetation change tracker (VCT; Huang et al. 2010), used a Z-score procedure to normalize each image in the time series by dividing by the standard deviation of reflectance values for a set of undisturbed forest pixels. Anomalies were then mapped based on significant, long-lasting excursions from the time series (Huang et al. 2010). Both the year of disturbance and the spectral magnitude were included in the final products (Figure 12.2). It should be noted that the annual timestep used in the algorithm allows partial disturbances (such as thinning, partial harvest, and mortality from storms and insects) to be tracked.

Overall, the sampling results indicate about 1.1% of forest area disturbed each year in the United States during the 1985–2005 period. Although

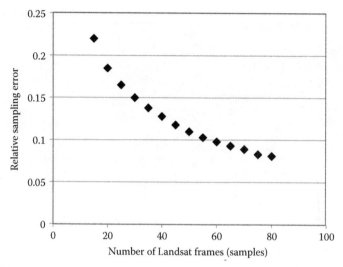

FIGURE 12.1
Relative error of U.S. national disturbance rate estimates as a function of the number of Landsat frames (path/row locations) used in the geographic sample, estimated from initial drafts of sample-level disturbance rate for all years from 1985 to 2005. Relative error is the proportional difference between the estimated value from the sample and the unknown true value, subject to a 90% confidence interval. (Analysis courtesy of Robert Kennedy, Oregon State University, Corvallis, OR.)

disturbance rates in the western United States are dominated by fire and insect damage, while rates in the east are dominated by harvest, overall disturbance rates were not significantly different between the west and east. However, there were significant year-to-year differences. For example, disturbance rates in the western United States increased to 1.5% per year during the early 2000s as a result of extremely active fire years. There were also significant geographic differences in disturbance rate within individual forest type strata.

The fact that disturbance rates vary significantly in both space and time raises doubts that sampling approaches can adequately character-ize the disturbance regime at continental scales. The assumption behind the NAFD sampling approach was that disturbance rate was fundamen-tally a function of forest type (or at least that forest type could act as a proxy for the controlling factors). This assumption has not been borne out by the scene-by-scene results. As a result, the latest phase of the NAFD project has abandoned the geographic sampling scheme and switched to an ambitious "wall-to-wall" characterization of annual disturbance rate for the entire conterminous United States. This effort will require pro-cessing in excess of 20,000 Landsat images and is taking advantage of

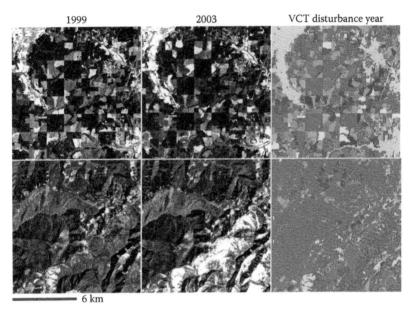

FIGURE 12.2
(See color insert.) Examples of NAFD disturbance mapping from southern Oregon (Landsat path 46, row 30). Top row: RGB imagery (bands 7, 5, 3) and VCT disturbance maps for an area of active harvest; bottom row: RGB imagery and disturbance map for the northern edge of the 2002 Biscuit Fire. The VCT maps shows permanent forest (green), permanent nonforest (gray), and the year of mapped disturbance from 1985 to 2009 (other colors).

the NASA Earth Exchange (NEX) parallel computing environment at the NASA Ames Research Center.

12.3 Operational Fire Monitoring: MTBS and LANDFIRE

Although wildfire is a primary disturbance agent within the United States, the area affected by forest fire has not been well characterized. The National Interagency Fire Center (NIFC) maintains a database of major wildfires, but does not consistently discriminate between forest fires and other wildfires (e.g., brushfire or grassfire). Furthermore, the area recorded is based on an external perimeter of each large fire, rather than the actual area affected by burning. Two operational projects, Monitoring Trends in Burn Severity (MTBS) and LANDFIRE, are using Landsat remote sensing to improve burned area and fire risk monitoring.

A collaboration between the USGS and the U.S. Forest Service (USFS), the MTBS project is seeking to supplement the NIFC database with accurate information on U.S. fire area and burn severity (Eidenshink et al. 2007). The primary goal of MTBS is to provide sufficient information to quantify interannual variability in U.S. burned area and to understand the extent to which forest management and environmental factors may be influencing longer term trends in fire.

MTBS has acquired preburn and postburn (1 year after fire) Landsat imagery for all major wildfires within the United States since 1984, as identified from government databases (Eidenshink et al. 2007). Fires larger than 1,000 acres in the western United States and larger than 500 acres in the eastern United States are considered in the project. The normalized burn ratio (NBR) spectral index is calculated for the image pair bracketing a major fire, and a difference (dNBR) image is generated by subtracting the pre- and postfire NBR values. The NBR metric takes advantage of the fact that recent fires leave considerable char, ash, and mineral soil, which tend to be relatively bright in the shortwave infrared compared to the near-infrared. While the NBR metric has been questioned as a suitable proxy for overall fire severity in Boreal ecosystems (Hoy et al. 2008), it has also been shown to be highly correlated with canopy damage (Hoy et al. 2008) and overall fire impact in temperate ecosystems (Cocke et al. 2005). MTBS data are available online (http://www.mtbs.gov) in a variety of formats, including geospatial products and statistical summaries of annual burned area by region and ecosystem.

LANDFIRE is a multipartner project producing 30 m Landsat-based maps of vegetation, fuel, fire regimes, and ecological departure from historical conditions across the United States (Rollins 2009). Leadership is shared by the wildland fire management programs of the USDA Forest Service and the U.S. Department of the Interior. LANDFIRE's maps are widely used for both fire management and ecological modeling. Circa-2000 imagery was used to produce LANDFIRE's original maps, and a combination of approaches is used to track subsequent disturbances so that maps may be kept up to date (Vogelmann et al. 2011).

The initial updating mechanism involved intersecting LANDFIRE maps with the fire events mapped by the MTBS project (described above). This approach has recently been augmented with management activities (conducted mostly on federal lands), which have been recorded in a spatial database. Because a more automated process was needed for incorporating the effects of disturbance events, LANDFIRE has recently done extensive work with the VCT algorithm described earlier under the activities of the NAFD project. An estimated 30,000 Landsat images will ultimately be used to map disturbance extent and magnitude across the conterminous United States (Vogelmann et al. 2011).

Because the cause, or type, of a disturbance plays an important role in its effect upon fuel conditions, cause attribution is underway for LANDFIRE's VCT maps. This process makes use of the MTBS data to some extent, but it also currently involves a good deal of manual classification. Current

LANDFIRE disturbance-mapping efforts focus on the Landsat TM era, but extension into the MSS era is a longer term goal, as is continued mapping into the future.

12.4   Forest Cover Conversion: Trends, NLCD, and C-CAP

While the emphasis of the preceding projects has been on characterizing forest disturbance rates, permanent conversion of forest cover remains an ongoing process within the United States. The trend during most of the twentieth century was toward increased forest cover via agricultural abandonment. However, increased urban and suburban growth during the last 50 years has altered this pattern in some areas. The projects discussed here have viewed forest change from the perspective of land cover conversion and have used Landsat-based change detection to separate gross forest change (including harvest and other disturbances) from the lower rates of long-term land cover and land use conversion.

The USGS Trends project began in the late-1990s using a random sample of Landsat subsets, stratified by EPA ecoregion, to characterize both regional and national trends in land cover (Loveland et al. 2002). Each Landsat subset was either 10,000 or 40,000 ha (e.g., 10 km × 10 km or 20 km × 20 km). For each subset, images were collected for the years 1973, 1980, 1986, 1992, and 2000 and manually classified into a series of land use classes, as well as two classes representing recent mechanical disturbance (harvest) and fire. As in the NAFD project, the sampling framework allowed sampling uncertainty to be quantified. The overall goals were to provide estimates of gross change with an uncertainty of <1% at an 85% confidence interval (Drummond and Loveland 2010).

The Trends data set has been used widely for studies of land use conversion (Drummond and Loveland 2010), ecosystem carbon (Liu et al. 2006), biodiversity, and surface energy balance (Barnes and Roy 2008). For the eastern United States, Drummond and Loveland (2010) assessed both gross and net forest cover change using the Trends data and concluded that eastern forests experienced 142,000 ha/year of net forest conversion during the 1973–2000 period due mostly to urbanization, surface mining, and reservoir construction. Gross forest change rates were 2.5 times higher and mostly reflected harvest activities.

The National Land Cover Database (NLCD) project, coordinated by the USGS EROS Data Center, has produced wall-to-wall U.S. maps of land cover for 1992, 2001, and 2006. While NLCD image selection criteria, classification methods, and target classes have evolved over the course of the project, significant efforts have been made to ensure interpretable maps of change among different land covers (Xian et al. 2009; Fry et al. 2011). Changes between the 1991 and 2001 products were identified through a "retrofitting" process, which involved standardization of classification schemes, and a sequence of decision tree–based operations that first identified and then labeled land cover transitions. Differences in Landsat ratio-based indices were primary predictors for this process.

Likewise, (different) Landsat-based indices were used in the 2001–2006 change identification process, which used complex heuristic-based threshold rules to identify changed pixels and to indicate whether changed pixels were losing or gaining biomass. This change-detection process, multi-index integrated change analysis (MIICA; Fry et al. 2011), produced change maps that were intersected with the land cover product from 2001 (considered to be the base year) in nondeveloped areas to generate the 2006 cover product.

Mapping projects such as the NLCD's are an important complement to inventory estimates of forest change. In the United States, the FIA provides ground-acquired estimates of land use not available from automated satellite processes, and it provides a design-based error structure for its estimates of net change of forest area. However, the FIA does not measure gross transitions to and from other cover types. NLCD can specify that its estimate of a net loss of 16,720 km^2 of evergreen forest cover, for example, is the result of a 36,000 km^2 gross loss and a 19,000 km^2 gross gain (Fry et al. 2011). As discussed earlier, maps also provide a picture of change at much more localized scales than is achievable with a simple random sample.

The NOAA's Coastal Change Analysis Program (C-CAP) maintains a nationally standardized database of landcover and landcover change in coastal regions of the country (Dobson et al. 1995). Thematic classes, including those for forests, are consistent with those used in the NLCD (described earlier), and C-CAP is actually the source of the NLCD data in coastal zones. Landsat has been the basis for classification and change detection for C-CAP national maps using imagery from 1996, 2001, 2006, and 2011 (in progress), as well as high-priority local analyses going back to the mid-1980s.

The 2001 cover map, produced from three dates of imagery collected by the MRLC (multiresolution land characteristics consortium), is considered the baseline product, and only those areas determined to have changed are reclassified in subsequent products (J. McCombs, NOAA, personal communication). For changes between 2001 and 2006, CCA (crosscorrelation analysis) was used to detect change using imagery from the two dates. Landcover transitions were estimated with classification and regression trees (CARTs). Change detection between 2006 and 2011 will be consistent with the multithreshold change vector analysis used by NLCD (Xian et al. 2009). Spatial data and customized summaries of CCAP maps are freely available from the CCAP Web site.*

---

* http://www.csc.noaa.gov/digitalcoast/data/ccapregional/index.html

12.5 Synthesis of U.S. Forest Dynamics

Given that several studies have used satellite data to quantify forest disturbance rates in the United States, how do these estimates compare with each other, and with inventory-based rates? Satellite-based estimates of disturbance rates are available in Masek et al. (2008), and via the MTBS data products (available online). In addition, Hansen et al. (2010) presented MODIS-based measures of gross forest loss that includes both disturbance and deforestation. Figure 12.3 shows a comparison among these estimates. We also show inventory-based estimates of harvest, fire, and insect damage from Smith et al. (2009) and U.S. EPA (2011) annualized for the 2000–2008 period. Finally, we also derived an annualized rate of "stand-clearing" disturbance from the FIA by taking the area of U.S. forests less than 20 years of age, and dividing by 20, under the assumption that a stand-clearing event should reset the measured stand age on the FIA plot.

The range of estimates shows an expected trend, with shorter remeasurement periods (e.g., the annual NAFD) and finer resolution (e.g., Landsat vs. MODIS),

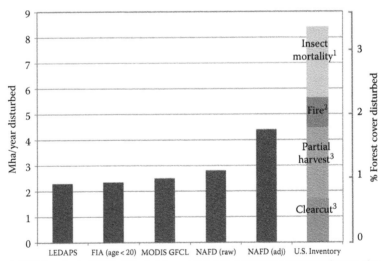

1. USDA Forest Service (2010) for USA—includes areas of mortality only, excluding defoliation without mortality.
2. U.S. EPA (2010)—includes Alaska fires.
3. Smith et al. (2009).

FIGURE 12.3
(See color insert.) Comparison of disturbance rates among satellite-based and inventory-based studies. LEDAPS (Masek et al. 2008) and NAFD (Kennedy et al. in preparation) are based on Landsat change detection. NAFD (adj) reflects compensation for net omission errors based on visual validation. MODIS GFCL is based on MODIS gross forest cover loss (GFCL) (Hansen et al. 2010). The FIA (age < 20) is based on equating the area of young forestland from the FIA with an annualized turnover rate. The percent forest cover values are based on the area of forest land in the "lower 48" conterminous United States (~250 Mha).

leading to more disturbances mapped and higher overall rates (Figure 12.3). The LEDAPS result of 0.9% per year corresponds closely to the FIA-derived stand-clearing rate (0.9% per year) and the gross loss derived from MODIS (1.0% per year). The ability of the NAFD annual data to capture some thinning and partial harvest likely explains the somewhat higher rates (1.1% per year). Further adjusting the NAFD rates for the net omission error of the products would increase these rates further, to about 1.5% per year.

All of these rates are significantly lower than what can be derived from inventory estimates alone (Smith et al. 2009; U.S. EPA 2010). This may reflect the difficulty in measuring minor disturbances using remote sensing, including selective harvest that does not significantly alter the forest canopy. Field- and satellite-based disturbance estimates may also differ in what they label "disturbance." The FIA's definition of disturbance includes "mortality and/or damage to 25 percent of all trees in a stand or 50 percent of an individual species' count" (FIA 2011). In addition to the canopy mortality targeted through remote sensing, this characterization certainly includes large areas affected by insects or storms, where sublethal damage may affect only a small fraction of the trees. Thus, discrepancies in Figure 12.3 may be due to both varying sensitivity and inconsistent definitions among data sources.

12.6 Looking Forward

The opening of the Landsat archive and advances in computing technology have paved the way for broader and more innovative applications of Landsat data for forest monitoring. These innovations include mapping at wider geographic scales (e.g., wall-to-wall national monitoring), the use of dense time series to better characterize intra- and interannual variability, and a greater sophistication in leveraging multiple data sources to attribute the origin of forest change as well as ecosystem consequences. Given that the Landsat archive is most complete within the United States, it is natural that many of these techniques are being pioneered for U.S. applications. However, given the increased global data collections implemented for Landsat-7 and the upcoming LDCM (Landsat Data Continuity Mission), these approaches could be applied to global monitoring as well.

12.6.1 Operational Monitoring of Forest Dynamics: LCMS

The Landscape Change Monitoring System (LCMS) is under development by a consortium of scientists, agencies, and projects engaged in remotely sensed change detection in the United States. Coordinated by the Forest Service and Department of Interior/USGS, LCMS is intended to be a hub around which existing national change products such as those from MTBS and NLCD may be integrated and extended. The ideal change monitoring system would provide up-to-date and consistent information about the location, magnitude, and nature of vegetation changes on all land cover types across the country. The research- and agency-based monitoring efforts described in this chapter address different aspects of this ideal system, and several steps are being taken under LCMS to promote both integration of existing products and cooperation on the development of new products.

First, LCMS is conducting an independent needs assessment of the land management community. Acquired information about the type, precision, and frequency of needed land change information will guide development of new monitoring strategies. Many of the needed answers are likely to come from the combination of current resources. For instance, LANDFIRE is anticipating a more meaningful update of their fuel maps as fire records from the MTBS are augmented with more general all-disturbance mapping achieved through the VCT (Vogelmann et al. 2011). Similar benefits of product integration likely extend into processes such as carbon accounting, where disturbance emissions are strongly influenced by event type and magnitude.

Any new products developed by the LCMS to meet identified needs will likely depend heavily upon the Landsat archive and will draw upon the experience of participating partners. Like the MTBS project, the LCMS will follow a collaborative multiagency business model, with an emphasis upon meeting operational monitoring needs by producing consistently updated and validated products. The LCMS is expected to be deployed during 2013.

12.6.2 Hypertemporal and Near-Real-Time Change Detection

The use of dense image time series has pushed the "epoch length" (i.e., the time between images used for monitoring change) to shorter and shorter periods. Not surprisingly, the range of forest dynamics that can be assessed has expanded as well. While semidecadal time series are useful for monitoring net land cover change and stand-clearing disturbance (Jin and Sader 2002; Drummond and Loveland 2010; Masek et al. 2008), more subtle disturbances require annual image acquisition. Thus the algorithms proposed by Kennedy et al. (2007) and Huang et al. (2010) are capable of detecting significant thinning, partial harvest, and selective mortality from insects and disease. However, even these algorithms may not record subtle and short-lived degradation of the forest canopy due to insect defoliation, storm damage, and selective cutting.

A variety of approaches are being prototyped to obtain seasonal or even submonthly information from Landsat. The WELD (Web-Enabled Landsat Dataset) project at the University of South Dakota is using MODIS-style compositing to generate monthly and seasonal gridded composites of Landsat-7 data. Zhu and Woodcock (in press) have proposed an approach to mapping forest disturbance by fitting per-pixel phenological curves using every available Landsat observation. These methods not only obtain greater sensitivity to short-term canopy changes, but by explicitly considering changes relative to observed phenology they minimize errors due to mismatches between annual image dates.

12.6.3 Integration of Landsat with Active RS for Biomass Change

Landsat data have proven robust for estimating the area affected by forest change processes. However, Landsat data have not proven as useful for forest volume or biomass estimation, except in conditions of sparse tree cover where volume can be directly related to canopy cover (Powell et al. 2010). There are, however, a variety of ways in which change area from Landsat can be combined with active remote sensing in order to better quantify "three-dimensional" changes in forest structure and biomass. Direct fusion between Light Detection and Ranging (LIDAR) and Landsat has been proposed to improve retrieval of biomass. Landsat imagery has also been proposed as a way to spatially interpolate LIDAR "samples" across the landscape using krigging techniques as well as a way to group LIDAR measurements based upon patterns of forest structure and disturbance. More recently, disturbance and age information derived from Landsat have been combined with LIDAR data to estimate postdisturbance carbon accumulation rates and to improve spatial interpolation of height (Li et al. 2011). In principle, similar work could be carried out using one-time biomass retrievals from radar (including interferometric SAR) combined with historical disturbance data from Landsat.

12.6.4 Ecological Impacts of Climate Change and Recovery Trajectories

The projects discussed here have focused mostly on quantifying the fraction of U.S. forest land disturbed and the fraction that reverts back to forest after disturbance. The spectral information of Landsat time-series data also offers important information on the *rate* at which ecosystems recover from disturbance. In one example, Schroeder et al. (2007) related postharvest Landsat spectral trajectories in the Pacific Northwest to increases in canopy cover deduced from air photos. The current phase of the NAFD project is extending this work by assessing rates of forest recovery for all recently disturbed patches in the United States.

One application for such approaches is to understand how ecosystem recovery may be responding to climate warming. A number of studies have suggested increased rates of forest decline in the southwestern United States due to prolonged drought (Williams et al. 2010), and van Mantgem et al. (2009) found evidence for increased rates of tree mortality throughout the western United States. Disturbance events (fire, insect outbreaks, and disease) may be accelerated in climate-stressed forests, and successional pathways may be altered or slowed. Ultimately the 40+ year Landsat record will prove valuable for understanding the long-term shifts in forest composition and mortality associated with climate warming in the United States.

12.7 Conclusions

The application of Landsat remote sensing to the monitoring of U.S. forests has accelerated during the last decade. This trend reflects both the development of new algorithmic and computational approaches for dealing with large volumes of data and the opening of the Landsat archive for free distribution by the USGS. The projects discussed here represent large-scale mapping efforts that have sought to characterize U.S. forest dynamics during the Landsat era, including disturbance, recovery, and conversion.

Although the U.S. forest inventory will continue to provide our most robust national estimates of forest attributes, remote sensing is increasingly being called on to perform operational monitoring of forest and land cover change. The appropriate integration of geospatial information from remote sensing with forest attribute available from the FIA remains one of the significant challenges for the future. The *k*-nearest neighbor approach of assigning suites of FIA attribute data based on spectral properties has found acceptance within the USFS as it allows the statistical variance of the FIA-reported attributes to be "imported" to the geospatial products (McRoberts et al. 2002). Alternative approaches have sought instead to use statistical models to predict attributes from Landsat spectral data using the FIA attributes as training data (e.g., Powell et al. 2010) or to use Landsat-derived harvest maps to spatially distribute the FIA-recorded harvest volumes (Healey et al. 2009). Ultimately the extent to which remote sensing can support operational needs depends on the trade-off between measurement error and sampling error. Landsat remote sensing can record wall-to-wall dynamics, and thus has no sampling error, but may exhibit significant errors of omission and commission (measurement errors) depending on the attribute of interest.

The launch of LDCM in early 2013 will continue the Landsat legacy while providing a greater density of global acquisitions compared to Landsat-7. In addition, the ESA Sentinel-2 satellites will be launched during 2013–2014. Like Landsat, the Sentinel program has committed to open access for its archive. Taken together, the LDCM and Sentinel missions will provide an extremely rich source of global observations for the next decade. It is anticipated that many of the advances in the use of Landsat data described here, including time-series methods, will soon find global use.

About the Contributors

Jeffrey G. Masek is a research scientist in the Biospheric Sciences Laboratory at the NASA Goddard Space Flight Center. His research interests include mapping landcover change in temperate environments, application of advanced computing to remote sensing, and satellite remote sensing techniques. Dr. Masek currently serves as the Landsat project scientist at NASA. He received a BA in geology from Haverford College (1989) and a PhD in geological sciences from Cornell University (1994).

Sean P. Healey is a research ecologist who works for the FIA program at the U.S. Forest Service's Rocky Mountain Research Station. His interests include linking Landsat data to inventory records for the purpose of tracking forest carbon storage over time.

References

Barnes, C.A. and Roy, D.P., Radiative forcing over the conterminous United States due to contemporary land cover land use albedo change. *Geophysical Research Letters*, 35, L09706, 2008.

Birdsey, R.A. and Lewis, G.M., Current and historical trends in use, management, and disturbance of U.S. forestlands. in *The Potential of U.S. Forest Soils to Sequester Carbon and Mitigate the Greenhouse Effect*, edited by J.M. Kimble et al., pp. 15–33, CRC Press, New York, 2003.

Cocke, A.E. et al., Comparison of burn severity assessments using differenced normalized burn ratio and ground data. *International Journal of Wildland Fire*, 14, 189–198, 2005.

Cohen, W.B. and Goward, S.N., Landsat's role in ecological applications of remote sensing. *BioScience*, 54(6), 535–545, 2004.

Dobson, J.E. et al., *NOAA Coastal Change Analysis Program: Guidance for Regional Implementation*, Version 1.0.NOAA Technical Report NMFS 123, Scientific Publications Office, National Marine Fisheries Service, NOAA, Seattle, WA, 92 p., 1995.

Drummond, M.A. and Loveland, T.R., Land-use pressure and a transition to forest-cover loss in the eastern United States. *BioScience*, 60(4), 286–298, 2010.

Eidenshink, J. et al., A project for monitoring trends in burn severity. *Fire Ecology*, 3, 3–21, 2007.

FIA (Forest Inventory and Analysis Program of the U.S. Forest Service). *Interior West Forest Inventory & Analysis P2 Field Procedures, V5.00*. USDA Forest Service Rocky Mountain Research Station. Ft. Collins, CO., 409 pp., 2011.

Fry, J. et al., Completion of the National Land Cover Database for the conterminous United States. *Photogrammetric Engineering and Remote Sensing*, 77, 858–864, 2011.

Goward, S.N. et al., Forest disturbance and North American carbon flux. *American Geophysical Union EOS Transactions*, 89, 105–116, 2008.

Hansen, M.C. et al., Quantification of global gross forest cover loss. *Proceedings of the National Academy of Sciences*, 107, 8650–8655, 2010.

Healey, S.P. et al., Changes in timber haul emissions in the context of shifting forest management and infrastructure. *Carbon Balance and Management*, 4, 9, 2009.

Hoy, E.E. et al., Evaluating the potential of Landsat TM/ETM+ imagery for assessing fire severity in Alaskan black spruce forests. *International Journal of Wildland Fire*, 17, 500–514, 2008.

Huang, C.Q. et al., An automated approach for reconstructing recent forest disturbance history using dense Landsat time series stacks. *Remote Sensing of Environment*, 114, 183–198, 2010.

Jin, S. and Sader, S.A., Comparison of time-series tasseled cap wetness and the normalized difference moisture index in detecting forest disturbances. *Remote Sensing of Environment*, 94, 364–372, 2005.

Kennedy, R.E. et al., Trajectory-based change detection for automated characterization of forest disturbance dynamics. *Remote Sensing of Environment*, 110, 370–386, 2007.

Li, A. et al., Modeling the height of young forests regenerating from recent disturbances in Mississippi using Landsat and Icesat data. *Remote Sensing of Environment*, 115, 1837–1849, 2011.

Liu, J. et al., Temporal evolution of carbon budgets of the Appalachian forests in the U.S. from 1972 to 2000. *Forest Ecology and Management*, 222, 191–201, 2006.

Loveland, T.R. et al., A strategy for estimating the rates of recent United States Land Cover changes. *Photogrammetric Engineering and Remote Sensing*, 68, 1091–1099, 2002.

Masek, J.G. et al., North American forest disturbance mapped from a decadal Landsat record: Methodology and initial results. *Remote Sensing of Environment*, 112, 2914–2926, 2008.

McRoberts, R.E. et al., Stratified estimation of forest area using satellite imagery, inventory data, and the k-nearest neighbors technique. *Remote Sensing of Environment*, 82, 457–468, 2002.

Powell, S.L. et al., Quantification of live above-ground forest biomass dynamics with Landsat time-series and field inventory data: a comparison of empirical modeling approaches. *Remote Sensing of Environment*, 114, 1053–1068, 2010.

Rollins, MG., LANDFIRE: A nationally consistent vegetation, wildland fire, and fuel assessment. *International Journal of Wildland Fire*, 18, 235–249, 2009.

Ruefenacht, B. et al., Conterminous U.S. and Alaska forest type mapping using forest inventory and analysis data. *Photogrammetric Engineering and Remote Sensing*, 74, 1379–1388, 2008.

Schroeder, T.A. et al., Patterns of forest regrowth following clearcutting in western Oregon as determined from a Landsat time-series. *Forest Ecology and Management*, 243, 259–273, 2007.

Smith, W.B. et al., *Forest Resources of the United States, 2007*, General Technical Report WO-78, Forest Service, U.S. Department of Agriculture, Washington, DC, 336 pp., 2009.

Tucker, C.J. et al., NASA's global orthorectified Landsat data set. *Photogrammetric Engineering and Remote Sensing*, 70, 313–322, 2004.

U.S. Environmental Protection Agency. *Inventory of U.S. Greenhouse Gas Emissions and Sinks: 1990–2009* (EPA 430-R-11-005). US EPA, Washington, DC, 2011.

van Mantgem, P.J. et al., Widespread increase of tree mortality rates in the western United States. *Science*, 323, 521–524, 2009.

Vogelmann, J.E. et al., Monitoring landscape change for LANDFIRE using multi-temporal satellite imagery and ancillary data. *IEEE Journal of Selected Topics in Applied Earth Observations and Remote Sensing*, 4, 252–264, 2011.

Williams, A.P. et al., Forest responses to increasing aridity and warmth in the southwestern United States. *Proceedings of the National Academy of Science*, 107, 21289–21294, 2010.

Xian, G. et al., Updating the 2001 National Land Cover Database classification to 2006 by using Landsat imagery change detection methods. *Remote Sensing of Environment*, 113, 1133–1147, 2009.

Zhu, Z. and Woodcock, C.E., Continuous monitoring of forest disturbance using all available Landsat imagery. *Remote Sensing of Environment* (in press).

13

Long-Term Monitoring of Australian Land Cover Change Using Landsat Data: Development, Implementation, and Operation

Peter Caccetta, Suzanne Furby, Jeremy Wallace, and Xiaoliang Wu
Commonwealth Scientific and Industrial Research Organisation

Gary Richards
Department of Climate Change and Energy Efficiency
and
Australian National University

Robert Waterworth
Department of Climate Change and Energy Efficiency

CONTENTS

13.1 Background

The need to monitor greenhouse gas (GHG) emissions accurately has become a task of major importance over the last decade. Emissions and removals of GHG in the land sector represent a large proportion of Australia's total GHG emissions. Following the signing of the Kyoto Protocol in 1997, Australia began developing a new system to account for emissions and removals from the land sector. The result, the National Carbon Accounting System (NCAS), is a fully integrated modeling system that utilizes data from a variety of sources to estimate emissions and removals for the purpose of reporting to the United Nations Framework Convention on Climate Change (UNFCCC 2001) and accounting under the Kyoto Protocol.

Under the Kyoto protocol, Australia was required to estimate emissions from land use and land use change in 1990 and from 2008 to 2012 (the first Kyoto Commitment period) while ensuring time-series consistency, limiting potential errors of omission and commission, allowing for annual updating at fine (subhectare) spatial resolution, and focusing on areas of change rather than total extent. The size of Australia (769 Mha) and the extent of its forests (110 Mha) required that robust and cost-effective methods that could be reliably operated into the foreseeable future be developed to estimate emissions and removals from the land sector. A key component of this system would be to track areas of land use change. As no such data existed in Australia that could meet all of these criteria, the NCAS needed to consider alternative options to traditional forest inventory and mapping.

The NCAS Land Cover Change Program (NCAS-LCCP) was developed by the Australian government in collaboration with the Commonwealth Scientific and Industrial Research Organisation (CSIRO) and other partners to meet the exacting requirements of the Kyoto Protocol. The NCAS-LCCP delivers the framework for fine-scale continental mapping and monitoring of the extent and change in perennial vegetation using Landsat satellite imagery, allowing for an effective estimation of the GHG emissions from land use and land use change (Brack et al. 2006; Caccetta et al. 2010). The program has been successively developed (see, for example, Furby 2002; Caccetta et al. 2003, 2007; Furby et al. 2008) over a number of years and currently uses over 7,000 Landsat MSS, TM, and ETM+ images at a resolution of 25 m for 18 time periods from 1972 to the present time (2011) and continues on an annual update cycle, making it one of the largest and most intensive land cover monitoring programs of its kind in the world.

While the remote sensing program was designed specifically for the purposes of GHG accounting, it has many additional benefits for bodies interested in monitoring land use change generally. The resultant products represent one of the few nationally consistent time-series data for the land sector.

Moving remote sensing from the realm of a technical research program to fully operational systems with ongoing update cycles was a considerable undertaking. Issues of scientific expertise, technical capacity, ongoing data supply and analysis, and accessing and processing large archives of data all needed to be considered. While many of these issues, in particular, those related to storage and compute capacity, have largely been removed through technological advancements, the operation of such a system still requires ongoing planning and management. The operational procedures adhere to a strict processing guideline: the output from each processing stage is checked against specific accuracy and consistency standards through a rigorous quality assurance process. Given the above operating environment, accuracy, interpretability (for outsourcing and QA), computational efficiency, the ability to incorporate "better" algorithms, and reliability when applied through space and time are important aspects for consideration during methodology development.

13.2   Materials and Methods

13.2.1   Method Selection

Although no national-scale remote sensing program for land use change existed at the start of the NCAS program, several operational broad-scale monitoring programs [for example, Land Monitor (Caccetta et al. 2000; Land Monitor 2008) and SLATS (Goulevitch et al. 1999)] did exist at the subnational scale. These had been implemented to serve the natural resource management needs of subnational agencies rather than for the specific purposes of tracking land use change for carbon accounting. To assess the suitability of the differing methods, a series of workshops and pilot projects were conducted from which the national Landsat-based forest monitoring program was established. The end product was not the whole-scale adoption of a single method but rather a selection of the best aspects of several different systems.

The approach adopted is based on:

- Long-term sequences of orthorectified and calibrated Landsat MSS, TM, ETM+ satellite data
- Discriminant analysis techniques to (spectrally) separate classes of interest

- Supervised and automated approaches to specify/estimate classifier parameters
- Spatial/temporal models to reduce errors

As the task included the analysis and processing of thousands of historical Landsat scenes, as well as the requirement that the information be updated annually during reporting periods, operational components of the methodology were vital to the success of the system. To do this required:

- Detailed specification of the application of the methods in operations manuals (Furby 2002)
- Training and subsequent processing of the data by third parties with documented quality assurance checks
- Independent review of the outputs by an independent third party to provide insight into the characteristics of errors (Jones et al. 2004) for use in method refinement through a continuous improvement exercise

13.2.2   Landsat Data

The initial step for the program was to develop specifications for the selection of Landsat scenes. Landsat has a return time of 16 days, resulting in around 22 images available per year for any specific area. To develop the annual maps of forest extent required by the system required selection of the optimal image. The selections were based on both preferred time sequence according to factors including reporting requirements, seasonality, greenness, sun angle, and other artifacts such as cloud, fire, and smoke. As the purpose of the program is to determine changes in forest cover, images that maximize the separation between tree and other cover (i.e., usually drier conditions) are generally selected.

13.2.3   Landsat Data Geometric Rectification

Accurate orthorectification of the Landsat data is vital to ensure that any change is due to real changes on the ground rather than edge effects due to image misalignment. In the NCAS-LCCP, this was achieved using a rigorous earth orbital model (PCI OrthoEngine software; Toutin 1994; Cheng and Toutin 1995), with a specification requiring subpixel accuracy. The first step was to establish a common orthorectified base mosaic of Landsat data. Once the orthorectified base was established, ground control points (GCPs) were automatically matched using a crosscorrelation technique and the temporal sequences of images orthorectified to the common base. This approach improves efficiency and accuracy of the results. For quality assurance, visual inspection and numerical summaries based on crosscorrelation feature matching are used to assess the accuracy of orthorectification of the time-series images.

13.2.4 Image Calibration/Normalization

Radiometrically calibrated images allow for comparisons between image scenes and the possibility of better extrapolation of a chosen classifier. We convert raw digital counts to be consistent with a chosen reference image.

The five main steps in the calibration and normalization (see Figure 13.1) of the Landsat data are:

- Top-of-atmosphere (TOA) reflectance calibration (as described by Vermote et al. (1997), which is to correct the reflectance differences caused by the solar distance and angle.

- Bidirectional reflectance distribution function (BRDF) calibration, described by Wu et al. (2001).

- Empirical correction for atmospheric and other affine effects via the use of invariant targets (Furby and Campbell 2001).

- Terrain illumination correction (Wu et al. 2004), which is based on the C-correction (Teillet et al. 1982). This step is required where there are significant terrain illumination effects, resulting in bright and dark sides of hills and mountains.

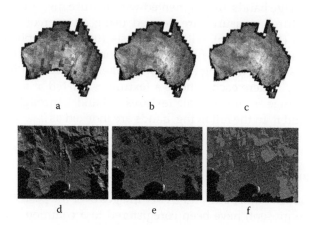

a b c

d e f

FIGURE 13.1
(See color insert.) Image calibration (top) and normalization (bottom). Calibration: Landsat mosaic of Australia showing (a) uncalibrated, (b) TOA correction, and (c) TOA + BRDF correction. Normalization (From Wu et. al., 2004.): (d) uncorrected, (e) terrain illumination correction, and (f) estimated occlusion mask overlaid and shown in gray. (From Wu, X., et al., An approach for terrain illumination correction. Australasian Remote Sensing and Photogrammetry Conference, Fremantle, Western Australia, 2004.)

- Occlusion detection (Wu et al. 2004) to identify terrain not observed due to the combination of terrain and the viewing geometry. This step identifies true shadow, which is labeled as missing data.

A relatively high-resolution digital elevation model (typically better than the 90 m SRTM) is required to achieve adequate occlusion detection and removal of terrain illumination effects.

13.2.5 Landsat-Derived Texture Measures

There are many natural and seminatural areas that have significant extents of heterogeneous perennial woody vegetation that do not meet the structural definition of forests or are at the lower limit of the definition of forests that is difficult to interpret and draw a line on a map so to speak. Here we refer to perennial woody vegetation having less than 20% canopy cover as sparse.

Seasonal weather changes and management effects may change the characteristics of these regions, and this in conjunction with the limited ability of remote sensing technology to distinguish this 20% canopy cover limit typically results in seasonal transitions between forests and sparse.

Based on observations that some sparse regions had a textured appearance, measures of texture were demonstrated to have useful information for distinguishing between forest, sparse, and nonforest classes and have been trialled at subnational scale (Caccetta and Furby 2004), progressively being incorporated into the work described here (Furby et al. 2007), where the Landsat image bands are augmented with texture measures in the analysis. The "texture" measures are derived using an overcomplete wavelet decomposition (Unser 1995), with Haar basis functions applied to forest/nonforest linear discriminant functions of the original Landsat bands. These measures are smoothed using an adaptive filter. This results in an n-band "texture image," where each band is a texture estimated at a coarser scale. The textures range from fine-scale textures in band 1 through coarse-scale textures in band n. In the following, bands are indexed as $h_0 \ldots h_n$ where h_0 is the finest scale texture and h_n the coarsest.

13.2.6 Comments

Some 7,000 Landsat MSS, TM, and ETM+ images over the past 39 years (from 1972 until the present) have been coregistered to a common orthorectified base mosaic using the above methods. The process is ongoing with an annual updating process. The program also periodically evaluates the potential for data from other sensors such as IRS, SPOT, and CBERS (Furby and Wu 2007, 2009; Wu et al. 2009) as possible candidates for operational use should data from the Landsat series no longer be available. To ensure access to those wishing to use data processed to this national standard, the data are then provided back to the Australian government agency responsible for remote sensing. These data are then made available for the cost of data transfer.

13.2.7 Forest Extent and Change Analysis

13.2.7.1 Geographic Stratification

Consistent with the experience of subnational programs, a stratified approach was adopted, allowing the local optimization of classifier parameters across the many different land cover–soil associations (Commonwealth of Australia 2005, 2009) that exist in such a large area. The stratification was adaptively derived, starting from boundaries based on (Landsat) spectral and other (such as topographic) consistency properties of strata during analysis. In all, about 400 strata were defined, as depicted in Figure 13.2.

13.2.7.2 Training and Validation Data

The process of classification requires that a quantitative assessment of the information in the available data is performed; the class labels, after having assessed the information in the data, are defined; a choice of model is made; and the accuracy of the results validated. Sample locations of known land cover are used to derive the classifier parameters or to train the classifier, and we refer to such data as training data. Similar sites independent of the training data are used to assess the accuracy of the results, and we refer to these data as validation data. The primary sources of training and validation data that have been used for the project include: about 800 historical aerial photographs whose locations are distributed across the continent;

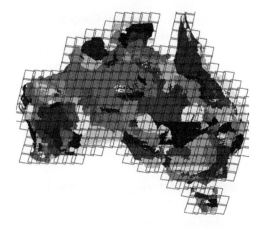

FIGURE 13.2
Stratification zones with Landsat scene boundaries overlaid used in analysis and subsequent processing. Within each zone, training data are used to estimate the parameters of the multitemporal classifier.

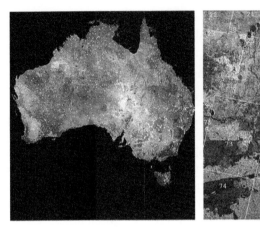

FIGURE 13.3
(See color insert.) (Left) Graphical depiction of the location of high-resolution IKONOS data used in the derivation of classifier training information. (Right) Typically, samples are required by intersection of zone and image, though well-calibrated data can reduce this requirement by allowing extrapolation across scene boundaries in many cases.

about 1,000 IKONOS images distributed across the continent (locations as depicted in Figure 13.3); and secondary less formal and generally available information such as regional expert knowledge, plantation location, and type information as provided by ground-based surveys and inventory information where it exists.

13.2.7.3   Multitemporal Model Used for Classification

Here we follow the approach described by Caccetta (1997) and Kiiveri and Caccetta (1998) for combining the multitemporal land cover information provided by the Landsat observations to form multitemporal classifications of land cover. The approach uses a probabilistic framework for combining data, with the view to classifying the data. Useful properties of the approach include:

- Propagation of uncertainties in inputs and calculation of uncertainties in outputs
- Production of hard and soft maps
- Handling of missing data by using all available information to make predictions
- Existence of well-developed statistical tools for parameter estimation

We note these characteristics are useful in practice as operational monitoring programs face issues such as availability of cloud-free imagery, variable (historic) atmospheric conditions, and changing sensor characteristics resulting in time-series data that vary in quality, completeness, and spectral discrimination.

13.2.8 Accuracy Assessment

The accuracy of the final forest presence/absence classification was independently validated, with initial results recorded by Lowell et al. (2003) and a subsequent update by Jones et al. (2004). Results from the latter are summarized below (see Section 13.3.3). Validation involved the comparison of classifications against "truth" obtained from aerial photo interpretation. The classes "forest," "nonforest," "regrowth," and "deforestation" were considered. As noted by Lowell et al. (2003) and Jones et al. (2004), the sampling strategy was constrained by the availability of (historical) aerial photography and was further constrained by the variable quality and scale of the photography. Routine collection of aerial photography resides with the states within Australia, with the collection being tailored by the states to individual state needs. This results in variable geographic, temporal, and spatial resolution when considering a national program.

Due to the variable availability and quality of aerial photography, Lowell et al. (2003) and Jones et al. (2004) adopted an approach that required the analyst to attach a degree of confidence to the cover class interpretations. Results were thus summarized as a "fuzzy" confusion matrix.

13.2.9 Attribution

Land cover change does not directly relate to land use change, in particular for deforestation and reforestation. Forest cover can change for a variety of reasons including clearing or establishment of trees, fire, pest attack, and drought. Further, there is a degree of error in any remote sensing analysis that need to be removed wherever possible, especially to remove false change due to the random errors in forest extent between years.

Attribution is a largely manual process that relies on expert judgment and experience. However, it can be greatly assisted by other products that allow for rules-based methods to be applied. For example, tenure can be used in many cases to separate forest cover loss due to forest management (such as clear felling) from that due to clearing for agriculture (deforestation). Mapping of fire scars can be used to separate change in forest cover from fire from areas of deforestation or forest management. Other mapping products, such as areas of known plantation establishment, allow for separation of areas of natural regrowth from human-induced reforestation.

The process of attribution is directly related to the policy, reporting, and accounting requirements. While the remote sensing sets the base for the system, it is the attribution that ensures that the final outputs of the system are policy relevant.

13.3 Results and Discussion

The key outputs from the system are raw and policy-relevant time-series data of forest cover, forest cover change, forest cover trend, and plantations identified as being either hardwood or softwood. From these analyses, it is possible to effectively age areas of forest accurately from ages 1 to 38, with a further class of 38 years or older.

The dense and extended time-series data developed through the NCAS-LCCP allows for analysis that has not previously been available. Such data provide detailed insight into the key processes in the land sector that drive emissions and removals.

13.3.1 Comparison to Existing Manual Mapping Products

A variety of other mapping products exist in Australia that were developed for a number of purposes, including biodiversity, conservation, and watershed management (Commonwealth of Australia 2008, 2009). For the purposes of change analysis, such mapping products are unable to track the change in forest extent due to human-induced activities. For example, Commonwealth of Australia (2008) uses manual methods that are not time-series consistent. Although these mapping products are constrained for change analysis, they still play a vital role in the estimation of emissions and removals from forests. This is an excellent example of using data that are fit for purpose.

13.3.2 Relationship to Modeling and Natural Resource Management

The remote sensing program has produced a rich source of spatial information for use in the emissions modeling framework, allowing Australia to report accurately on emissions and removals from the land sectors (Figure 13.4) as well as being used to report on rates of forest conversion. As the program expands, new information is being derived and progressively incorporated into the framework. We briefly describe the progress of the land cover information derived to date.

The forest presence/absence information has been derived for each of the Landsat epochs in the time series. Based on spatial and temporal rules, areas most likely to be plantations are identified and classified as being either hardwood or softwood (Chia et al. 2006).

Spectral indices providing an ordination from forest to nonforest have been derived for each Landsat TM epoch for 1989 onward. The perennial vegetation cover trend information provides subtle information on historic changes within forest (and ultimately sparse) areas and offers a surveillance tool for forest managers (Wallace et al. 2006). See Lehmann et al. (2011) for details. These indices are used with an "ever forest" mask, which is derived

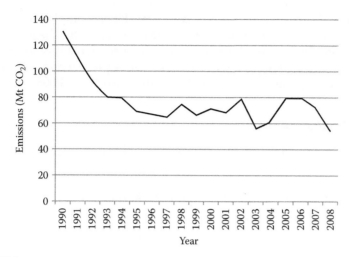

FIGURE 13.4
Emissions from forest land converted to cropland and grassland in Australia, 1990–2008. (From Australia National Inventory Report 2008.)

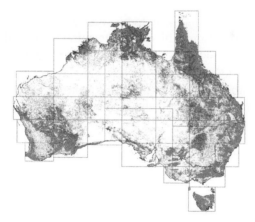

FIGURE 13.5
(See color insert.) Map of Australia showing NCAS forest extent (green) and sparse extent (red).

from the union of any area identified as forests at any point in the forest presence/absence time series. Together they provide temporal information on the trajectory of a pixel.

Sparse cover presence/absence classification (see Figure 13.5), which relies on image-derived spatial texture measures for discrimination, has been derived from the Landsat TM epochs in the time series 1989 onward (Furby et al. 2007). At the time of writing, the sparse cover information was being prepared and was not temporally complete. Upon completion, the trend information will also be derived for this class similar as for the forest class.

The land cover change information currently used for emissions modeling is the forest change (derived from the time-series presence/absence classifications) and the plantation type classification (Furby et al. 2008).

13.3.3 Accuracy Assessment

The overall forest presence/absence accuracy statement, as summarized by Jones et al. (2004), p. 8 of the report, is:

- Nationwide, the NCAS definite error rate was ~3%.
- Combined NCAS definite and probable error rate was ~12%.
- Nationwide, the forest definite error rate was ~2%.
- Nationwide, the nonforest definite error rate was ~4%.
- Nationwide, the forest combined definite and probable error rate was ~6%.
- Nationwide, the nonforest combined definite and probable error rate was ~15%.
- The amount of forest is likely to be underestimated continent wide, but the exact amount is difficult to determine because the CIVP sampling scheme was not a stratified or random sample.
- Regrowth and deforestation have considerably higher levels of errors associated with them, but are much rarer classes (only occurring ~2% and 1% of the time, respectively).

For the sparse covers, plantation, and trend information, validation is yet to be performed and will be in the scope of future works.

Acknowledgments

The authors thank all those who have also contributed to the larger NCAS system. This includes the number of contractors, scientists, and public servants involved in all stages of the NCAS development. In particular, the authors acknowledge the foresight of those who supported and promoted such a system in the late 1990s, including Gwen Andrews, Howard Bamsey, and Ian Carruthers, and for early scientific guidance from Dr Norm Campbell. The role of such dedicated public servants in ensuring that a vision reaches reality was a key factor in such success.

About the Contributors

Peter Caccetta received his PhD in computing science from the Curtin University of Technology, Perth, Australia. He is currently an adjunct professor with the Curtin University of Technology. He joined the Division of Mathematics, Informatics and Statistics, CSIRO, Wembley, Australia, in 1991, and worked on the development of broad-scale land cover mapping and monitoring technologies. Since 1999, he has been the principal scientist and leader of the Mapping and Monitoring Group, CSIRO. His current research interests include the application of remote sensing technologies for monitoring for natural resource management, including satellite and high-resolution monitoring obtained from aerial photography of urban and periurban environments.

Suzanne Furby is a member of the Mapping and Monitoring Research Group in the CSIRO Division of Mathematics, Informatics and Statistics, Perth, Australia. She has developed image calibration methodology and has been a key member of the team that developed methods for deriving quantitative information on land cover and condition change from remotely sensed and other spatial data, including the mapping and monitoring of dryland salinity and trends in perennial vegetation cover. She coordinated the remote sensing program to monitor land use changes across the continent. Results from this program data are used in the Australian National Carbon Accounting System (NCAS).

Xiaoliang Wu received his PhD in photogrammetry and remote sensing from the Wuhan Technical University of Surveying and Mapping, Wuhan, China, in 1993. In 1996, he joined the CSIRO, Perth, Australia, as a research scientist, where he contributed to the development and delivery of a satellite-based monitoring system—Australian NCAS. He is currently a principal research scientist with the CSIRO Division of Mathematics, Informatics and Statistics, developing an airborne-based monitoring system called "urban monitor."

Jeremy Wallace is a member of the CSIRO's Division of Mathematics for Mapping and Monitoring Group, based in Perth, Western Australia. This group has focused on the operational development of monitoring systems for land and vegetation using remote sensing. He has worked on Australia's National Carbon Accounting System Land Cover Change Program (NCAS-LCCP) since its inception and more recently in the training and implementation of remote sensing monitoring in Indonesia. He also has interests in the applications of remote sensing and related data for ecological and natural resource management applications.

Robert Waterworth has a PhD from the Fenner School of Resources, Environment and Society, Australian National University, Canberra, Australia. He is currently the director and senior research scientist of the International Forest Monitoring Section. Previous to this, Waterworth was a senior research scientist with Australia's NCAS and was responsible for the development of models and inventory systems. Prior to this, he worked in forest inventory systems and developed field measurement procedures for carbon accounting. His main area of expertise is system framework design, specifically implementing technical systems that meet policy and reporting requirements, forest growth modeling, and carbon accounting.

Gary Richards has a PhD in forestry and is an adjunct professor at the Fenner School of Resources, Environment and Society, Australian National University, Canberra, Australia. Richards is also cochair of the Clinton Climate Initiative, Carbon Measurement Collaborative, is an active participant on the work of the Intergovernmental Panel on Climate Change, is a national colead for the intergovernmental Group on Earth Observations (GEO) Forest Carbon Tracking Task, and convenes the GEO Global Forest Observations Initiative.

He is currently assistant secretary for the Energy Branch of the Department of Climate Change and Energy Efficiency. Richards was formerly principal advisor (International Forest Monitoring) for the Australian Government Department of Climate Change and Energy Efficiency. Richards was responsible for establishing Australia's NCAS that pioneered the combination of remotely sensed data, ground data, and carbon cycle models as the basis for national emissions reporting from land sectors.

References

Brack, C. Richards, G., and Waterworth, R., Integrated and comprehensive estimation of greenhouse gas emissions from land systems. *Sustainability Science*, 1, 91, 2006.

Caccetta, P.A., Allen, A., and Watson, I., The Land monitor project. The 10th Australasian Remote Sensing and Photogrammetry Conference, Adelaide, Australia, August 21–25, Remote Sensing and Photogrammetry Association of Australasia Ltd., Australia, 2000.

Caccetta, P.A. Remote sensing, GIS and Bayesian knowledge-based methods for monitoring land condition. PhD thesis, School of Computing, Curtin University of Technology, 1997.

Caccetta, P.A., et al., Notes on mapping and monitoring forest change in Australia using remote sensing and other data. Proceedings 30th ISRSE Honolulu, Hawaii, November 10–14, 2003.

Caccetta, P.A. and Furby, S.L., Monitoring sparse perennial vegetation cover. Proceedings 12th Australasian Remote Sensing and Photogrammetry Conference, Fremantle, Western Australia, October, 18–22, 2004.

Caccetta, P., et al., Continental monitoring: 34 years of land cover change using Landsat imagery. International Symposium on Remote Sensing of Environment, San José, Costa Rica, 2007.

Caccetta, P., et al., Monitoring Australian continental land cover changes using Landsat Imagery as a component of assessing the role of vegetation dynamics on Terrestrial Carbon Cycling. ESA Living Planet Symposium, Bergen Norway, June 28–July 2, 2010.

Cheng, P. and Toutin, Th. High accuracy data fusion of satellite and airphoto images. *Proceedings of ACSM/ASPRS Annual Convention*, Charlotte, North Carolina, American Congress on Surveying and Mapping (ACSM) and American Society for Photogrammetry and Remote Sensing (ASPRS), Bethesda, Maryland, Vol. 2, 453–464, 1995.

Chia, J., et al., Derivation of a perennial vegetation density map for the Australian continent. Australasian Remote Sensing and Photogrammetry Conference, Canberra, ACT, Australia, 2006.

Commonwealth of Australia, Assessment of Australia's terrestrial biodiversity 2008. (Available online at http://www.environment.gov.au/biodiversity/publications/terrestrial-assessment/index.html. Last accessed 11 October 2011), 2009.

Commonwealth of Australia, Interim biogeographical regions of Australia. (Available online at http://www.environment.gov.au/parks/nrs/science/pubs/regions.pdf. Last accessed 11 October 2011), 2005.

Commonwealth of Australia, Australia's State of the Forests Report: Five-yearly report 2008. (Available online at http://adl.brs.gov.au/forestsaustralia/_pubs/sofr2008reduced.pdf. Last accessed 11 October, 2011), 2008.

Furby, S. and Campbell, N., Calibrating images from different dates to 'like-value' digital counts. *Remote Sensing of Environment*, 77, 186, 2001.

Furby, S.L., Land cover change: Specification for remote sensing analysis. National Carbon Accounting System technical report no. 9, Australian Greenhouse Office, Canberra, 2002.

Furby, S.L, Wallace, J.F., and Caccetta, P.A., Monitoring Sparse Perennial Vegetation Cover over Australia using Sequences of Landsat Imagery. 6th International Conference on Environmental Informatics, Bangkok, Thailand, November 2007.

Furby, S., et al., Continental scale land cover change monitoring in Australia using Landsat imagery. International Earth Conference: Studying, Modeling and Sense Making of Planet Earth, Mytilene, Lesvos, Greece, 2008.

Furby, S. and Wu, X., Evaluation of IRS P6 LISS-III and AWiFS image data for forest cover mapping. CSIRO Mathematical and Information Sciences CMIS Technical Report 06/199, 2007.

Furby, S.L. and Wu, X., Evaluation of alternative sensors for a Landsat-based monitoring program. *Innovations in Remote Sensing and Photogrammetry*, 75, 2009.

Goulevitch, B.M., Danaher, T.J., and Walls J.W., The statewide landcover and trees study (SLATS)-monitoring land cover change and greenhouse gas emissions in Queensland. Proceedings of IEEE International Geoscience and Remote Sensing Symposium, Hamburg, June 1999.

Jones, S., et al., Update on the NCAS continuous improvement and verification methodology. NCAS Technical Report No. 46, 2004.

Kiiveri, H.T. and Caccetta, P.A., Image fusion with conditional probability networks for monitoring salinisation of farmland. *Digital Signal Processing*, 8(4), 225–230, 1998.

Kiiveri, H.T., et al., Environmental monitoring using a time series of satellite images and other spatial data sets. In *Nonlinear Estimation and Classification* by D.D. Denison (Ed.), M.H. Hansen, C. Holmes, B. Mallick, B. Yu, Lecture Notes In Statistics, New York, Springer Verlag, 49, 2003.

Land Monitor, Land monitor: A project of the Western Australian salinity action plan. (Available at http://www.landmonitor.wa.gov.au, verified September 25, 2009). Western Australian Land Information Authority, Perth, WA, Australia, 2008.

Lehmann, E.A., et al., Forest cover trends from time series Landsat data for the Australian continent (submitted for publication), 2011.

Lowell, K., et al., (2003), Continuous Improvement of the national carbon accounting system land cover change mapping. National Carbon Accounting System technical report No. 39. 2003. (http://www.greenhouse.gov.au/ncas/reports/pubs/tr39final.pdf).

Teillet, P.M., Guindon, B., and Goodeonugh, D.G., On the slope-aspect correction of multispectral scanner data, *Canadian Journal of Remote Sensing*, 8(2):84–106, 1982.

Toutin, Th., Rigorous geometric processing of airborne and spaceborne data. Proceedings of EUROPTO Symposium on Image and Signal Processing for Remote Sensing, Rome, Italy, Vol SPIE 2315, 825–832, 1994.

UNFCCC, Report of the conference of the parties on its seventh session held in Marrakesh, Morocco – Addendum. Part 2: action taken by the conference of the parties, FCCC/CP/2001/13/Add.1, Bonn, Germany, 2001.

Unser, M., Multigrid adaptive image processing, In: *Proceedings of IEEE International Conference on Image Processing (ICIP)*, Vol. 1. Washington DC, USA, 49–52, 1995.

Vermote, E.F., et al., Second simulation of the satellite signal in the solar spectrum, 6S: an overview. *IEEE Transactions on Geoscience and Remote Sensing*, 35, 675, 1997.

Wallace, J.F., Behn, G., and Furby, S.L., Vegetation condition assessment and monitoring from sequences of satellite imagery. *Ecological Management and Restoration*, 7, 31, 2006.

Wu, X., et al., A BRDF-corrected Landsat 7 mosaic of the Australian continent. IEEE International Geoscience and Remote Sensing Symposium, Sydney, Australia, 3274, 2001.

Wu, X. Furby, S., and Wallace, J., An approach for terrain illumination correction. Australasian Remote Sensing and Photogrammetry Conference, Fremantle, Western Australia, 2004.

Wu, X. et al., Evaluation of CBERS image data: Geometric and radiometric aspects. In *Innovations in Remote Sensing and Photogrammetry*, Lecture Notes in Geoinformation and Cartography, Springer, Berlin Heidelberg, p. 91, 2009.

14

Assessment of Burned Forest Areas over the Russian Federation from MODIS and Landsat-TM/ETM+ Imagery

Sergey Bartalev, Vyacheslav Egorov, Victor Efremov,
Evgeny Flitman, Evgeny Loupian, and Fedor Stytsenko
Russian Academy of Sciences

CONTENTS

14.1 Introduction

14.1.1 Background of Burned Area Mapping

Wildfires are one of the most important drivers of land cover changes in Russia. They affect annually millions of hectares of forests and other terrestrial ecosystems, such as tundra, grasslands, and peatlands (Korovin 1996). Earth observation allows characterizing the distribution and impact of wildfires from individual events up to the country level. Burned area mapping is a critical input for both fire management actions' planning and postfire impact assessment, including economic and environmental aspects.

There is a wide range of requirements for the delivery time and accuracy of burned area estimates in relation to the range of applications. Fire-fighting and suppression activities require information to be updated as frequently as possible and to be delivered to users as rapidly as possible. The fire-fighters require fire information to be updated very frequently, up to several times a day. However, they do not have stringent requirements for information accuracy. On another hand, the applications related to postfire impact assessment are highly dependent on the accuracy of burned area mapping, but do not have strong requirements for data delivery speed as postfire impact assessment data can be delivered a few weeks or even a few months after the fires. Postfire assessment is used in particular for forest inventories, forest management, biodiversity conservation, and carbon emissions reporting.

For more than two decades, earth observation techniques have demonstrated their capacities to provide various types of information related to vegetation fires, including active fire detection and monitoring, burned area mapping, and characterization. A number of methods have been developed for active fire detection based on the radiation temperature characteristics of fires. These methods are based on the use of a few main satellite remote sensing instruments: NOAA-AVHRR (Li et al. 2001), ERS-ATSR2 and Envisat-AATSR (Arino et al. 2005), as well as Terra-MODIS (Giglio et al. 2003). In spite of attempts made to assess the extent of burned area directly from the detection of active fire pixels, such approaches are not considered very robust and are reported with large ranges of uncertainties. Eva and Lambin (1998) did not find any significant correlation between estimates of active fire pixels (derived from NOAA-AVHRR sensor) and assessment of burned areas in Central Africa. By contrast, Loboda and Csiszar (2004) reported a very high correlation ($R^2 = 0.99$) between the number of active fire pixels (estimated from MODIS sensors) and burned areas (derived from Landsat-ETM+ imagery) in Russia, with only about 10% underestimation. The fundamental shortcomings of such approaches are due to the combination of a few factors (Giglio et al. 2006), mainly:

- Masking of active fires by clouds and smoke
- Limited temporal frequency of satellite observations
- Spatial and temporal heterogeneity of fires, related in particular to a large range of propagation speed, fuel contents, meteorological conditions, and temperature daily dynamics
- Coarse spatial resolution of the satellite sensors used for active fire detection

On the one hand, as some of these factors are stochastic in nature, a consistent assessment of burned area is difficult from these active fire detection approaches, and the accuracy of results may significantly vary between regions and time periods. On the other hand, such approaches are considered the most appropriate for fire-fighting activities for which data delivery time is the most critical factor.

A number of burned area mapping approaches are based on the detection of intra or interannual changes in land cover spectral properties using time series of coarse-resolution satellite imagery (500 m–1 km) mainly from NOAA-AVHRR (Sukhinin et al. 2005), SPOT-Vegetation (Grégoire et al. 2003), and Terra/Aqua-MODIS (Roy et al. 2008) instruments. These methods are usually based on surface reflectance measurements in the NIR or SWIR (near or short-wave infrared, respectively) spectral channels of these instruments. The NIR and SWIR channels are either used as direct inputs into change-detection algorithms or through spectral vegetation indexes (such as NDVI, SWVI, or NBI) with high discrimination power to separate burned areas from green vegetation. Other research studies (Fraser et al. 2000a,b; Bartalev et al. 2007) have demonstrated the efficiency of the combined use of both approaches, i.e., the combination of (1) active fire detection and (2) burned area assessment from changes in land cover spectral properties.

These latest approaches usually demonstrate higher accuracies for burned area estimates compared to methods based on active fire pixel detection only. Burned area products can also be produced on a regular basis, e.g., monthly (Zhang et al. 2003), decadal (Bartalev et al. 2007), or daily (Tansey et al. 2008) time frames. These products consist of multiannual time series of burned area data over large territories that are valuable inputs for the geosciences and for environmental assessments. However, they have very limited use for forest inventory and management applications because finer spatial resolution and higher accuracy are required by foresters. Moreover, so far such methods do not allow for rapid data delivery in an operational manner. The information is usually made available to users with a substantial delay.

There is also an extensive experience for burned area mapping from moderate spatial resolution (10 m–30 m) satellite optical imagery, such as Landsat-TM/ETM+ (Isaev et al. 2002). In spite of the existence of a number of methods, these methods have been mostly applied to episodic and local level assessments. Mapping of burned areas from moderate-resolution satellite imagery over large areas and at regular time intervals has been restricted mainly by data availability until recently. This restriction has been reduced drastically through the recent open data distribution policy and online access to the global multiannual Landsat-TM/ETM+ data archive (see Section II.2).

14.1.2 Forest Fire Monitoring Information System (FFMIS)

Mapping of burned area is one key feature of the FFMIS, developed by a consortium of institutes belonging to the Russian Academy of Sciences. The FFMIS constitutes an essential component of several environmental monitoring services, such as the VEGA service (Loupian et al. 2011), which is publicly available, and the forest monitoring information system (called in Russian *ISDM-Rosleshoz*) operated by the Russian Federal Forest Agency (Loupian et al. 2006; Bartalev et al. 2008). The FFMIS covers the full territory of Russia and provides information to a range of forestry services, from the local forestry districts up to the federal forest agency. The FFMIS focuses on daily information support for activities related to fire management and for environmental and economic impact assessment. Considering the size of the Russian territory and the users' requirements for information delivery speed and frequency, satellite remote sensing technology has been considered as the main source of data in the system. The FFMIS uses as main inputs the multiannual and daily updated archives of data acquired by the Terra-MODIS and the Landsat-TM/ETM+ instruments (since year 2000). The system considers three sources of input data for burned area assessment over Russia, as follows:

1. Locations of active fires detected using the MOD14 standard algorithm (Justice et al. 2006) and MODIS Level 1B data (Toller et al. 2006) collected via a network of satellite data-receiving stations distributed across Russia. As a backup data source, the Fire Information for Resource Management System (FIRMS) Web site is also used for the daily download of MOD14 products (http://firefly.geog.umd.edu/firms).

2. MODIS surface reflectance daily data including information on solar illumination and instrument viewing geometry (MOD09 standard products; http://lpdaac.usgs.gov/main.asp).

3. Landsat-TM/ETM+ data downloaded from USGS GLOVIS (http://glovis.usgs.gov). By the end of the year 2011, the FFMIS archive of Landsat-TM/ETM+ data contained more than 122,000 scenes over the Russian territory including about 23,000 scenes acquired during the year 2011 only.

A new approach for burned area assessment based on the integration of this large database has been developed by the Space Research Institute of the Russian Academy of Sciences. This new approach is aimed at benefiting from the complementarities of the different data sources and includes highly automatic satellite data processing. The system creates three different burned area products:

1. *AFBA product*: Burned area polygons at 1 km spatial resolution. This product is based on the spatiotemporal clustering of active fire pixels derived from MODIS data with the use of individual satellite passes.

2. *SRBA product*: Burned area at 250 m spatial resolution. This product is derived from MODIS data using land cover surface reflectance change combined with active fire detection.

3. *HRBA product*: Burned area at 30 m spatial resolution. This product is derived from Landsat-TM/ETM+ data.

The integrated burned area assessment approach produces continuous information during the fire season. All available satellite imagery from the three potential data sources are used for any date and fire event. In cases of more than one burned area product being available for a given fire event, the following priority ranking is used to select the potentially most accurate product: (1) HRBA, (2) SRBA, and (3) AFBA.

The AFBA product provides the most rapid assessment of burned areas. This product can then be complemented with one of both SRBA and HRBA products depending on their availability. The SRBA product is produced on a regular basis from daily MODIS data a few weeks after the AFBA product and is usually available before the HRBA product. However, when burned areas are too small to be retrieved from the SRBA product, only the HRBA product is used. In the following sections of this chapter, all three mentioned burned area products are described in more detail including the methods used to produce them and some results for Russia (national burned area estimates with accuracy assessment) are discussed.

14.2 Description of Three Burned Area Products

14.2.1 AFBA Product: Rapid Burned Area Mapping Based on Active Fire Detection from MODIS Sensor

The *AFBA* burned area product is generated from MODIS data. The raw MODIS data are acquired in the direct broadcast mode via a network of receiving stations located in Moscow, Pushkino (Moscow region), Khanty-Mansiysk, Novosibirsk, Krasnoyarsk, and Khabarovsk. The MODIS data are first preprocessed up to level 1B standard (MOD02 product) and are then used as inputs for the MOD14 active fire detection algorithm (Justice et al. 2006) in order to produce so-called hot spots. Each hot spot is characterized by a number of attributes: (1) geographical coordinates, (2) on-the-ground pixel size (including both pixel widths along and across the sensor scanning directions), and (3) brightness temperature derived from two MODIS spectral channels (with wavelength intervals centered at 4 μm and 11 μm). Then the hot spots detected from the acquired multitemporal MODIS imagery are used to generate burned area polygons and to monitor their temporal dynamics.

The FIRMS (http://firefly.geog.umd.edu) serves as an archive of hot spots detected with the MOD14 algorithm. All detected hot spots are automatically recorded into the FFMIS database with their attributes. The main role of the hot spot archive is to fill potential gaps resulting from accidental MODIS data-receiving stations' failures or data delivery delays.

The rapid burned area assessment includes the analysis of hot spot time series for the monitoring of fire temporal dynamics. An important step in this analysis includes the generation of active fire polygons from spatially scattered hot spot pixels. This polygon generation process is carried out for each new satellite image using available historical hot spot dynamics (i.e., hot spots detected on earlier imagery). The data-processing chain has been developed to provide both near-real-time burned area mapping and postfire season burned area assessment. The main steps of the near-real-time burned area mapping method are described hereafter in more detail.

Step 1: Retrieval of hot spot timing. The hot spots detected from MODIS imagery are first characterized with their satellite observation times. This timing information is incorporated into the FFMIS database in order to build a consistent data time series for further analysis. The hot spot observation time is assigned as the MODIS data-receiving time at the local receiving station or, in the case of FIRMS data, as the MODIS data granule time.

Step 2: Generation of hot spot polygons. The generation of polygons around individual hot spots is an intermediate step. This step uses the MODIS pixel dimensions along and across the sensor scanning directions. The MODIS pixel dimensions are approximated by using geographical directions (along parallels and meridians).

Step 3: Generation of active fire polygons. In order to generate active fire polygons for each satellite image, the corresponding hot spot polygons have to be merged considering a spatial proximity criteria. Two hot spot polygons are merged into one single fire event if their areas are overlapping or the distance between them is less than 0.3 km. For each MODIS image, an individual fire event polygon corresponds to a burned area estimate for the date of the satellite observation. By considering a full time series of such fire polygons, an exhaustive burned area assessment can be carried out.

Step 4: Generation of burned area polygons. This data-processing step is the most complex step. It consists in monitoring the fires dynamic and in aggregating all individual fire polygons detected at different dates into one single burned area polygon (corresponding to a single fire event). The FFMIS database includes full time series of all active fire polygons that have been detected from the beginning of the fire season. One essential step of the burned area polygon generation procedure is the decision to take for each newly generated fire polygon: either (i) to be aggregated to an existing registered fire event or (ii) to create a new fire event in the database. The fire polygon identification procedure aims to check if derived from of last satellite pass data active fire polygon overlapped with or close (distance is less than 1 km) to one of already existing burned area polygon. In case if outcome from such test was positive, the last active fire polygon is geometrically merged to one of existing burned area polygon, otherwise it is considered as new event to be recorded in the database. The algorithm for active fire polygon identification includes also the consideration of particular cases or outliers which can significantly impact the burned area mapping results. One main particular case relates to new active fire polygons which overlap with more than one previously detected burned area polygon. In such case we assume that at the date of the new active fire polygon, these burned area polygons get connected and have to be considered later on as a joint single event.

The hot spot pixels detected with 1 km spatial resolution MODIS data are used as input data for the generation of burned area polygons. These hot spot pixels are based on thresholds of radiation temperature within the sensor's field of view and can obviously include unburned area. Assuming that the burned area error reaches a maximum at the fire border and declines toward the center of the fire, we use a heuristic formula to correct directly the burned area estimates:

$$S_C = \begin{cases} \left(1 - \dfrac{k \times \Delta \times (1-\sigma)}{\sqrt{S_G}}\right) \times S_G \ \forall \ S_G > (k \times \Delta)^2 \\[3mm] \sigma \times S_G \ \forall \ S_G \leq (k \times \Delta)^2 \end{cases} \tag{14.1}$$

where
S_G is the area of burned area polygons in km^2;
S_C is the corrected burned area in km^2;
$\Delta = 1.1$ is the nominal pixel size in km;
$\sigma = 0.25$ is the coefficient of correction;
$k = 4$ is a constant value.

Equation 14.1 assumes that a higher relative error corresponds to smaller areas (and vice versa) due to a larger proportion of boundary pixels. According to Equation 14.1, the correction procedure reduces the burned area estimates with a maximum factor of 4 for fires smaller than k^2 pixels. As fire size grows ($S_G \to \infty$), the correction coefficient decreases up to a value of 1, and thus for very large fires the correction does not change significantly the area estimates.

This burned area mapping method is implemented as an automatic processing chain within the FFMIS. Each new MODIS imagery is processed automatically when acquired in the system. The system provides burned area updates with a frequency of up to six times a day. The full data-processing cycle takes from 20–70 min depending on the number and area of active fires and on available computing resources. In case of the FIRMS being used as the source for hot spots, the data delivery extra time is at least 50 min and is usually about 2–3 hours after the satellite pass.

14.2.2 SBRA Product: Burned Area Mapping Based on Land Cover Change Detection from MODIS Sensor

This product is aimed at providing wall-to-wall assessments of burned areas during the fire season with higher accuracy and reliability than AFBA product. The method has been designed using existing approaches that combine two types of information derived from satellite imagery: surface reflectance changes and thermal anomalies (Fraser et al. 2000a,b; Bartalev et al. 2007). In such approaches, the thermal anomalies are used to separate fire-related land cover changes from other types of vegetation changes (due to other disturbance factors). The SBRA product includes a step of comparison to historical spectral dynamics. Historical multiannual satellite data time series are used to derive optimized land cover change-detection thresholds for any geographical location.

The *SRBA* burned area product is generated at 250 m spatial resolution based on the use of two MODIS data standard products, namely:

- The multiannual daily surface reflectance MOD09 data;
- The active fire (hot spots) MOD14 data for a single year.

The burned area mapping method includes several data-processing steps as follows:

- Detection of pixels contaminated by clouds and cloud shadows, sensor failures, and seasonal snow cover
- Building of multiannual time series of SWVI (short-wave vegetation index) daily composites from uncontaminated pixels
- Generation of the SWVI multiannual "reference" based on SWVI annual time series along a reference period
- Land cover change detection through detection of seasonal anomalies by comparison to SWVI reference
- Burned area mapping using a consistency criteria between detected land cover changes and active fires

The MODIS data preprocessing aims at detecting contaminated pixels and consists of following steps:

- Masking-out pixels with satellite observation and sun elimination angles above certain thresholds
- Detection of clouds, cloud shadows, and snow cover–related pixels
- Detection of residually contaminated pixels through statistical filtering of time-series data

The threshold criteria, such as view zenith angle $\theta > 40^0$ and sun zenith angle $\delta > 80^0$, are applied to mask-out pixels which are not suitable due to extreme geometrical observation and illumination conditions.

Clouds and snow-cover detection involves surface reflectance data as measured in the blue (459–479 nm) R_3 and SWIR (1,628–1,652 nm) R_6 MODIS channels, as well as normalized difference snow index (NDSI) (Hall et al. 1995), which is calculated using Formula 14.2:

$$NDSI = \frac{R_3 - R_6}{R_3 + R_6} \tag{14.2}$$

Assuming that any pixel can be assigned to one of four classes (clouds, semitransparent clouds, snow, and "clear surface"), the R_3-NDSI bidimensional space (Figure 14.1) can be subdivided as follows:

- «Snow» if $R_3 > 0.05$ and NDSI > 0.1
- «Clouds» if $R_3 > 0.05$ and $-0.2 <$ NDSI < 0.1 (14.3)
- «Semitransparent clouds» if $R_3 > 0.05$ and $-0.35 <$ NDSI < -0.2
- «Clear surface» in all other cases

Pixels that are located in surroundings of «clouds» and «semitransparent clouds» areas are also classified as «clouds» or «semitransparent clouds» if their R_3 value is equal or higher than 0.05.

Assuming a maximum clouds' height as $H = 12$ km and considering the measured sun and view zenith angles, we can reconstruct the potential

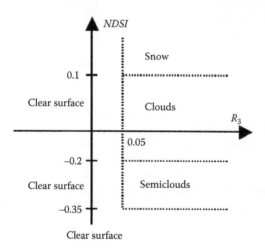

FIGURE 14.1
Discrimination of the classes of clouds, snow, and clear surface in the R_3-NDSI space.

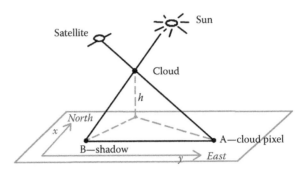

FIGURE 14.2
Geometrical modeling of the cloud-shadow position on the Earth surface (AB line).

cloud-shadow areas (Figure 14.2). If we consider an orthogonal coordinate system with origin O in a given cloud pixel with height H and axes Ox and Oy directed to geographical North and East, spatial shift of cloud shadow on the ground is estimated using Formula 14.4:

$$x = H\big(\cos(\psi)\,\mathrm{tg}(\theta) - \cos(\beta)\,\mathrm{tg}(\delta)\big)$$
$$y = H\big(\sin(\psi)\,\mathrm{tg}(\theta) - \sin(\beta)\,\mathrm{tg}(\delta)\big),$$

(14.4)

where
  ψ—view azimuth angle
  θ—view zenith angle
  β—sun azimuth angle
  δ—sun zenith angle

In general the geometrically modeled cloud-shadow areas include also "clear surface" pixels, which are removed from contaminated pixels through an additional spatial analysis step. The MODIS NIR channel R_2 (841–876 nm) image profile is analyzed along the cloud-shadow line (Figure 14.3) to identify the correct shadow segments.

The next analysis step is aimed at removing further false shadow pixels due to possible misclassification as clouds or snow-covered area with relatively low NDSI. The shadow pixel is considered as false detection if during a monthly period it has never been classified as "clear surface" and the following expression is true for the potential cloud-shadow pixels:

$$R_1(\Theta^*, t) > M_{R_1}(\Theta^*, t) + \sigma_{R_1}(\Theta^*, t),$$

(14.5)

where
  $M_{R_1}(\Theta^*, t)$ is the mean estimate of surface reflectance data in red (620–679 nm) channel centered at day t during a 31-day period

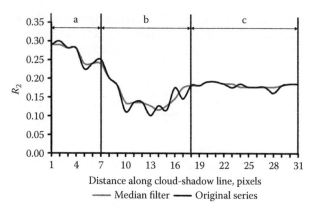

FIGURE 14.3
Example of spectral profile (MODIS NIR channel) with indication of sections corresponding to residual clouds (a), cloud shadows (b), and clear sky surface (c).

$\sigma_{R_1}(\Theta^*, t)$ is the standard deviation from mean
$R_1(\Theta^*, t)$ is the red surface reflectance data for given pixel with coordinates Θ^*

The additional statistical data filtering is aimed at reducing residual noise through the use of a monthly moving time window. The pixel with surface reflectance R_6 is considered as contaminated if the Expression 14.6 is true:

$$\left| R_6(\Theta^*, t) - M_{R_6}(\Theta^*, t) \right| \geq 2\sigma_{R_6}(\Theta^*, t) \tag{14.6}$$

From these preprocessing steps, the masks of different types of contaminated pixels are generated at 500 m spatial resolution.

Our main criteria to detect fires which are causing vegetation cover changes (i.e., which are burning the vegetation) is based on daily time series of the normalized SWVI (Fraser et al. 2000a)

$$SWVI = \frac{R_2 - R_6}{R_2 + R_6}, \tag{14.7}$$

SWSI is calculated using MODIS surface reflectance data (R_6) resampled from 500 to 250 m. The contaminated pixels (detected during preprocessing) are reconstructed based on a moving time-window polynomial algorithm to retrieve SWVI. The burned area mapping method uses the SWVI multiannual seasonal reference which is derived from MODIS time-series data acquired during previous years—so-called reference period. An experimental justification of the optimal reference period duration is given at the end of Section 14.2.2.

The assessment against the SWVI reference involves the estimation of mean $M_{SWVI}^N(\Theta^*, t)$ and standard deviation $\sigma_{SWVI}^N(\Theta^*, t)$ for every pixel with geographical coordinates Θ^* and image date t(DOY):

$$M_{SWVI}^N(\Theta^*, t) = \sum_{y=1}^{Y} \sum_{t-\Delta t}^{t+\Delta t} SWVI(\Theta^*, t, y) \tag{14.8}$$

$$\sigma_{SWVI}^N(\Theta^*, t) = \frac{1}{N} \left(\sum_{y=1}^{Y} \sum_{t-\Delta t}^{t+\Delta t} (SWVI(\Theta^*, t, y) - M_{SWVI}^N(\Theta^*, t))^2 \right)^{1/2} \tag{14.9}$$

$$\forall t(t = \overline{1, 365}) \quad \text{and} \quad \forall y(y = \overline{1, Y})$$

where

y is the year within the reference period with duration of Y years

Δt is the moving time-window length parameter for the SWVI intrayear statistical assessment

$N = Y(2\Delta t + 1)$ is the total number of measurements involved in the SWVI assessment for given pixel and DOY

The detection of pixels likely affected by fire is based on pixel-to-pixel and day-to-day differences between M_{SWVI}^N and the SWIR vegetation index time series for a given year $SWVI^C$. The detection of seasonal dynamic anomalies is based on following Formula 14.10:

$$SWVI^C(\Theta^*, t) - M_{SWVI}^N(\Theta^*, t) < -k\sigma_{SWVI}^N(\Theta^*, t), \tag{14.10}$$

where k is an experimentally tuned constant that allows to define the range of the SWVI reference interannual dynamics. A pixel is considered as abnormal and likely affected by a fire causing a land cover change if its $SWVI^C$ value is lower than reference SWVI seasonal values as presented in Figure 14.4. Such approach uses automatic thresholds $M_{SWVI}^N - k\sigma_{SWVI}^N$ which are calculated for any pixel location and date.

At this stage, the detected pixels include pixels affected by land cover changes caused by fire and by other disturbance factors such as, for example, flooding, insect outbreaks, and extreme weather conditions. They include also false changes due to particular atmospheric and angular conditions of observations and residual effects of interannual differences in phenological vegetation dynamics. A contextual spatial filtering is applied to remove such false detections. The mean $M_{SWVI}^W(\Theta^*, t)$ and standard deviation $\sigma_{SWVI}^W(\Theta^*, t)$ of $SWVI^C$ are computed using an increasing window size W for each given date t from five or more pixels surrounding a potential change pixel, with

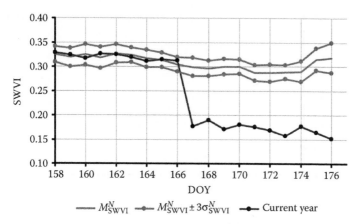

FIGURE 14.4

Example of fire detection from anomaly in the SWVI dynamics at pixel level. SWVI, short-wave vegetation index; DOY, day of the year.

geographical coordinates Θ^* excluding those pixels which have been detected at the previous stage. The pixel is considered as changed if its SWVI^C value is lower than $(M^W_{\text{SWVI}} - \sigma^W_{\text{SWVI}})$.

Finally, a clumping procedure is applied to group pixels detected as changed vegetation into spatially connected regions for each day. The resulting clumped areas are then compared to MODIS-derived active fire data to separate burned areas from areas that were subject to land cover changes resulting from other disturbances. The clumped area is considered as burned area if more than 1% of its total surface is spatially and temporally (within a 20-day time window) consistent with MODIS active fire data. This 1% area threshold has been determined empirically through visual tests and is aimed at elimination of false burned area detection such as crop harvesting.

This burned area mapping method requires setting values for a few main parameters. Two of them such as the reference period duration Y and the moving time-window length parameter Δt are aimed to determine the most appropriate SWVI reference parameters $M^N_{\text{SWVI}}(\Theta^*, t)$ and $\sigma^N_{\text{SWVI}}(\Theta^*, t)$. A third one, namely the scaling constant k, is used to define the reference range of SWVI for interannual variations. The most appropriate parameters' values have been estimated through a number of tests performed with MODIS data over the European part of Russia, which experienced an extreme fire season in 2010 due to exceptional heat wave and drought.

Figure 14.5 shows burned area as estimated by different combinations of reference period durations $Y(Y = 3,6)$ and time-window length parameter values ($\Delta t = 3$ and $\Delta t = 5$). Following these experiments, $Y = 5$ and $\Delta t = 3$ were considered as most appropriate parameters values. The shorter reference

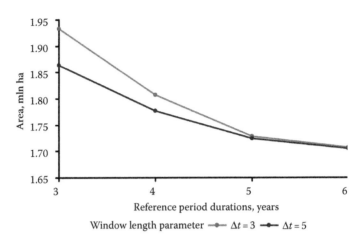

FIGURE 14.5
Burned area estimates for a MODIS tile (H20V03 granule) and for the year 2010 using different reference period durations and time-window lengths (SBRA product).

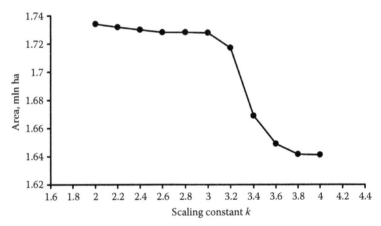

FIGURE 14.6
Burned area estimates with different scaling constant k values.

period ($Y < 5$) led to more false burned area pixels (controlled through visual interpretation), while an increase of the reference period to 6 years resulted in a negligible increase of burned area.

A number of burned area mapping tests have been also performed using different values of the scaling factor k. Figure 14.6 shows that when varying the scaling factor value between 2 and 3 it does not lead to significant changes in burned area estimates. The abrupt decline at $k = 3$ leads to the conclusion that this value can be considered as the most appropriate.

This method allows generating daily SRBA products over the entire Russian Federation in a routine manner within 20–30 days depending on the availability of uncontaminated satellite data.

14.2.3 HRBA Product: Burned Area Mapping from Landsat-TM/ETM+ Sensor

From the three burned area products of FFMIS, the HRBA is the product which takes most time to be delivered but which is potentially the most accurate at the level of individual fires. This product is derived from Landsat-TM/ETM images at 30 m spatial resolution. The main methodological difference with the two other burned area products stands is the involvement of human visual expertise for burned area control and delineation from Landsat-TM/ETM imagery. Another important characteristic of the HRBA product is that it cannot be used alone for an assessment at country level: the HRBA product can only complement the national estimates derived from AFBR and SRBA. This is due to the potential gaps in suitable quality Landsat-TM/ETM imagery over the country during the fire season (missing data). The national yearly completeness of the HRBA product may also significantly differ from year to year due to availability of human resources for assessment and interpretation of the imagery.

The HRBA approach has been developed from the FFMIS web-service interface which provides access to remote sensing data and products along with mapping tools (Figure 14.7). The web-service interface is based on the GEOSMIS system which includes generic GIS and dedicated vegetation analysis tools (Tolpin et al. 2011). The information available from the web interface includes imagery from both Landsat-TM/ETM+ and MODIS sensors as well as data on land cover, fires and meteorological conditions.

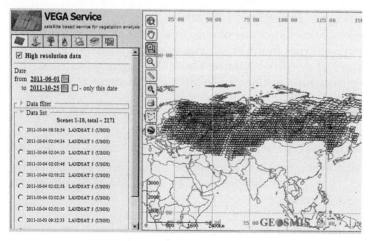

FIGURE 14.7
Display of the Web-service map interface with selected Landsat-TM frames.

In more details the web-service data analysis tools provide the following tools:

- Joint analysis of Landsat-TM/ETM+ and MODIS data, along with thematic maps and data
- Analysis of long-term time series of spectral vegetation indices to assess land cover changes and driving factors. The web-service allows to select an area of interest and to derive instantaneously a multiannual temporal profile of spatially averaged vegetation index
- Management of the database of burned areas (contours and characteristics)
- GIS analysis of satellite data and derived products

The web-service allows for an easy and quick access to all products derived from Landsat-TM/ETM+ and MODIS satellite data, which are automatically and continuously downloaded from the USGS data archive. Daily MODIS data are automatically processed. First, pixels contaminated by clouds and other noise are eliminated, and then weekly composite images are generated. These weekly composite images are used to produce normalized difference vegetation indices (NDVIs). Gaps in weekly time-series data are filled in though an interpolation procedure. Time-series data are then smoothed to reduce remaining noise due to cloud-contaminated pixels. The MODIS-derived NDVI time series are recorded into the database and used to create multiannual vegetation index profiles for each MODIS pixel. The Landsat-TM/ETM+ data are first preprocessed (radiometric and geometric correction), and then color composite images are made available trough the web interface. The MODIS data-derived land cover map of Russia for year 2010 at 250 m spatial resolution (Bartalev et al. 2011) is also made accessible through the web system. This map is the most up-to-date country-level map of forest type distribution.

The web-service allows mapping burned areas at 30 m resolution when an ABFA polygon exists in the FFMIS database and a corresponding appropriate postfire and cloud-free Landsat-TM/ETM+ image is available. The HRBA mapping procedure includes the following main steps:

- Selection of one MODIS-derived AFBA polygon and search of the best available Landsat-TM or ETM+ image
- Selection of the option: automatic or visual burned area delineation
- Visual evaluation of the automatic burned area delineation results and, in case of insufficient quality, replacement by visual burned area delineation

The automatic burned area delineation method is based on a multispectral image segmentation algorithm (Zlatopolskyy 1985) combined with automatic segment labeling and merging steps (Bartalev et al. 2009). The labeling

FIGURE 14.8
(See color insert.) Example of burned area polygons derived from the three methods: red polygon, AFBA product; black polygon, SRBA product; yellow polygon, HRBA product. The results are displayed in the Web-service user interface with the Landsat-TM scene used for the HRBA product as a background image.

and merging steps are based on distance criteria from corresponding AFBA and SRBA polygons along with the brightness histogram analysis of Landsat-TM/ETM+ image.

Figure 14.8 provides an example of burned area polygons derived from the three different methods available from the FFMIS web-service user interface.

14.3   Integrated Burned Area Assessment

The integrated burned area assessment approach consists in the combination of the three burned area products, *AFBA, SRBA* and *HRBA*, which are updated continuously during a fire season. This combination is aimed at producing best estimates of burned areas from all the products available in the FFMIS database. The fire events recorded from these three products are linked through an identification process which initiates from the MODIS hot spots–based polygons and related fire events in the AFBA database. The approach links these AFBA events to the fire events from the SRBA and HRBA databases. In case of a fire event existing only in the AFBA database (i.e., with no related event in the SRBA and HRBA databases), the related AFBA burned area is taken into account for the compilation of burned area estimates at national and regional levels and at daily frequency during the full fire season. When an event appears in the SRBA or HRBA databases the AFBA estimate is replaced by the estimate derived from the SRBA or HRBA event, with priority to HRBA product as it is considered to have higher accuracy.

However the practical implementation of this approach is complex due to the difficulty to set unique correspondences between fire events from the three burned area products. It happens often that several AFBA polygons overlap with one unique HRBA polygon, due to a complex spatial pattern or to limitations in data availability. The relationship between AFBA and SRBA polygons is in general even more complex as one single fire event on the ground can correspond to several polygons with no spatial correspondence between the two set of polygons. The impact of this lack of coherence on burned area estimates is limited over large territories, e.g., at national or regional scale. However this issue has to be taken into account for assessment at the level of individual fire events.

In order to address such issue, the integration procedure includes a step of polygon clustering. The clustering procedure is aimed at identifying polygons that are likely to be related to the same fire event. The SRBA and HRBA polygons are grouped within clusters which are linked to the AFBA polygons. An analysis of interlinkages between polygon pairs is carried out for the AFBA–SRBA and AFBA–HRBA combinations. The clustering procedure subdivides the total set of polygons into subsets of polygons which are considered to be related to single fire events. This clustering procedure provides for more detailed intercomparison and analysis at individual fire event level.

14.4 Burned Area Assessment over Russian Federation for Year 2011

The three burned area mapping methods, AFBA, SRBA and HRBA, and the integrated assessment approach have been applied over the full territory of the Russian Federation to estimate the extent and impact of the fire season of year 2011.

From the AFBA product the estimate of total burned areas for year 2011 over the entire country is 10.27 million ha, including 5.06 million ha of burned forest areas. For the same year the SRBA method leads to an estimate of 10.41 and 4.38 million ha of total burn areas and burned forest areas respectively. When looking at the burn area estimates derived from the SRBA method during the last 7 years, it appears that the 2011 fire season was obviously of not exceptional magnitude in Russia (Figure 14.9).

From the FFMIS database of available Landsat-TM/ETM+ imagery acquired during year 2011, the HRBA product has resulted in 3,609 polygons with a total burned area of 5.94 million ha, including 4.00 million ha of burned forests. The burned area sizes of individual fire events show a wide range as presented by their distribution histogram (Figure 14.10). Being not comprehensive enough to provide an accurate estimate at country level, this data set can be used as reference for and evaluation of the accuracy of burned area products derived from MODIS data. This product is also critical for the integrated burned area assessment.

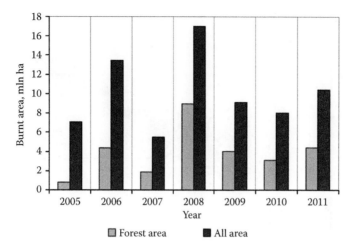

FIGURE 14.9
Annual estimates of burned areas (derived from the SRBA product) over the Russian Federation from 2005 to 2011.

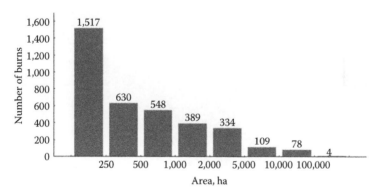

FIGURE 14.10
Burned area distribution of fire events from the HRBA product derived from Landsat-TM/ETM+ imagery.

The cross-comparison of Landsat-TM/ETM+ and MODIS hot spot–derived burned area products including both initial and corrected estimates (correction using Formula 14.1) demonstrates a bias reduction for corrected estimates in particular for small burned areas (<1,000 ha) burns (Figure 14.11).

The accuracy of forest burned area estimates derived from the AFBA product and corrected with Formula 14.1 has been assessed from the comparison to the HRBA product with the following results:

$$RMSE = \pm2.43\% \text{ and bias} = -14.1\%$$

a

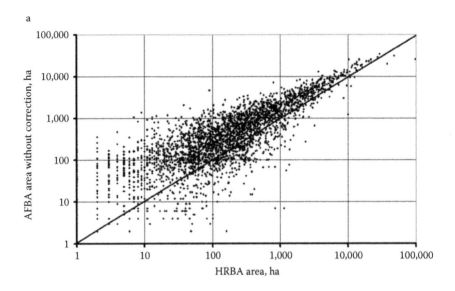

b

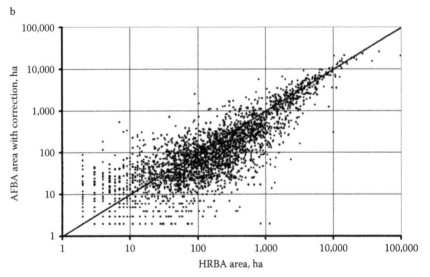

FIGURE 14.11
Correlation between burned areas from the AFBA product (MODIS hot spots) and burned areas from the HRBA product (Landsat-TM/ETM+): (a) before correction and (b) after correction for bias using Equation 14.1.

A similar assessment has been made for the SRBA product (Figure 14.12):

$$\text{RMSE} = \pm 1.52\% \text{ and bias} = -8.7\%$$

Figure 14.13 shows a decrease of RMSE as a function of the area size for small burns (<1,000 ha) considering either all wildfires together or only forest fires.

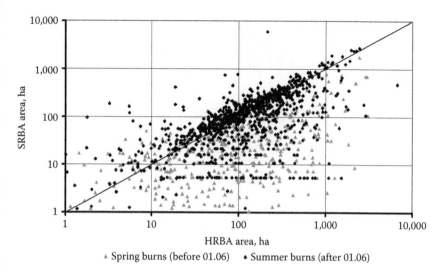

FIGURE 14.12
Correlation between burned areas from the SRBA product (MODIS surface reflectance change) and burned areas from the HRBA product (Landsat-TM/ETM+) with identification of spring and summer fires (before or after June 1).

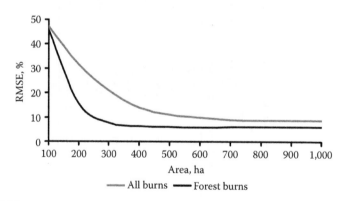

FIGURE 14.13
RMSE (root mean square error) of burned area estimates from the SRBA product in relation to burned area size for total burned areas and forest burned areas.

The combined use of all three burned area products into an integrated assessment leads to an estimate of 14.32 million ha of total burned area including 5.79 million ha in forest domain only. Table 14.1 shows the burned area estimates from the three burned area products and from the integrated assessment at the levels of entire country and federal districts of Russian Federation. Figure 14.14 shows spatial distribution of the burned areas of the year 2011 across the territory of Russia.

TABLE 14.1

Burned Area in Russia for the Year 2011 as Estimated Using MODIS and Landsat-TM/ETM+ Data with Three Different
Methods and Integrated Assessment Scheme

Federal District	AFBA Product (ha)		SRBA Product (ha)		HRBA Product (ha)		Integrated Assessment (ha)	
	Total Area	Forest Burns	Total Area	Forest Burns	Total Area	Forest Burns	Total Area	Forest Burns
Central	190,923	11,957	837,272	11,321	36,215	6,066	921,400	24,845
Southern	519,866	3,726	855,424	3,930	9,745	2,639	1,053,002	3,702
Northwestern	163,879	96,794	267,839	142,798	155,947	112,242	247,852	138,478
Far Eastern	5,144,318	3,320,603	4,023,000	2,855,103	3,924,097	2,818,567	6,194,216	3,900,744
Siberian	3,231,180	1,335,362	2,660,764	1,109,667	1,356,684	817,199	3,794,757	1,375,317
Urals	622,398	281,640	808,306	250,709	445,511	242,796	1,002,686	333,100
Volga	177,963	7,204	707,892	6,893	8,435	1,414	767,978	9,211
North Caucasian	215,802	1,711	247,488	159	800	512	340,224	1,458
Russian Federation	10,266,329	5,058,997	10,407,985	4,380,580	5,937,433	4,001,436	14,322,117	5,786,855

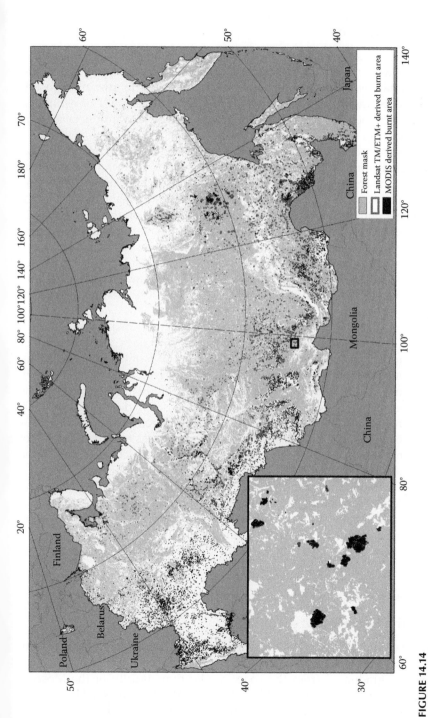

FIGURE 14.14
(See color insert.) Burned areas for the year 2011 over Russian Federation as depicted by the SRBA product (black areas) and the HRBA product (red polygons).

14.5   Conclusions

This chapter presents an approach for the mapping of burned forest area that combines two remote sensing data sources: MODIS and Landsat-TM/ETM+ satellite imagery. This approach is aimed to answer to two main users' requirements: rapidity of information delivery and accuracy of estimates. The approach includes three complementary burned area products at 1 km, 250 m, and 30 m spatial resolution, respectively.

The first burned area product (AFBA) is based on temperature anomalies detected from MODIS data at 1 km resolution. This product is generated with the most rapid but least accurate method. It allows providing burned area estimates several times a day with an acceptable level of accuracy. The second burned area product (SRBA) is based on the use of MODIS data-detected surface reflectance changes combined with radiation temperature anomalies. This method is less rapid (daily assessments are produced within 20–30 days delay) but leads to more accurate results at 250 m resolution. These two methods are fully automated and allow producing regular updated wall-to-wall burned assessments for the entire Russian Federation during a fire season.

The most spatially accurate burned area product (HRBA) is derived from Landsat-TM/ETM+ imagery. This product on its own does not allow providing a comprehensive burned area assessment at the country level but is considered as complementary to the MODIS data-derived estimates. However, this approach was applied operationally over the Russian Federation for the fire season of year 2011 and has resulted in the detailed mapping of 3,609 burned area events. The total burned areas derived from Landsat-TM/ETM+ imagery correspond to about 57% of total burned areas and to about 91% of total burned forest areas derived from the 250 m product for entire Russia during the year 2011.

The MODIS-derived products have different levels of RMSE (14.1% and 8.7% for AFBA and SRBA, respectively) with an underestimation of burned areas. Our integrated burned area assessment approach allows providing more comprehensive and accurate estimates.

About the Contributors

Sergey Bartalev is head of the Terrestrial Ecosystems Monitoring Laboratory at the Space Research Institute of the Russian Academy of Sciences, Moscow, Russia. He graduated from the space applications faculty of the Moscow State University of Geodesy and Cartography, Moscow, Russia. He received his PhD in remote sensing and a doctor of sciences in experimental physics.

His research interests are focused on the development of highly automated methods for land cover mapping and monitoring over large areas using time series of earth observation data. He has coauthored more than 80 peer-reviewed scientific publications.

Vyacheslav Egorov is a senior scientist at the Space Research Institute of the Russian Academy of Sciences, Moscow, Russia. His current research interests include the development of remote sensing techniques for forest mapping and change detection in boreal Eurasia. Vyacheslav Egorov graduated from the Moscow Institute of Physics and Technology, Moscow, Russia, and received his PhD in remote sensing from the Moscow State University of Geodesy and Cartography, Moscow, Russia. He has coauthored about 20 peer-reviewed publications in scientific journals.

Victor Efremov is a scientist at the Space Research Institute of the Russian Academy of Sciences, Moscow, Russia. He graduated from the Moscow Institute of Physics and Technology, Moscow, Russia, and his work is focused on the development of large data archives and user Web interfaces for various environmental monitoring systems based on remote sensing techniques. He has coauthored about 40 peer-reviewed publications in scientific journals.

Evgeny Flitman is a scientist at the Space Research Institute of the Russian Academy of Sciences. His primary professional activity is the development of software for the highly automated processing of persistent data streams. He is currently responsible for the development of an automatic data-processing and archiving system for monitoring forest fires. He has a masters degree in computing methods for civil engineering. He has coauthored about 70 peer-reviewed publications in scientific journals.

Evgeny Loupian graduated from the Moscow Institute of Physics and Technology, Moscow, Russia. He has a doctor of science diploma and works as a deputy director at the Space Research Institute of the Russian Academy of Science, supervising earth remote sensing research. He works at developing methods and techniques of automated satellite data processing. He participated in many projects concerning satellite data accumulation, processing, and distribution. He has coauthored more than 130 peer-reviewed publications in scientific journals.

Fedor Stytsenko is a PhD student at the Space Research Institute of the Russian Academy of Sciences, Moscow, Russia. He has a master of science degree from the Moscow State University of Geodesy and Cartography, Moscow, Russia. His current research is focused at continental-scale forest monitoring using optical remote sensing data and developing systems and methods for satellite data processing.

References

Arino, O., Plummer, S., and Defrenne, D., Fire disturbance: The ten years time series of the ATSR World Fire Atlas. *Proceedings of the MERIS-AATSR Symposium*, ESA publication SP-597, 2005.

Bartalev, S.A. et al., The vegetation mapping over Russia using MODIS spectrora-diometer satellite data. *Contemporary Earth Remote Sensing from Space*, 8, 285, 2011. [in Russian]

Bartalev, S.A. et al., Multi-year circumpolar assessment the area burned in boreal ecosystems using SPOT-Vegetation. *International Journal of Remote Sensing*, 28, 1397, 2007.

Bartalev, S.A. et al., The forest fire satellite monitoring information system of Russian Federation's forest agency (actual status and development perspectives). *Contemporary Earth Remote Sensing from Space*, 5, 419, 2008. [in Russian]

Bartalev, S.A., et al., Automatised forest burned area delineations based on HRV and HRVIR satellite data. *Contemporary Earth Remote Sensing from Space*, 6, 335, 2009. [in Russian]

Eva, H.D. and Lambin, E.F., Remote sensing of biomass burning in tropical regions: sampling issues and multisensor approach. *Remote Sensing and Environment*, 64, 292, 1998.

Fraser, R.H., Li, Z., and Landry, R., SPOT-VEGETATION for characterising boreal forest fires. *International Journal of Remote Sensing*, 21, 3525, 2000a.

Fraser, R.H., Li, Z., and Cihlar, J., Hotspot and NDVI Differencing Synergy (HANDS): A new technique for burned area mapping over boreal forest. *Remote Sensing and Environment*, 74, 362, 2000b.

Giglio, L., et al., An enhanced contextual fire detection algorithm for MODIS. *Remote Sensing and Environment*, 87, 273, 2003.

Giglio, L., et al., Global estimation of burned area using MODIS active fire observa-tions. *Atmospheric Chemistry and Physics*, 6, 957, 2006.

Grégoire, J.M., Tansey, K., and Silva, J.M.N., The GBA2000 initiative: Developing a global burned area database from SPOT-VEGETATION imagery. *International Journal of Remote Sensing*, 24, 1369, 2003.

Hall, D.K., Riggs, G.A., and Salomonson, V.V., Development of methods for mapping global snow cover using moderate resolution imaging spectroradiometer data. *Remote Sensing and Environment*, 54, 127, 1995.

Isaev, A.S., et al., Using remote sensing to assess Russian forest fire carbon emissions. *Climate Change*, 55, 235, 2002.

Justice, C., et al., *MODIS Fire Products Algorithm Technical Background Document. Version 2.3*, 2006. http://modis.gsfc.nasa.gov/data/atbd/atbd_mod14.pdf. Accessed on March 30, 2012.

Korovin, G.N., Analysis of the distribution of forest fires in Russia, in fire in ecosys-tems of boreal Eurasia. *Forestry Sciences*, 48, 112, 1996.

Li, Z., et al., A review of AVHRR-based fire active fire detection algorithm: Principles, limitations, and recommendations. In *Global and Regional Vegetation Fire Monitoring from Space, Planning and Coordinated International Effort*, SPB Academic Publishing, The Hague, The Netherlands, 199–225, 2001.

Loboda, T. and Csiszar, I., Estimating burned area from AVHRR and MODIS: Validation results and sources of error. *Contemporary Earth Remote Sensing from Space*, 2, 415, 2005.

Loupian, E.A., et al., Satellite monitoring of forest fires in Russia at federal and regional levels. *Mitigation and Adaptation Strategies for Global Change*, 11, 113, 2006.

Loupian, et al., Satellite service VEGA for vegetation monitoring. *Contemporary Earth Remote Sensing from Space*, 6, 190, 2011. [in Russian]

Roy, D.P., et al., The collection 5 MODIS burned area product: Global evaluation by comparison with the MODIS active fire product. *Remote Sensing and Environment*, 112, 3690, 2008.

Sukhinin, A.I., et al., Satellite-based mapping of fires in Russia: New products for fire management and carbon cycle studies. *Remote Sensing and Environment*, 94, 428, 2005.

Tansey, K., et al., A new, global, multi-annual (2000–2007) burned area product at 1 km resolution and daily intervals. *Geophysical Research Letters*, 35, L01401, 2008.

Toller, G.N., et al., *MODIS Level 1B Product User's Guide for Level 1B Version 6.1.0 (Terra) and Version 6.1.1 (Aqua)*, NASA Goddard Space Flight Center, 2006. http://mcst.gsfc.nasa.gov/uploads/files/M1054.pdf.

Tolpin, V.A., et al., Development of user interfaces for satellite data based environment monitoring systems (GEOSMIS system). *Contemporary Earth Remote Sensing from Space*, 8, 93, 2011. [in Russian]

Zhang, Y.H., et al., Monthly burned area and forest fire carbon emission estimates for the Russian Federation from SPOT VGT. *Remote Sensing and Environment*, 87, 1, 2003.

Zlatopolskyy, A.A., Image segmentation for objects with discontinued limits. *Automatic and Telemechanics*, 9, 109, 1985. [in Russian]

15

Global Forest Monitoring with Synthetic Aperture Radar (SAR) Data

Richard Lucas and Daniel Clewley
Aberystwyth University

Ake Rosenqvist
solo Earth Observation

Josef Kellndorfer and Wayne Walker
Woods Hole Research Center

Dirk Hoekman
Wageningen University

Masanobu Shimada
Japan Aerospace Exploration Agency

Humberto Navarro de Mesquita, Jr.
Brazilian Forest Service

CONTENTS

15.1   Introduction

Remote sensing data acquired by synthetic aperture radar (SAR) provide unique opportunities for forest characterization, mapping, and monitoring, largely because of sensitivity of the radar signal to vegetation physiognomic structure and the provision of observations that are largely independent of atmospheric (e.g., cloud and smoke haze) and solar illumination conditions. Spaceborne SAR have been operating at a near global level since the 1990s, and the wide range of frequencies, polarizations, and observation strategies provide numerous opportunities for retrieving information on the past and current state of forests and surrounding landscapes and changes associated with natural and anthropogenic change, including climatic fluctuation. The development of systems and algorithms for characterizing, mapping, and monitoring forests, however, has been informed by studies using data acquired by SAR onboard airborne and spaceborne systems (e.g., the Shuttle Imaging Radar) and through dedicated missions.

This chapter reviews the use of spaceborne SAR for forest monitoring at regional to global scales. Particular focus is on the use of single- and dual-polarization backscatter data acquired at X-, C-, and L-bands, as these are the most widely available to those charged with forest monitoring. However, examples of how polarimetric SAR (POLSAR) and inteferometric SAR (InSAR) data can be used to improve monitoring are considered. The chapter provides essential background information on SAR and an overview of how key change processes of deforestation, degradation, and regeneration/afforestation can be detected using these data. Case studies relating to SAR-based monitoring of tropical rainforests in the Brazilian Amazon and Borneo, tree–grass savannas in Australia, and boreal forests in Siberia are then presented. Advantages of SAR for forest monitoring, either singularly or in combination with other sensors, are conveyed. The future of SAR for forest monitoring is discussed, particularly as this type of data is now increasingly used in support of local, national, and regional to global forest monitoring frameworks.

15.2   Suitability of SAR for Forest Monitoring

15.2.1   Forest Structural Diversity and Radar Modes

A wide range of forest types exist globally, with distinct formations occupying large areas including tropical rainforests, boreal and temperate forests, and tree–grass savannas. In all biomes, forests can be broadly categorized into evergreen, semi-deciduous, or deciduous. Common leaf types include broad-leaf, needle-leaf, and palm-like. Canopy cover ranges from sparse to closed and primarily as a consequence of prevailing environmental conditions (e.g., precipitation, evapotranspiration, soil types). As well as cover, forests are often distinguished on the basis of height and the number of canopy layers which, when distinct, can range from single layer (with no understory) to multilayer. The plants themselves also vary in their moisture content, canopy form and orientation, density, and size of their foliar and woody components. The substrate underlying forests may range from dry to wet and be smooth or rough, depending on the soil and geology and levels of inundation. Forest structure in all regions is highly variable and depends on growth stage, management practices, and natural and human-induced events and processes.

These different characteristics of forests are primary determinants of the variability in the SAR response at different frequencies and polarizations and over time. Hence, an understanding of microwave interactions with different components of the vegetation and the underlying surface is essential if these data are to be used for monitoring. A recent overview of imaging radar principles is provided in Kellndorfer and McDonald (2009), but information specific to forest monitoring is conveyed in the following sections.

15.2.2   SAR Frequencies and Polarisations

Spaceborne SAR, which provide capacity for monitoring over large areas, operate at X- (~9.6 GHz, 3.1 cm), C- (~5.3 GHz, 5.7 cm), and L-bands (~1.275 GHz, 23.5 cm). Within closed-canopy and taller forests, the shorter X- and C-band waves interact primarily with the foliage and smaller branches in the upper layers of the canopy and allow discrimination of forest types primarily as a function of differences in their leaf and small branch dimensions, orientations, and densities (Mayaux et al. 2002). However, in more open forests (e.g., tree–grass savannas), interactions may occur with the ground and woody components of the vegetation (Lucas et al. 2004). In all forests, microwaves emitted at lower frequency (L- and P-bands) generally penetrate through the smaller elements of the canopy and interact with the larger woody branches and trunks and ground surface.

At all frequencies, single- and double-bounce scattering result in a large amount of reflected energy returning to the sensor in the same polarization as that transmitted (i.e., HH or VV, with H and V representing the horizontal and vertical polarizations, respectively). The strongest returns are often at HH polarization, where double-bounce scattering between the ground surface and vertical structures (e.g., plant stems) occur, enhanced when forests are inundated by water. Volume scattering leads to depolarization of the transmitted signal and is caused by multiple interactions with structures (e.g., branches, leaves) that have multiple angles of orientation. Returns are comparatively lower from the cross-polarized wave (i.e., HV or VH) and are typically minimal for bare areas, including water. However, the HV backscatter generally increases asymptotically with the amount of plant material in the canopy and has been related to the above ground biomass (AGB) of forests at lower frequencies.

15.2.3   Interferometry

Spaceborne X-, C-, and L-band interferometric data have been used to map forest extent, the distribution of plant components in the forest volume, and canopy height. With *single pass interferometry*, one antenna is used to emit and receive a wave (in a single polarization), while a second detects the same polarization component of the reflected wave. In other words, both antennas measure the backscatter in the same polarizations but, as they are separated in range direction over a certain baseline, this causes a very small time lapse between the reception of reflection. This can be associated with the angle of the observed scatterer while the total elapsed time corresponds with the distance of the scatterer. Consequently, the position and height of the so-called scattering phase center can be determined. In areas without vegetation, the height of the terrain can be determined, while in areas with high vegetation in a single-resolution cell, scatterers over a range of heights are present. This range of phases is expressed by a parameter called interferometric phase difference, and the total correlation (normalized similarity) between the two data is commonly referred to as coherence. Most current SAR systems allow *repeat pass interferometry*, which is based on the use of only one antenna where the second measurement is undertaken within a short time period (from hours to weeks) and from a slightly different position (thereby forming the baseline). Coherence is high (approaches 1) when the same interaction with objects on the ground occurs between two images and decreases as a result of temporal decorrelation (e.g., because of changes in environmental conditions including surface moisture and wind) and volume decorrelation (because of variable scattering within volumes, including forests, as a function of observation parameters). Interferometric coherence is typically lower over forests, although it depends upon the season of observation.

15.3 Development of SAR for Forest Monitoring

15.3.1 Sensors Available for Monitoring

The benefits of using SAR for forest monitoring were recognized by a number of early studies, commencing with the 1970 Brazilian RADAMBRASIL project, in which airborne X-band SAR data were acquired over the entire Brazilian Amazon Basin, and followed by those making use of the shuttle imaging radar (SIR-A/B, SIR-C) SAR missions and other airborne datasets (e.g., NASA's AIRSAR). The Japanese Earth Resources Satellite (JERS-1) SAR provided the first L-band observations globally over the years 1992 to 1998 while the Canadian RADARSAT SAR and SAR on board the European Remote Sensing (ERS-1 and -2) satellites provided C-band observations and interferometric capability. From the mid-2000s and onwards, Italy and Germany launched X-band satellite missions, the COSMO SkyMed constellation and TerraSAR-X. Fully polarimetric observations were provided by the advanced land observing satellite phased array L-band SAR (ALOS PALSAR) (L-band), RADARSAT-2 (C-band), and TerraSAR-X (X-band) instruments at a near global level. The 2000 Shuttle Radar Topographic Mission (SRTM) and the TanDEM-X mission from 2010 provided unique capability for generating digital surface models (DSMs) at global scales, allowing retrieval of canopy height in more densely vegetated areas and the topographic ground surface.

The practical use of SAR for forest monitoring has followed developments in the technology and observation capability. The RADAMBRASIL project was the first to provide a baseline of the extent of forest cover in the Brazilian Amazon without interference from cloud or smoke haze. Focusing on more local areas, the SIR-C missions (X-, C-, and L-bands) allowed researchers to identify the benefits of using different radar wavelengths and polarizations for detecting forest extent, characterizing areas cleared of forest, and retrieving forest biomass and structural attributes (Kellndorfer et al. 1998). The capacity of interferometric SAR for retrieving forest height across larger areas was demonstrated using SRTM (Kellndorfer et al. 2004). The JERS-1 mission provided the first consistent pan-tropical and pan-boreal observations, from which regional-scale mosaics of the boreal and tropical zones were generated as part of the global rain forest mapping (GRFM) and global boreal forest mapping (GBFM) projects (Rosenqvist et al. 2000). The long-wavelength L-band SAR data proved useful for the classification of forest/nonforest areas and identification of secondary growth (Sgrenzaroli et al. 2002), particularly when time-series data were used. The L-band HH data also facilitated temporal mapping of standing water below closed-canopy forests, and hence differentiation of floodplain and swamp forests, and better understanding of the seasonal dynamics of inundation across large river catchments such as the Amazon and Congo (Hess et al. 1995). The successor of the JERS-1 SAR, the ALOS PALSAR, provided the first global systematic observations at a global level between 2006 and 2011. The ALOS mission highlighted the potential of SAR for operational forest monitoring, with the HV data providing better detection of deforestation across many regions compared to HH data. As the data accumulated into a time series, the benefits of using these and also JERS-1 SAR data for identifying events or processes that might lead ultimately to expansion of the area deforested or degraded or tracking histories of land use became apparent. As well as changes in the backscattering coefficient, interferometric observations proved useful for detecting disturbances within the canopy and suggested capacity for mapping degraded forest or identifying specific events (e.g., selective logging). In the boreal regions in particular, the advantages of using coherence data derived from combinations of spaceborne C- and L-band SAR for forest characterization, mapping, and monitoring became apparent. The advantages of integrating data from multifrequency SAR, optical sensors, and light detection and ranging (LiDAR) were also recognized.

15.3.2 SAR Observation Strategies

The use of satellite data for forest monitoring is currently moving from local studies on a limited number of satellite scenes, to regional or national scales where whole countries are to be monitored on a regular basis. Many countries have or are establishing operational national forest monitoring systems to meet their national reporting obligations in support of international conventions, with a key driver being the UN Framework Convention on Climate Change (UNFCCC) Reduction of Emissions from Deforestation and Degradation (REDD+). However, national-scale monitoring requires the availability of satellite sensor data that are consistent over countries, in terms of both coverage (no gaps) and temporal frequency (all acquisitions within a limited time period). A major strength of remote sensing technology is that long-term, systematic, and repetitive observations can be provided over large areas, particularly as SAR is not limited by low sun angles or persistent cloud cover. However, many moderate (10–30 m) spatial resolution sensors have not acquired data uniformly and regularly across large areas but have, instead, focused on areas where specific requests have been submitted. Consequently, some areas have received systematic coverage over long periods of time while neighboring areas have been totally neglected. Many satellite missions have also followed gap-filling background mission objectives as and when operational resources permit. However, the data have often been acquired without consideration given to the impacts of temporal effects, and the heterogeneous archives are of limited use.

Optical missions, and notably Landsat, have generally been more successful than their microwave counterparts, particularly over countries that have their own ground receiving stations. However, where regions are associated with frequent cloud cover, obtaining a full national coverage on an annual basis remains a major challenge. While more or less continuous observations are available through coarser spatial resolution MODIS or AVHRR data, the significantly higher data rates associated with moderate or fine spatial resolution sensors require a higher degree of planning if regional fragmentation is to be avoided. Therefore, a systematic observation strategy is needed for moderate and also fine (<10 m) spatial resolution datasets in order to meet the requirements of a remote sensing–based national forest monitoring system. In particular, the following should be taken into consideration, as highlighted by Rosenqvist et al. (2003):

- Spatially and temporally consistent observations over large areas to avoid gaps in acquisitions and minimize backscatter variations caused by seasonal differences in surface conditions between passes
- Adequate repetition frequency to facilitate detection of temporal changes as a result of, for example, flooding or land use
- Appropriate timing such that long-term repetitive observations are taken over the same time frame each year and ideally targeted to seasons where backscatter conditions are more stable
- Consistency in sensor observation modes such that acquisitions are limited to a small number of "best trade-off" sensor modes, thereby maximizing data homogeneity and minimizing programming conflicts
- Long-term continuity such that observations can be continued from sensors that are preceding or launched in the future

The first radar-based systematic observation strategy dates back to experiences gained with the JERS-1 SAR which, during the last 3 years of its lifetime (1995–1998), was used to acquire data in a consistent manner over the entire tropical and boreal zones of the Earth (Rosenqvist et al. 2000). For the first time, the utility and feasibility of acquiring moderate spatial resolution data systematically and repetitively at continental scales was demonstrated. The global acquisition strategy concept was implemented, in full, for the ALOS satellite, with the PALSAR programmed to achieve at least one gap-free coverage of all land areas every 6 months (Rosenqvist et al. 2007), as illustrated in Figure 15.1.

The importance of systematic acquisition strategies is becoming increasingly recognized, and a number of near-future satellite missions are planning similar global observation plans. Of particular note was the joint effort made from 2012 to establish a coordinated multimission acquisition strategy by a number of national space agencies under the framework of the global forest observation initiative (GFOI) of the group on Earth observations (GEO), comprising moderate-to-fine (<10 m) spatial resolution optical and X-, C-, and L-band SAR.

15.3.3 Synergistic Use of SAR and Optical Data

The development of a forest monitoring program that integrates both SAR and optical data acquired across a range of frequencies is ideal and the benefits are being increasingly realized, with demonstration in a few cases

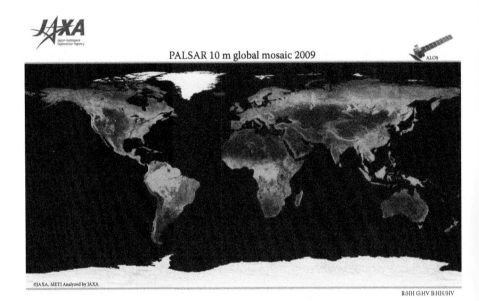

FIGURE 15.1
(See color insert.) Global ALOS PALSAR color composite mosaic at 10 m pixel spacing (R: HH, G: HV, B: HH/HV). 95% of the data—a total of approximately 70,000 scenes—were acquired within the time period June–October 2009. (Courtesy of JAXA EORC, Tsukuba, Japan.)

(e.g., Queensland, Australia). The benefits include the provision of complementary information on the foliage/canopy and woody components of vegetation, which can assist mapping of forest types (e.g., regrowth, mangroves) and retrieval of linked biophysical properties (e.g., canopy cover and AGB). SAR data can also be used to "in fill" gaps in time series of optical remote sensing data where cloud or haze cover prevents acquisition of the latter or the revisit frequency or timing of acquisition is suboptimal. ScanSAR data, in particular, have proven to be particularly useful for this purpose. Where the timing of SAR and optical data acquisitions is not coincident, more comprehensive time-series datasets detailing changes in forest cover can be generated. SAR data may also prove to be the workhorse of operational monitoring programs in the event of failure by one or more sensors (e.g., Landsat).

A multi-sensor and multi-scale approach to monitoring also allows better detection of hotspots of change (e.g., through observations of fire activity from, for example, MODIS) or areas that are vulnerable to future change. For example, fire activity detected by sensing in the middle or thermal infrared wavelengths from coarse spatial resolution sensors of high temporal frequency can be followed up by SAR observations of the area affected (Siegert and Hoffman 2000). Adverse changes in the long-term trends in measures of vegetation productivity (e.g., the normalized difference vegetation index [NDVI] or enhanced vegetation index [EVI]), as derived from coarse spatial resolution optical data, may indicate areas of regrowth or degradation involving accumulation or loss of plant material, which can potentially be characterized through time-series comparison of SAR data. In both cases, fine spatial resolution and programmable data, such as that provided by the Tandem-X mission or very high-resolution (VHR) optical sensors (e.g., Worldview, Quickbird), can then be used to associate observed changes with an actual or likely cause such that measures can be put in place to prevent further loss or degradation of forests. Long-term and regular wall-to-wall observations at a regional level are critical as approaches that sample the landscape often omit changes in forest cover because of their restricted extent (e.g., along road networks or the borders between lowlands and uplands). In all cases, the combination of optical and SAR data provides enhanced benefits for forest monitoring and also understanding the processes of change.

15.4   Processes of Forest Change

Changes in forest cover are typically associated with specific events (e.g., clearcutting), long-term degradation, natural succession, or human-induced regeneration following clearance or disturbance. In each case, SAR can play a role in mapping and monitoring change and also estimating the magnitude of changes in structure and AGB, as outlined in the following sections.

15.4.1   Deforestation

Deforestation is defined as a conversion of forest to nonforest. However, establishing the boundary between forests and nonforest or the magnitude of change that constitutes a deforestation event is often compromised by factors including the nature of forest loss in terms of structural components removed and the methods of clearing. Using SAR data, such definitions are compromised by prevailing climatic (e.g., rainfall, freeze–thaw cycles) and background conditions (e.g., surface roughness, soil water-holding capacity).

Deforestation is ordinarily associated with complete removal of woody vegetation and hence a change in the dominance of volume and double-bounce scattering to surface scattering. However, in some cases, cut stumps, fallen woody material, and individual trees (e.g., palms, nonproductive timbers) are often remaining following clearance events. In the tropics and at L-band, an increase in the backscattering coefficient at HH polarization is typically observed because of double-bounce interactions with woody debris (slash; Almeida-Filho et al. 2009), which can be greater than that of the original forest, as shown in the example from Riau Province in Indonesia (Figure 15.2a). However, this is typically followed by a rapid decline because of loss of woody

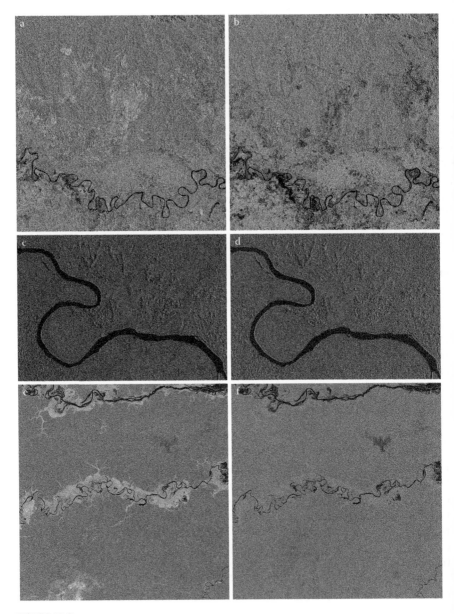

FIGURE 15.2
For a site in Riau Province, Indonesia, ALOS PALSAR image highlighting the difference between forest and nonforest at (a) HH polarization and (b) HV polarization. For a tropical rainforest site in Guyana, open gold mining is less evident within Cosmo-SkyMed X-band (c) HH polarization data compared to (d) **(See color insert.)** a composite of HH data from two dates (September 12 and 15, 2011) and coherence (in RGB respectively; blue areas indicate deforested areas). Due to double-bounce scattering between tree stems and the water surface at L-band, inundated forest in the Central Amazon Basin is clearly visible (bright) at (e) HH polarization, while barely visible at (f) HV polarization.

debris such that deforested areas often become indistinguishable for a short period because of similarity in backscatter with adjacent forest. Over time, however, these become more separable, exhibiting a lower backscatter than primary forest because of the dominance of specular scattering. In some cases, woody material can be piled into rows, which leads to a high backscatter at HH polarization. Trees can also be left standing and exhibit a high return at L-band HH but a lower return if observed using, for example, C-band SAR or optical data (Lucas et al. 2008). Areas of open ground may also exhibit a similar backscatter as areas with dead standing trees. For detecting deforested areas, greater contrast with undisturbed forests is generally obtained at L-band at HV polarization (Figure 15.2b). However, the distinction between forest and nonforest is often compromised using higher frequency (C-band/X-band) SAR single-polarization data because of similarities in backscatter with herbaceous vegetation (e.g., pastures). Differences are greater where interferometric coherence data are used, as illustrated in the X-band example in Figure 15.2c and d, allowing detection of the deforested area (appearing blue in Figure 15.2d). The environmental conditions prevailing at the time of the SAR image acquisition also have implications for mapping and monitoring the extent of forest cover. For example, a reduction in SAR backscatter may occur as a consequence of thawing or snowmelt, reductions in precipitation, or lowering/raising of the water level beneath a forest canopy. In the latter case, the signature can be similar to recently cleared forest, and hence misinterpretations may occur when mapping deforested areas (Figure 15.2e and f).

Methods for defining the forest/nonforest boundary have ranged from simple thresholding to more complex classifications, but, in each case, compromises have been necessary or errors are introduced for the reasons mentioned above. However, the decision as to what constitutes the boundary has significant implications for countries reporting on the extent of forest cover and hence the detection of change. An alternative option is to retrieve biophysical attributes of forests (e.g., area, height, and cover) that are used in standard definitions of forest cover and can be more easily interpreted, although this has rarely been undertaken to date.

15.4.2 Forest Degradation and Natural Disturbances

Forest degradation typically involves the removal of individuals or groups of trees through processes such as selective logging and fuel wood collection as well as dieback as a consequence of, for example, burning or drought. Typically, degradation results in a loss of trees and hence canopy material, with a corresponding reduction in backscattering coefficient and a change in texture typically evident within the SAR image depending on frequency and polarization.

Many studies have highlighted how forest degradation can be observed from SAR data. One example is in the peat swamp forests of Indonesia where drainage of the central dome contributed to underground peat fires, which eventually led to the collapse of the forest. This sequence of events was captured in a time series of JERS-1 SAR data acquired between 1995 and 1998 (Figure 15.3). Until 1996, the dome was still hydrologically intact, but the construction of a very wide canal through the dome was visible in the JERS-1 SAR image of 1997. In the third image of the sequence (September 1997), the canal was filled with water leading to specular scattering away from the sensor and hence its black appearance in the image. A small but bright area is also evident, which then grew in area, becoming brighter until the collapse of the forest, as observed in January 1998. For many peat swamp areas in Borneo and Sumatra, large series of historical JERS-1 images collected during the period 1992–1998 (as many as 30 scenes) and ALOS PALSAR data from 2007 to 2010 show evidence of degradation of the peat swamp forests. The sequence illustrated highlights the benefits of using time series of SAR data from L-band, although data from other sensors can also indicate degradation.

From single-date SAR imagery, selective logging is often difficult to discern because of the relatively coarse spatial resolution, although multitemporal datasets can be used to better identify such areas. SAR coherence measurements can also indicate disturbance. As examples, interferometric ERS-2 SAR data have been used to detect losses of canopy in Kalimantan following large wildfires, while ALOS PALSAR coherence data have proved useful for detecting the impacts of severe fires in Victorian forests in Australia in 2009. These data also have the potential for detecting natural disturbances associated with, for example, downdrafts and lightning strikes as well as long-term declines in the condition of forests as a consequence of drought or flooding.

15.4.3 Secondary Forests and Woody Thickening

Secondary forests often establish following deforestation or degradation while in some intact forests, thickening of the vegetation may occur as a consequence of rainforest expansion or lack of burning over long time periods. A limitation of using SAR data, particularly in tropical regions, is that the rapid increase in woody material renders them indistinguishable from primary forest within a few years. Therefore, most information relating to different stages of regrowth is gathered in the early years of regrowth.

Where forests regenerate on land used previously for agriculture or clear felled of trees, these can be identified through time-series comparison of SAR data, although the point at which regrowing woody vegetation can be considered to be forest can be contentious. Temporal datasets can, however, be used to track the progression of regrowth as the SAR backscatter typically increases over time up to the level of saturation. As an example, the recovery of mangroves in Perak, Malaysia, that had been cleared in rotation can be readily detected using time series of JERS-1 SAR and ALOS PALSAR data through changes in backscattering coefficient. Time-series comparisons of remote sensing data classifications assume that forests of the same age are

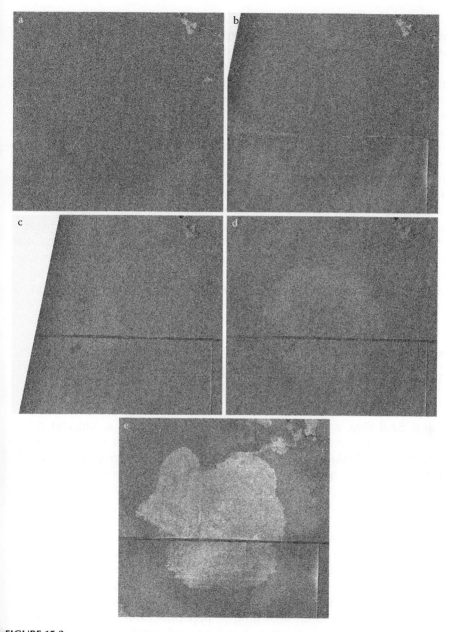

FIGURE 15.3
The collapse of a forest on top of a peat dome in Central Kalimantan, Indonesia, as observed using time series of JERS-1 SAR data acquired on (a) July 12, 1995, (b) March 19, 1997, (c) September 11, 1997, (d) October 25, 1997, and (e) January 21, 1998. The image width is ~21 km.

similar in terms of their structure, species composition, and biomass. SAR data can, however, be used to differentiate forests that are of similar age but may differ in terms of their structure or accumulated AGB.

The use of multifrequency SAR and optical data provides unique opportunities to map the extent of regrowth and differentiate growth stages based on structural development rather than age. At C-band, early stages of regrowth are often indistinguishable from herbaceous vegetation while at L-band and particularly at P-band, these may be unable to be distinguished from nonforested areas as the stem size and density may be insufficient to evoke a response. Forests at more advanced stages of growth are, however, able to be detected. Hence, the use of X- and/or C-band or optical data to establish the presence of plant material and lower frequency L- and P-band SAR to determine whether woody components exist and their relative sizes are useful for determining the nature of regrowth as a function of its structural development. Applying thresholds to SAR data to discriminate regrowth forests from nonforest is often problematic because of confusion with other land and water surfaces and hence the application of thresholds to layers representing retrieved biophysical attributes (e.g., AGB retrieved from SAR data) may be more appropriate.

15.5 Forest Monitoring

15.5.1 Overview

Using SAR data, a large amount of information on the extent and nature of deforestation and degradation associated with human activities, natural disturbances through specific events and long-term processes, and the patterns and dynamics of regrowth can be quantified. Such knowledge can be used to inform subsequent use of the land, in planning for conservation and sustainable management of the existing forest area, and for restoring forests on land that had been previously cleared or degraded. In many cases, forested landscapes have been classified into thematic categories (forest, nonforest, regrowth stages, logged forest), with thresholds or specific algorithms used. Forests can, however, be described on the basis of retrieved biophysical attributes (e.g., height from interferometric SAR or AGB obtained from lower frequency SAR), with continuous surfaces produced which are subsequently subdivided according to predefined intervals. Nevertheless, complications in the description, mapping, and monitoring of change have arisen from variations in environmental conditions (e.g., surface moisture). The following sections provide several case studies from boreal and tropical forests as well as wooded savannas in which SAR data have been used for monitoring forests at local to regional levels and demonstrate how some of the issues presented above have been addressed.

15.5.2 Xingu Watershed, Mato Grosso, Brazil

The Xingu watershed headwaters in southeastern Brazil is representative of many areas along the Amazon's "arc of deforestation." The native vegetation within the 387,000 km^2 of the headwaters includes tall evergreen (25–45 m) and transitional semideciduous (10–30 m) and riparian forests as well as savannas, with these encompassing cerrado woodland, grassland mosaics, thickets, and gallery forest. In the late 2000s, the headwaters contained more area in dense humid forest (~221,000 km^2) than 90% of the world's tropical nations. However, annual deforestation in the forest biome from 2000 to 2007 ranged between 649 km and 3,170 km^2, with an average annual rate over the 7-year period of 1951 km^2. This represents between 5% and 13% of all deforestation in the Brazilian Amazon (Stickler et al. 2008).

Using a mosaic of ALOS PALSAR data generated using spatially and temporally consistent images acquired during the period June to August 2007 (Figure 15.4a), areas of forest and nonforest (Figure 15.4b) were differentiated by applying a random forest algorithm to objects generated using eCognition and aggregating classes at several levels (Walker et al. 2010). This resulted in an accuracy of 92.4% when ancillary spatial/topographic predictor variables were included. A similar approach applied to a Landsat sensor mosaic (Figure 15.4c), comprised of data acquired over the same timeframe, resulted in an accuracy of 94.8%. The overall agreement between the PALSAR and Landsat-based forest cover products varied from 89.7% (1 pixel window) to 93.8% (11 pixel window), with minor discrepancies in some class boundaries (e.g., in the agreement of forests/field edges) being the primary source of spatial dissimilarity between the maps.

The observation strategy developed for the ALOS PALSAR allowed the generation of both dual and single polarimetric data mosaics from quasi-identical periods between 2007 and 2009. Change detected through comparisons of backscatter included clearcutting, logging, and wildfires. For the detection of change, the use of objects rather than pixels was preferred, as management activities or natural degradation typically occurred in areas larger than single pixels. Furthermore, the reduction in image speckle associated with averaging pixels within objects increased signal stability. Change gradients, based on regression tree approaches (e.g., random forest or support vector machines), were preferred over simple comparisons of forest cover maps at several time steps, because of reductions in errors.

A composite of HV data acquired in 2007, 2008, and 2009 over the Xingu watershed is shown in Figure 15.5. Forests clearcuts with slash removed between the 2007 and 2008 acquisitions appear in bright red because of significantly lower backscatter from clearcut areas (darker green and blue tones due after logging). Areas with the same treatment imposed between the 2008 and 2009 acquisitions appear in bright yellow. Other land covers include lower biomass cerrado with no significant change (medium gray), grassland, and bare soil (dark gray to black). Agricultural fields are generally darker,

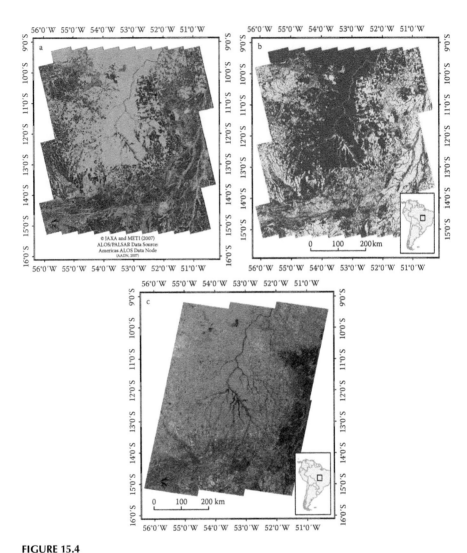

FIGURE 15.4
(See color insert.) Satellite image mosaics produced for the Xingu River headwaters region.
(a) ALOS PALSAR mosaic consisting of 116 individual Level 1.1 (single-look complex) fine
beam, dual-polarimetric scenes (R/G/B = polarizations HH/HV/HH–HV difference). (b) Map
of forest (green) and nonforest (beige) generated with an overall classification accuracy of
92.4% ± 1.8%. (c) Landsat 5 mosaic consisting of 12 individual Level 1G (Geocover) scenes
(R/G/B = bands 5/4/3).

with the variability in backscatter associated with changes in crop phenology and also surface (soil and vegetation) moisture. While algorithms for
change detection can be based on general principals of change, several ambi-
guities need to be taken into account when mapping change. In particular,
moisture changes from rain events can enhance the backscatter signal but are

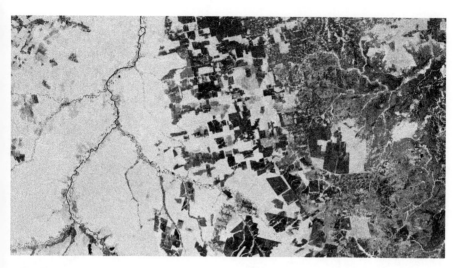

FIGURE 15.5
(See color insert.) Multitemporal ALOS PALSAR L-band HV image generated from data acquired in 2007 (red), 2008 (green), and 2009 (blue) for a part of the Xingu watershed. Closed forest (white) is interspersed with fire scars (red tones) along the main stem of the Xingu River and tributaries (black).

often readily identified at a swath scale and bias corrections can sometimes be applied in mosaic generation. In general, backscatter levels of standing (primary) forest are around –7 to –9 dB in HH and –13 to –15 dB in HV, and losses in AGB associated with logging, clearcutting, or fire reduce the ALOS PALSAR backscatter by 5–7 dB. Smaller changes are mostly related to agricultural activities and changes in both phenology and surface moisture.

15.5.3   Detecting Forest Degradation in Borneo

Borneo is the third largest island in the world and covers an area of approximately 750,000 km². Almost three quarters of the island is part of Indonesia (Kalimantan) while Sarawak and Sabah are territories of Malaysia, and the Sultanate of Brunei Darussalam occupies a small area. Until the 1950s, Borneo was almost entirely covered by tropical evergreen broadleaved forest, with other major natural vegetation covers including peat swamp forests along the coastal and subcoastal lowlands, freshwater swamps along the inland rivers, and mangrove forests on the coastal plains. However, intensive logging of predominantly commercial dipterocarp species and conversion to cropland, oil palm, and timber plantations have reduced forest cover significantly.

The establishment of baseline maps of forest cover and type against which to quantify and determine the nature and impact of change is essential. Using ALOS PALSAR fine beam single (FBS) and dual (FBD) polarization (path) image pairs acquired in 2007, a map of forest and land cover types was generated (18 classes in total). These maps have been used subsequently to assist government agencies in their spatial planning and reporting on the status of the environment, thereby allowing compliance with international environmental treaties (Hoekman et al. 2010). Furthermore, maps were generated annually using data acquired in 2008 and 2009, with these highlighting areas of forest degradation through selective logging (Figure 15.6).

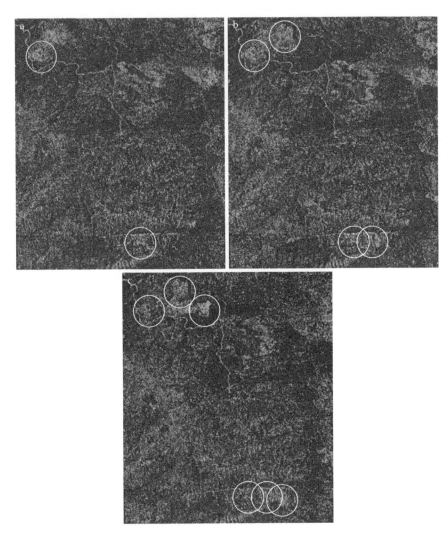

FIGURE 15.6
(See color insert.) Forest degradation in Sarawak through selective logging observed through comparison of forest maps generated using ALOS PALSAR data for the years (a–c) 2007 through to 2009.

15.5.4   Rapid Detection of Deforestation by ScanSAR

Using optical imagery, deforestation in the Brazilian Amazon is monitored and reported on an annual basis by the Brazilian Institute of Space Research (INPE). The majority of data is acquired during the dry season (July to October), although smoke haze and cloud reduce acquisition rates as this season progresses and during the wet season. While INPE provides deforestation alerts every 15 days, the Brazilian Institute for Environment and Natural Renewable Resources (IBAMA) is charged with implementing measures that prevent deforestation before it occurs. For this purpose, the Japan Aerospace Exploration Agency (JAXA) operated the ScanSAR routinely over Brazil every 3 days and provided the processed ScanSAR images to the IBAMA within 5 days from the acquisition date. Provided with information on deforestation events from IBAMA's Remote Sensing Centre, environmental law enforcement agents visit affected sites through ground or helicopter transportation. The imagery also assists the agents to define the logistics and strategies for subsequent field actions. While optical imagery is used, the wide-swath ScanSAR mode of the ALOS PALSAR has allowed detection of early deforestation. Each area identified as indicating a change is delineated within the image and the area is classified as being in the initial processes of deforestation or is a consequence of ongoing clearcutting of the forest. The information is assembled into a deforestation indication document enabling the law enforcement agents to respond rapidly to deforestation events, with particular focus on halting those that are illegal.

15.5.5   Change Detection in Boreal Forests

Boreal forests are extensive throughout the northern hemisphere and are located primarily in Siberia and North America. The SIBERIA project aimed to generate baseline maps of boreal forest cover across Siberia by using a combination of ERS-1 and ERS-2 SAR tandem coherence data and JERS-1 SAR backscatter data for 1997 to 1998. Mapping of forest cover was informed by relationships established between growing stock volume and both C-band coherence and L-band backscatter. The classification was undertaken using a maximum likelihood algorithm based on class statistics generated from training data.

Within the boreal zone, the ability to detect change depends upon the timing of observation. During the winter months, extensive snow cover and frozen conditions limit detection of forest cover using backscatter data. However, using interferometric pairs of ALOS PALSAR data, Thiel et al. (2009) established that temporal decorrelation was low during the winter months, and areas of forest and nonforest could be separated using a combination of winter-coherence data and PALSAR summer backscattered intensities. Operational delineation of forest cover was suggested, with accuracies exceeding 90% when an object-based classification was applied.

15.5.6 Quantifying Regrowth Dynamics in Amazonia and Australia

Australia supports a diversity of vegetation, with the greatest expanse associated with sparse to open tree–grass savannas. Temperature, sub-tropical, and tropical closed forests occur towards the coast and often at higher elevations. Within Queensland, Australia, vegetation monitoring is undertaken through the Statewide Landcover And Trees Study (SLATS) (Danaher et al. 2010) and primarily using time series of Landsat sensor data. The extent of woody vegetation is mapped on an annual basis using Landsat-derived foliage projective cover (FPC). The type and ecological importance of vegetation cleared is determined by intersecting mapped areas of deforestation with regional ecosystem (RE) mapping of vegetation types. Through this approach, changes in vegetation cover are tracked and ameliorative measures taken where appropriate.

While SAR data have not yet been used for operational monitoring in Queensland, potential exists for refining maps of woody vegetation and forest growth stage, thereby increasing the reliability of estimates of deforestation and regenerating forest areas. For example, confusion between herbaceous and woody vegetation occurring within Landsat sensor data is largely overcome by integrating ALOS PALSAR data because of the lack of interaction with the former although confusion with rough ground can occur, particularly with increasing amounts of surface moisture. Integration of the ALOS PALSAR with Landsat FPC data also allows the detection of the early stages of woody regrowth, which typically exhibit an FPC equivalent to forest (i.e., >12%, equivalent to a canopy cover of 20%) but an L-band backscatter more characteristic of nonforest (Lucas et al. 2006). Using such an approach, the dynamics of regrowth can be tracked, including the progression of regrowth through different stages.

15.5.7 Wider Use and Future Sensors

The studies outlined above have highlighted the benefits of using SAR data for monitoring deforestation, degradation, regrowth dynamics, and natural disturbances. In each case, the benefits for better understanding the cycling of carbon through landscapes, conserving biodiversity, and contributing to a range of national policy and international conventions are evident. However, in many cases, such datasets have not been effectively exploited nor recognized.

In the future, a number of SARs are planned, which are anticipated to provide significant advances in forest characterization, mapping, and monitoring at a global scale. These include the European Space Agency (ESA) Sentinel satellites, which are anticipated to provide interferometric and polarimetric observations at C-band (two satellites). The ALOS-2 and the Argentinian SAOCOM satellites are expected to provide L-band SAR observations while the ESA BIOMASS mission will be the first to provide P-band observations, specifically for the retrieval of forest biomass. The NASA DESDynI mission is also intended to provide a dedicated L-band SAR. The challenge will be the full integration of data from these sensors into forest monitoring systems and the use of data acquired in different modes and following different acquisition strategies.

15.6   Conclusions

While optical remote sensing data are the workhorse of many forest monitoring systems, SAR are able to acquire data regardless of clouds and haze, and are increasingly providing opportunities to uniquely detect deforestation activity as well as degradation and regeneration in a consistent and repetitive manner. These data can also inform on the conditions imposed through clearance operations or during subsequent use of the land.

The benefits of providing routine and consistent observations have been demonstrated through the JERS-1 SAR and ALOS PALSAR and while the archives only span over limited number of years, comparison of these data has allowed long-term trends in the amount and type of woody vegetation to be quantified in some cases.

While the relative benefits of SAR and optical data have been debated in the remote sensing community for some time, the integration of these datasets provides the greatest potential for monitoring systems. In particular, SAR data can fill in gaps where cloud cover or smoke haze prevents observations from optical sensor data (for periods covering several years) or can be integrated to provide better mapping of, for example, regeneration stages.

The increasing diversity of observation modes is expected to enhance the use of SAR into the future. The continued and future provision of global single, dual, and fully polarimetric data at X-, C-, and L-bands and interferometric capability together with a greater understanding of the information content of these data is anticipated to lead to increased use of SAR in many forest monitoring activities across a range of biomes and scales. The key challenge is to optimize the development and use of these data such that they ultimately contribute to not only halting the relentless loss of forest but also restoration through better understanding of the dynamics of the forest ecosystems in response to human activities.

About the Contributors

Richard M. Lucas obtained his PhD in 1989 from the University of Bristol, Bristol, United Kingdom. He currently leads the Earth Observation Group at Aberystwyth University, Aberystwyth, United Kingdom, where his research focuses on understanding change in terrestrial ecosystems through integration of ground, airborne, and spaceborne remote sensing data. Following a research position at Swansea University, Swansea, United Kingdom, focusing on mapping and understanding regrowth dynamics in Amazonia and Central Africa, he joined the University of New South Wales, Sydney, Australia (1998), where he initiated the long-term Injune Landscape Collaborative Project in central Queensland, which is aimed at understanding the implications of natural and human-induced change on woodland dynamics. He has been actively involved with the Japan Aerospace Exploration Agency (JAXA) Kyoto & Carbon Initiative, with research focusing on forest (including mangrove) characterization, mapping, and monitoring in the tropics and subtropics.

Ake Rosenqvist received his Dr Eng degree in microwave remote sensing from the University of Tokyo, Tokyo, Japan, in 1997. He joined the Swedish Space Corporation in 1990, where he was engaged in the French/Swedish SPOT program. He was invited to the National Space Development Agency of Japan, Tokyo, in 1993, where he became involved in the development of the JERS-1 application program and the establishment of the Japanese Earth Resources Satellite (JERS-1) SAR Global Forest Mapping project. He spent 6 years at the European Commission Joint Research Centre, Ispra, Italy, and another 5 years as a senior scientist at the Japan Aerospace Exploration Agency (JAXA), where he developed the concept of systematic data observations, which was adopted by JAXA and implemented for the advanced land-observing satellite (ALOS) mission. He founded solo Earth Observation (soloEO), Tokyo, in 2009, and is currently involved in the group on Earth observations (GEO) forest carbon tracking task, coordination of the Kyoto and Carbon (K&C) Initiative, and supporting the development of systematic acquisition strategies for the ALOS-2 and Argentinean SAOCOM-1 missions. Rosenqvist was awarded the International Society for Photogrammetry and Remote Sensing (ISPRS) President's Honorary Citation in 2000.

Josef M. Kellndorfer received his PhD in geosciences from the Ludwig–Maximilians University in Munich, Germany. His research focuses on the monitoring and assessment of terrestrial and aquatic ecosystems, and the dissemination of earth observation findings to policy makers through education and capacity building. Using geographic information systems, remote sensing, and modeling, he studies land use, land cover, and climate change on a regional and global scale. His projects include carbon and biomass mapping of the United States and mapping forest cover across the tropical forested regions of Latin America, Africa, and Asia through the generation of consistent data sets of high-resolution, cloud-free radar imagery. Before joining the Woods Hole Research Center, Falmouth, Massachusetts, he was a research scientist with the radiation laboratory in the Department of Electrical Engineering and Computer Science at the University of Michigan, Ann Arbor, Michigan.

Dirk H. Hoekman received his PhD degree from Wageningen University, Wageningen, The Netherlands, in 1990. He is employed at Wageningen University since 1981 where he currently holds the position of Associate Professor in remote sensing with main interests in the physical aspects of remote sensing, microwave remote sensing, and applications of remote sensing in forestry, agriculture (agro), hydrology, climate studies, and environmental change. He was scientific coordinator for the European Space Agency/Brazilian Institute of Space Research (ESA/INPE) Amazonian expedition (1986); for tropical rain forest sites in Colombia and Guyana (1992–1993); co-coordinating investigator for tropical rain forest sites in Indonesia, including NASA's Pacific Rim (PACRIM)-2 campaign (2000), and ESA's INDREX-1 and -2 campaigns (1996, 2004). He is currently coordinating research programs to support the realization of operational wide-area radar monitoring systems for sustainable forest management and the United Nations Reduction of Emissions from Deforestation and Degradation (UN REDD), and exploring the scientific use of advanced land-observing satellite phased array L-band synthetic aperture radar (ALOS PALSAR) for tropical peat swamp forest hydrology and forest cover change monitoring in Indonesia and the Amazon in the framework of the Japan Aerospace Exploration Agency (JAXA) Kyoto & Carbon Initiative.

Masanobu Shimada (M′97–SM′04–F′11) received his PhD degree in electrical engineering from the University of Tokyo, Tokyo, Japan, in 1999. Since 1979, he has been with the National Space Development Agency of Japan (NASDA), Tokyo, which is currently the Japan Aerospace Exploration Agency (JAXA). He has been assigned duties at the Earth Observation Research Center, where he serves as the advanced land-observing satellite (ALOS) science manager, is responsible for ALOS calibration/validation activities, global forest mapping, and SAR interferometry projects since 1995, and is currently a principal researcher. With JAXA, he designed a NASDA scatterometer. From 1985 to 1995, he developed data-processing subsystems for optical and SAR data (i.e., MOS-1, SPOT, and JERS-1) at the Earth Observation Center. He was a visiting scientist at the Jet Propulsion Laboratory, Los Angeles, California, in 1990.

Daniel Clewley received his BSc (Physics), MSc (remote sensing and GIS), and PhD from Aberystwyth University, Aberystwyth, United Kingdom, in 2006, 2007, and 2011, respectively. His work had focused on modeling and inversion of microwave interactions in tree–grass systems and mapping of regrowth stages in Queensland, Australia, using combinations of SAR and optical data. He is currently based at the University of Southern California, Los Angeles, California.

Wayne S. Walker received a PhD in remote sensing from the University of Michigan's School of Natural Resources and Environment, Ann Arbor, Michigan. He is an ecologist and remote sensing specialist interested in applications of satellite imagery to the assessment and monitoring of temperate and tropical forests at regional to global scales. His research focuses on measuring and mapping forest structural attributes, land cover/land use change, and terrestrial carbon stocks in support of habitat management, ecosystem conservation, and carbon cycle science. He is committed to building institutional capacity in the tools and techniques used to measure and monitor forests, working in collaboration with governments, NGOs, and indigenous communities across the tropics.

Humberto Navarro de Mesquita Jr. received his MSc in ecology at the University of São Paulo (USP), São Paulo, Brazil, and a PhD from the São Paulo Landscape Ecology Laboratory in conjunction with the Department of Geography, University of Edinburgh, Edinburgh, United Kingdom. From 2000 to 2002, he was involved in the GRFM and GBFM initiatives and has extensive involvement with the Japan Aerospace Exploration Agency (JAXA) Kyoto and Carbon (K&C) Initiative. He was head of the Remote Sensing Centre (CSR) of the Brazilian Institute of Environment and Natural Renewable Resources (IBAMA) until June 2010.

References

Almeida-Filho, R. et al., Using dual polarized ALOS PALSAR data for detecting new fronts of deforestation in the Brazilian Amazônia. *International Journal Of Remote Sensing*, 30, 3735–3743, 2009.

Danaher, T., Scarth, P., Armston, J., Collett, L., Kitchen, J. and Gillingham, S., Remote Sensing of Tree-Grass Systems: the Eastern Australian Woodlands. In: *Ecosystem Function in Savannas: Measurement and Modeling at Landscape to Global Scales* (Eds., Michael J. Hill and Niall P. Hanan), CRC Press, Boca Raton, Florida, US, 175–194, 2010.

Hess, L.L. et al., Delineation of inundated area and vegetation along the Amazon floodplain with the SIR-C synthetic aperture radar. *IEEE Transactions on Geoscience and Remote Sensing*, 33, 896–904, 1995.

Hoekman, D.H., M.A.M. Vissers, and N.J. Wielaard, PALSAR wide-area mapping methodology and map validation of Borneo. *IEEE Journal of Selected Topics in Applied Earth Observations and Remote Sensing (J-STARS)*, 3(4), 605–617, 2010.

Kellndorfer, J. and MacDonald, K., Microwave remote sensors. in T. Warner, M.D. Nellis, and G.M. Foody (Eds.), *The SAGE Handbook of Remote Sensing* (586 pages), Sage, London, 13, 179–198, 2009.

Kellndorfer, J.M. et al., Toward consistent regional-to-global-scale vegetation characterization using orbital SAR systems. *IEEE Transactions on Geoscience and Remote Sensing*, 36, 1396–1411, 1998.

Kellndorfer, J.M. et al., Vegetation height estimation from Shuttle Radar Topography Mission and National Elevation Datasets. *Remote Sensing of Environment*, 93, 339–358, 2004.

Lucas, R.M., Moghaddam, M., and Cronin, N., Microwave scattering from mixed species woodlands, central Queensland, Australia. *IEEE Transactions on Geoscience and Remote Sensing*, 42(10), 2142–2159, 2004.

Lucas, R.M. et al., Integration of radar and Landsat-derived foliage projected cover for woody regrowth mapping, Queensland, Australia. *Remote Sensing of Environment*, 100, 407–425, 2006.

Lucas, R.M. et al., Assessing human impacts on Australian forests through integration of airborne/spaceborne remote sensing data. in R. Lafortezza, J. Chen, G. Sanesi, and T.R. Crow (Eds.). *Patterns and Processes in Forest Landscapes: Multiple Uses and Sustainable Management*. Springer, Dordrecht, pp. 213–240, 2008.

Mayaux, P. et al., Large scale vegetation maps derived from the combined L-band GRFM and C-band CAMP wide area radar mosaics of Central Africa. *International Journal of Remote Sensing*, 23, 1261–1282, 2002.

Rosenqvist A., Milne, A.K., and Zimmermann, R., Systematic data acquisitions: A pre-requisite for meaningful biophysical parameter retrieval? *IEEE Transactions on Geoscience and Remote Sensing* and *Communication*, 41(7), 1709–1711, 2003.

Rosenqvist, A., Shimada, M., and Watanabe, M., ALOS PALSAR: A pathfinder mission for global-scale monitoring of the environment. *IEEE Transactions on Geoscience and Remote Sensing*, 45(11), 3307–3316, 2007.

Rosenqvist, A. et al., The global rain forest mapping project: A review, *International Journal of Remote Sensing*, 21, 1375–1387, 2000.

Sgrenzaroli, M. et al. Tropical forest cover monitoring: estimates from the GRFM JERS-1 radar mosaics using wavelets zooming techniques and validation. *International Journal of Remote Sensing*, 23, 1329–1355, 2002.

Siegert, F. and Hoffmann, A.A., The 1998 forest fires in East Kalimantan (Indonesia): A quantitative evaluation using high resolution, multi-temporal ERS-2 SAR images and NOAA-AVHRR hotspot data. *Remote Sensing of Environment*, 72, 64–77, 2000.

Stickler, C.M. et al., The opportunity costs of reducing carbon emissions in an Amazonian agroindustrial region: The Xingu River headwaters. in Berlin Conference on the Human Dimensions of Global Environmental Change. Berlin, Germany, 2008.

Thiel, C., Thiel, C., and Schmullius, C., Operational large-area forest monitoring in Siberia using ALOS PALSAR summer intensities and winter coherence. *IEEE Transactions Geoscience and Remote Sensing*, 47(12), 3993–4000, 2009.

Walker, W.S., Large-area classification and mapping of forest and land cover in the Brazilian Amazon: A comparative analysis of ALOS/PALSAR and Landsat data sources. *IEEE Journal of Selected Topics in Applied Earth Observations and Remote Sensing*, 3, 594–604, 2010.

16

Future Perspectives (Way Forward)

Alan Belward and Frédéric Achard

Joint Research Centre of the European Commission

Matthew C. Hansen

University of Maryland

Olivier Arino

European Space Agency

CONTENTS

16.1 Introduction

Satellites in polar orbits, like Landsat, image the entire planet's surface every day or every couple of weeks, depending on the swath of the satellite overpass; images with detailed spatial measurements (1–30 m) are usually only available once or twice a month—for example Landsat 5 and 7 (image every 16 days at 30 m resolution)—while coarser resolution imagery (e.g., the MODIS sensor on Terra at 250 m or the SPOT satellites' Vegetation sensor at 1 km) are provided nearly daily. Because the information is captured digitally, computers can be used to process, store, analyze, and distribute the data in a systematic manner. And because the same sensor on the same platform is gathering images for all points on the planet's surface, these measurements are globally consistent and independent—a synoptic record of earth observations ready-made for monitoring, reporting, and verification systems linked to multilateral environmental agreements as well as individual government policies.

Forty years ago, the United States of America was the only source of earth observation imagery—today there are more than 25 space-faring nations

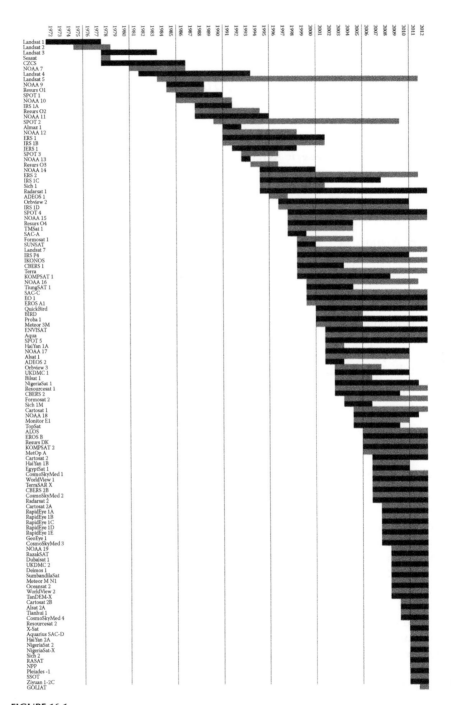

FIGURE 16.1
Polar orbiting satellites with imaging capability launched since 1972. The horizontal bars show period of operation.

flying imaging systems. In 1972, Landsat 1 was the only civilian satellite capable of imaging Earth at a level of spatial detail appropriate for measuring any sort of quantitative changes in forests; today there are more than 60 satellites flying that can provide suitable imagery (or at least they could, *if* they had a suitable data acquisition, archiving, processing, access, and distribution policy). Figure 16.1 lists the polar orbiting imaging satellites, in chronological order according to launch date and shows period of operation. Earth observations from space are becoming more widely employed and increasingly sophisticated. The latest systems launched, such as the Franco–Italian Pleiades system (the first of which was launched December 17, 2011), combine very high spatial resolution (70 cm) with a highly maneuverable platform, capable of providing an image of any point on the surface (cloud cover permitting) within a 24 h period. Concurrent to these technological advances is an increasing appropriation of the land surface in the production of food, fiber, and fuel at the global scale. Forests in particular are under increasing pressure from humankind. Earth observations are critical in assessing and balancing the immediate economic drivers of forest change with the equally important, but less appreciated ecosystem services forests provide.

The previous chapters of this compilation show that recent developments in regional to global monitoring of forests from earth observations have profited immensely from changes made to data policies and access (Woodcock et al. 2008). We now have an unbroken record of global observations stretching back over four decades, all freely available. This chapter provides some perspectives on future earth observation technology for monitoring forests at the global scale.

16.2 Future Earth Observation Technology

Monitoring forest areas over anything greater than local or regional scales would be a major challenge without the use of satellite imagery, in particular for large and remote regions. Satellite remote sensing combined with a set of ground measurements for verification plays a key role in determining rates of forest cover loss and gain. Technical capabilities and statistical tools have advanced since the early 1990s, and operational forest monitoring systems at the national level are now a feasible goal for most countries of the world.

The use of medium spatial resolution satellite imagery for historical assessment of deforestation has been boosted by changes to the policy determining access and distribution of data from the U.S. Landsat archive. In December 2008, the U.S. government released the entire Landsat archive at no charge (Woodcock et al. 2008). This open access data policy means that anyone interested in global forest monitoring now has access to an archive of data spanning four decades. Current plans for the Landsat Data Continuity

Mission (LDCM), with a launch scheduled for early 2013, and the European Sentinel-2 mission (Martimort et al. 2007), with a launch date of mid-2014, will both adopt global data acquisition strategies and will both (at least at the time of writing) provide free and open access to acquired imagery.

LDCM, to be christened Landsat 8 upon reaching orbit, will have a swath width of 185 km and feature a 15 m spatial resolution panchromatic band, nine 30 m multispectral bands (six of which will correspond to heritage Landsat bandwidths), and two 120 m thermal bands. The 16-day revisit rate will match that of past Landsat sensors but with an increased acquisition rate of at least 400 images per 24 h period. The open and free data policy will continue.

The Sentinel-2 satellite will have a swath width of 290 km and carry onboard a multispectral sensor having four bands with a spatial resolution of 10 m, five bands at 20 m, and three bands at 60 m. The Sentinel-2 mission comprises two identical satellites (the second has a tentative launch date for 2015) in identical orbits, but spaced 180° apart. This mission configuration gives a revisit time of 10 days for one satellite and 5 days when both satellites are operational. The Sentinel-2 mission will include a systematic acquisition plan of satellite imagery over all terrestrial land areas of the world between −56° and +83° latitude. The envisaged data policy will allow full and open access to Sentinel-2 data, aiming for maximum availability of earth observation data in support of environmental and climate change policy implementation.

In the near future, the practical utility of radar data is also expected to be enhanced from better data access, processing, and scientific advances. In particular, future space missions will provide complementary Synthetic Aperture Radar (SAR) imagery systems for the monitoring of forest area and biomass. The Sentinel-1 mission (Attema et al. 2007) is a pair of two C-band SAR sensors, the first is planned for launch in 2013 to be followed by a second satellite a few years later. This system is designed to provide biweekly global coverage of radar data at a fine spatial resolution (10 m × 10 m) with a revisit time of 6 days (a swath width of 240 km).

The finer spatial resolution of data from the Sentinel satellites (from 10 m × 10 m) can be expected to allow for more precise forest area estimates and canopy cover assessments, and therefore more reliable statistical information on forest area change, in particular for estimating forest degradation and forest regrowth.

16.3 Perspectives

The basic fact is that natural resources, such as natural forests, are becoming increasingly scarce. There is considerably more pressure on our natural resource base, and establishing a balanced use of forest resources is required. Do you use a forest as a carbon sink? Do you use it as a protected area for biodiversity? Or do you use it for fuel wood or agroindustrial development? To make sensible decisions on the trade-offs between different uses, information on where different forest resources are, what condition they are in, and how they are changing is required. In the framework of the UNFCCC REDD+ activities, the extension of the analysis of tropical deforestation to degradation and forest regrowth will be a crucial requirement (Asner et al. 2009). There are also strong incentives to reduce uncertainty in the estimation of carbon fluxes arising from deforestation by using better data on forest aboveground biomass or carbon stocks (Saatchi et al. 2011, Baccini et al. 2012) in combination with improved satellite-derived estimates of deforestation (Harris et al. 2012).

Mature forest monitoring methods need to be ported to operational settings. Monitoring systems such as Brazil's PRODES deforestation mapping program need to be replicated in other countries where results can be directly incorporated into policy and governance settings. Effective technology transfer of mature, proven methods to developing world institutions needs to be advocated and implemented. This can be envisaged as a leapfrog technology where agencies with little or no past technical capacity may advance in one step to the state of the art.

Researchers will be responsible for developing new capabilities by testing new data sets, processing methods, and thematic outputs. Future satellite image technology, including radar and optical imagery at finer spatial resolutions (10 m finer) and higher temporal frequencies, will require both improved scientific approaches, but also advanced processing systems, including cloud-computing environments (Nemani 2011). The ongoing methodological advances will narrow the gap between the demand for more accurate estimation of the global carbon budget and the limitations of current monitoring approaches.

The adoption of progressive data policies, such as those of NASA, USGS, ESA, and INPE, should be promoted. International coordination between space agencies and implementing institutions (e.g., through the Committee on Earth Observation Satellites—CEOS—or the Group on Earth Observations—GEO) is key to this prospect. Such international cooperation will ensure repeated coverage of the world's forests with varying observation types, all with easy access at low or no cost (GEO 2010). Progress will be measured by how quickly the methods reported here are made obsolete.

About the Contributors

Alan S. Belward works for the European Commission at the Joint Research Centre, where he heads the Land Resource Management Unit. Belward received a BSc degree in Plant Biology from the University of Newcastle upon

Tyne, United Kingdom, in 1981, and MPhil and PhD degrees in remote sensing studies of vegetation, both from Cranfield University's (Bedfordshire, United Kingdom) School of Agriculture Food and Environment in 1986 and 1993, respectively. In the 1990s, he cochaired the International Geosphere Biosphere Programme's Land Cover Working Group and chaired the Committee for Earth Observing Satellites (CEOS) Working Group on Calibration and Validation. From 2002 to 2006, he chaired the Global Climate Observing System's (GCOS) Terrestrial Panel, and in 2009 he was appointed to the GCOS Steering Committee. He is a member of the NASA and USGS Landsat Data Continuity Mission Science Team and the European Space Agency's Sentinel-2 Mission Advisory Group and is also a visiting lecturer at the Technical University of Vienna, Vienna, Austria, where he teaches environmental technologies and international affairs.

Olivier Arino received his PhD degree in physics specializing in remote sensing of surface albedo from the Institut National Polytechnique, Toulouse, France. He performed his research at the Laboratoire d'Etudes et de Recherche en Teledetection Spatiale, Toulouse, France. After two years of postdoctoral fellowship for the International Geosphere Biosphere Program and the European Commission, he joined the European Space Agency in 1991. Since 1999, as head of the projects section in the science applications and future technologies department of the Earth Observation Programme Directorate, he has managed the data user programme in close collaboration with user communities (e.g., 400+ user organizations involved). He was nominated senior advisor in the Earth Observation Programme Directorate in 2008 by a board of ESA directors. He has coauthored more than 20 peer-reviewed scientific papers in the fields of albedo, fire, vegetation, land cover, and sea surface temperature.

References

Asner, G.P. et al., A contemporary assessment of change in humid tropical forests. *Conservation Biology*, 23, 1386, 2009.

Attema, E. et al., Sentinel-1 the radar mission for GMES operational land and sea services. *European Space Agency Bulletin*, 131, 11, 2007.

Baccini, A. et al., Estimated carbon dioxide emissions from tropical deforestation improved by carbon-density maps. *Nature Climate Change*, 2, 182, 2012.

GEO, *Report on the Concept Phase for developing a Global Forest Observations Initiative*. The Group on Earth Observations, Geneva, Switzerland. 2010.

Harris, N.L. et al., Baseline map of carbon emissions from deforestation in tropical regions. *Science*, 336, 1573, 2012.

Martimort, P. et al., Sentinel-2 the optical high-resolution mission for GMES operational services. *European Space Agency Bulletin*, 131, 19, 2007.

Nemani, R. et al., Collaborative supercomputing for global change science. *EOS Transactions*, 92, 109, 2011.

Saatchi, S.S. et al., Benchmark map of forest carbon stocks in tropical regions across three continents. *Proceedings of National Academy of Science USA*, 108, 9899, 2011.

Woodcock, C.E., et al., Free access to Landsat imagery. *Science*, 320, 1011, 2008.

Index

For Product Safety Concerns and Information please contact our EU representative GPSR@taylorandfrancis.com Taylor & Francis Verlag GmbH, Kaufingerstraße 24, 80331 München, Germany

Printed and bound by CPI Group (UK) Ltd, Croydon, CR0 4YY

01/05/2025

01858493-0001